Ecology

From Ecosystem to Biosphere

Ecology

From Ecosystem to Biosphere

Christian Lévêque
Director of Research
Institut de Recherches pour le Développement (IRD)
Paris
FRANCE

CRC Press
Taylor & Francis Group
Boca Raton London New York

CRC Press is an imprint of the
Taylor & Francis Group, an **informa** business

A SCIENCE PUBLISHERS BOOK

First published 2003 by Science Publishers
Published by arrangement with Dunod, Paris

Published 2019 by CRC Press
Taylor & Francis Group
6000 Broken Sound Parkway NW, Suite 300
Boca Raton, FL 33487-2742

© 2003, Copyright reserved
CRC Press is an imprint of Taylor & Francis Group, an Informa business

First issued in paperback 2019

No claim to original U.S. Government works

ISBN-13: 978-0-367-44686-4 (pbk)
ISBN-13: 978-1-57808-294-0 (hbk)

Visit the Taylor & Francis Web site at
http://www.taylorandfrancis.com

and the CRC Press Web site at
http://www.crcpress.com

Library of Congress cataloging-in-Publication Data

Lévêque, C.
   [Écologie de lécosystème à la biosphère, English]
   Ecology from ecosystem to biosphere/Christian Lévêque
      p.cm
   Includes bibliographical references (p).
   ISBN 1-57808-294-3
      1. Ecology. I. Title

QH541.L4513 2003
577--dc21

                                                          2003052800

Translation of: *Écologie de l'écosystème à la biosphère*, Dunod, Paris, 2001.
French edition: © Dunod, Paris, 2001

Preface

The average person is often confused between political and scientific ecology. There is, moreover, a confusion between environmental sciences and ecology, which is sometimes called on to answer questions on topics beyond its expertise.

The lack of visibility of scientific ecology lies also in the very wide field it is supposed to cover. It ranges from evolutionary sciences to population biology and now to biosphere dynamics. A number of other scientific disciplines are also involved, such as zoology, botany, geology, geomorphology, climatology, hydrology, and chemistry. In such a context, what is the status of ecology as compared to more established sciences? Is it a metadiscipline whose main goal is to synthesize the scientific information provided by other disciplines? Or is it a full discipline that has developed its own concepts and theories?

One concrete response to the above questions is to provide a general overview including the origins, methods, and concepts, in other words, the paradigms, of ecosystem ecology. This holistic science was built as a systemic approach in order to develop theories and methods to understand the temporal and spatial dynamics of ecosystem components.

Two main ideas are developed in this book:

1) Ecosystem ecology, which originated historically in the life science domain, now claims a double filiation: life sciences and earth sciences. Its ultimate goal is the understanding of the joint dynamics of physical, chemical, and biological components. It is a framework within which the interactions between the dynamics of the biosphere and the atmosphere can be studied.

In the name of the unity of ecology, some ecologists are in favour of integrating the two main ecological paradigms of population biology and ecosystem ecology. In fact, however, these are two different domains that are well identified at the operational and conceptual levels. We have to recognize this difference and that ecosystem ecology is a well-identified interdisciplinary science among the natural sciences. It is based on a strong conceptual framework, and it involves analytical, experimental, and comparative methods using sophisticated field and laboratory techniques. New tools such as remote sensing, modelling, isotopes, and molecular biology have opened the way to the study of new processes at new spatial and temporal scales.

To achieve this goal we need to do two things. One is to develop a truly systemic approach in order to answer the complex questions that are inherent in the very nature of ecosystems. Scientists cannot spontaneously undertake interdisciplinary research. Such research has to be organized and encouraged by institutions. The second is to develop long-term research observatories given the spatial and temporal scales involved in ecosystem functioning and the existence of delayed feedback and responses, and to validate simulation models. Such organization is already operational for climate studies or astronomy.

2) There is a strong tendency to look at ecology as an operational environmental science. Under the term *ecosystem management* or *ecosystem approach*, there is an attempt to conceptualize such a practice. In many countries, social scientists often accuse ecologists of not giving a primary focus to man in their studies.

The situation is confusing. It is not reasonable to expect ecology to provide answers to questions whose origins, and solutions, have to be sought in social behaviour. Yet, such an attitude is encouraged by some ecologists who try to promote their discipline by emphasizing the social issues. This insistence on applicability is surprising because chemists or biologists, for example, do not need to stress the applicability of their science. And sociologists are clearly reluctant to be involved in decision-making.

It seems, therefore, that scientific ecology seeks for a social legitimacy. This attitude is irrelevant for two reasons. The first is that transformation of the scientific and naturalistic knowledge of ecosystems into recommendations for management with strong socio-economic components and political feedback requires expertise that most ecologists are not prepared to assume. Therefore, most advice provided by ecologists is not practically useful and does not really meet social needs. In other words, ecology goes beyond its field of expertise and it is irrational to perpetuate this confusion. The second reason is that if ecosystem ecology becomes demand-driven, it cannot devote enough time to elaborate paradigms. The risk is that it will become just an operational science mobilizing knowledge for short-term purposes, that is, conservation biology. If so, ecosystem ecology will no longer be considered a cognitive science, and there is a serious risk that it will lose its scientific legacy.

That ecology, like hydrology and chemistry, contributes to solving environmental problems is another story. No ecologist can ignore the fact that human beings are very active in ecosystems. Any member of the public may raise questions about the relationships between man and nature. European landscapes have been shaped by human activities and result from a co-evolution between ecological systems and social systems. The concept of anthroposystems has been devised by the CNRS Environmental Research Programme to characterize such interactive systems made up of ecosystems, resources, and the human communities that live on them.

Anthroposystem science has a strong human dimension: the sustainable use of ecosystems. Science provides knowledge but choices belong to people. Management is, in a way, the art of compromise according to the needs and perceptions of society. This is far from ecosystem ecology, in a domain where ethical, economical, and philosophical considerations are the driving factors of the dynamics of natural systems.

In short, each to his trade. Ecosystem ecology has to develop further without setting up man as the focus of its concern. On the other hand, like other scientific disciplines (e.g., hydrology, economy, sociology), ecology provides decision-makers with concepts and knowledge.

I would like to add that my experience in aquatic ecology has influenced the overall structure of the book. Most often, ecology books are written by terrestrial ecologists, while aquatic ecology appears as limnology or oceanography. However, lakes and oceans are very good examples of ecosystems and probably played a major role in the growth of the ecosystem concept, as shown in the historical part of this book. It is fair, therefore, to provide a unified view of ecosystem ecology.

Christian Lévêque

Contents

Part II

STRUCTURE AND ORGANIZATION OF ECOSYSTEMS

Part III
FUNCTIONING OF ECOSYSTEMS

Part IV
GLOBAL ECOLOGY

Chapter 1

The "Natures" of Ecology

If you come to look at the Earth from the perspective of space, everything seems silent and calm. The convulsions of the earth's crust are too slow to be perceived in the scale of our short lives. ... But if you travel from this rocky crust up to the atmosphere, you will find only noise and excitement compared to the eternal silence of immobile space. Not only does the air swirl and the water boil, but an extraordinary network of living things murmur and act on the face of the earth. These beings are distributed on a thin layer at which the atmospheric film is in contact with the rocky crust. They share the energy that is received on the earth, as they share the surface of the earth and the immense three-dimensional space of the oceans. They mingle with each other and accommodate each other, living and letting live, always adapted to the mode of life they must lead, a life that takes highly diverse forms.

The researchers who study the functioning of all these phenomena are ecologists—in the primary sense of the word.

translated from Paul Colinvaux (1993)

"Science has been called 'the art of the soluble' because science succeeds by answering questions, not simply posing them. Unfortunately, ecology has often asked intractable questions."

Peters (1991)

1.1 THE BIRTH CERTIFICATE

The term *ecology* (from the Greek *oikos*, meaning house or habitat, and *logos*, meaning discourse) was created in 1866 by a follower of Darwin, the German biologist Ernst Haeckel. By *ecology* he referred to the science of relationships of organisms with the external world, in which we can recognize the factors of the struggle for existence, in the widest sense (Dajoz, 1984). These factors are partly inorganic, such as physical and chemical characteristics of the habitat, the climate, water quality, and the nature of the soil. However, the conditions of existence also involve the relationships that organisms have with each other, knowing that each has friends and enemies among the other organisms. In 1870, Haeckel evoked a body of knowledge concerning the economy of nature, that is, a study of all the relationships of the animal with its inorganic as well as organic environment, especially the inimical or hostile relationships with those animals and plants it comes into direct or indirect contact with.

Haeckel himself did not contribute to the development of the concept he created. But he was one of its most visible philosophical inspirers, as an active member of the Monistic League, whose philosophy affirmed the unity of the state of inert things and living things, plants, animals, and human societies (Deleage, 1993). Like other members of this league, Haeckel believed in political reform based on scientific knowledge of the relationships between man and the world and on a fundamental respect for the beauty and order of nature.

In reality, many science historians see in Darwin's *Origin of Species* one of the major sources of modern ecology. Others dispute this filiation.

From the end of the 19th century, the term *ecology* was used, notably by the Danish botanist E. Warming, to designate the part of botanical geography that studies the relationships of plants with their environment. From the 1930s onward, it was commonly used by scientists in England and the United States and began to be known in France. It appeared in French dictionaries only in 1956, defined as the part of biology that studies the relationships of living things with the natural environment.

Definition of the term *ecology*

By ecology we mean the science of the set of relationships that organisms have with each other and with the external world (Haeckel, 1866).

Ecology is considered the science of relationships between living things and their environment. Since living things are closely integrated with their environment, ecology is the science of complex, functional biological systems called *ecosystems*. It also encompasses the study of relationships between living things (Duvigneaud, 1974).

Defined as the study of relationships of organisms with their environment, or as the study of interactions that determine the distribution and abundance of organisms (Krebs, 1972), or even as the study of ecosystems, ecology opens on to a wide field, ranging from physiology to biogeography. Seen from this angle, it is a sort of general biology of organisms, a naturalistic approach to the living world (Barbault, 1983).

"Ecological theory today spans a large range of topics, from the physiology and behavior of individuals or groups of organisms, through population dynamics and community ecology, to the ecology of ecosystems and the biochemical cycles of the entire biosphere. Ecology theory also embraces large parts of evolutionary biology, including paleontology and systematics, and of the earth sciences, especially oceanography and tectonics" (Roughgarden et al., 1989).

Ecology is the study of interactions between living organisms and the environment, and among living organisms themselves in natural conditions (Frontier and Pichod-Viale, 1991).

At present, many definitions of the term *ecology* have been proposed, and it is difficult to cover them all. Some definitions are proposed in this work. From the definition of Krebs, which is concise but very general, to that proposed by

Roughgarden, there are notable differences. The former emphasizes mostly the species-environment interactions, the distribution of species being determined by the quality of habitats offered by the environment. The latter draws up the makings of a catalogue of disciplines potentially involved and widens the field covered by ecology. The definition proposed by Frontier and Pichod-Viale is not, in essence, far different from that of Haeckel.

1.2. IS ECOLOGY A SCIENCE?

Well before ecology emerged as a scientific discipline, people were already concerned with questions that arose about the functioning of so-called natural environments. For a long time, in fact, nature was transformed by activities linked to the use and exploitation of resources. Foresters, agronomists, and geographers, as well as farmers, woodcutters, hunters, and fishermen, who had travelled and/or shaped many landscapes, had accumulated an important empirical ecological knowledge. Historically, studies of applied biology, such as agronomy, were the source of discoveries and concepts in the physiology of living species as well as the flow of nutrients and energy (photosynthesis). Because of this, the nascent field of ecology was strongly marked by operational research.

The science of ecology was also subject to the influence of the cultural and social context in which it developed. For example, Worster (1992) showed that there existed a relationship between the pioneer traditions in North America and the ideas developed by the avant-garde American ecologists. Similarly, Wiener (1968) saw a connection between the organicist concepts of the first Russian ecologists and the traditions of a people imbued with a particularly rich imagination centring around water and the forest.

According to Worster (1992), there are two extreme attitudes towards ecology, which he calls *arcadian ecology* and *imperialist ecology*. The first, represented by G. White, considers nature in a "humble, romantic" fashion and leads to a search for a mode of life in harmony with creation, from a perspective that is more sentimental than scientific and rational. For believers in arcadian ecology, who originated movements for the conservation of nature, man is an integral part of nature, which has an intrinsic value. The imperialist ecologist, on the other hand, considers that nature has no value except what we attribute to it in order to exploit it. Such an ecologist adopts the concept of ecosystem as a central dogma and studies its functioning from a productivity point of view. In this view we find the values of modern economy. Its ambition is to establish the dominion of man over nature.

For all these reasons, ecology has not been perceived in the same way as molecular biology or particle physics. Ecology is more of an everyday science for people, from fishermen to hikers. People have the impression that their understanding underlies the legitimacy of ecological knowledge. It is here that we find a great ambiguity. To the extent that this science appears accessible to all, there is a risk of lack of scientific rigour in the approach of questions, because they are insufficiently formulated.

We are thus well-advised to question (as many do) the processes and methods of ecology that are said to be scientific. Is ecology really a science and not just a current of thought? In asking this rather provocative question, J.M. Drouin (1991), in the introduction to his work *L'ecologie et son histoire*, launched a debate that he himself promptly settled. Arguing from the fact that the history of ecology was marked by many research studies, theoretical reflections, and controversies well before ecology appeared as a sociological phenomenon in relation to the protection of the framework of human life, Drouin states that it was an entirely separate scientific discipline. Still, in the eyes of some people and decision-makers, and even in the eyes of scientists, ecology is not generally viewed as a science in the same way as nuclear physics or molecular biology.

Deleage (1992) poses the question in a different way: what science is involved here? He adds, what best defines its epistemological status is the point of view it adopts. Ecology is especially the field of heuristic principles, that is, hypotheses that we do not seek to prove as true or false, but that are adopted provisionally as directive ideas in the search for facts. Unlike physics or genetics, for example, ecology has not created a significant construct of organized laws. In this sense, many propositions of ecology, such as the concepts of niche, climax, or even biosphere, cannot be tested in the sense that Popper defines testing. But one of the reasons why we sometimes question the status of ecology as science is that it is often difficult to eliminate the particular point of view of the observer and to eliminate all value judgements on the object of study (Dealeage, 1993).

In reality, behind this rather academic debate lies the major question of recognition of scientific domains that do not raise the same paradigms as those of physics, which have dominated science till recently. There are domains in which the elaboration of universal and deterministic laws, and the experimental process, are much more difficult to implement, taking into account their nature and complexity. Some even cast doubt on the fundamentals of scientific discourse and search for alternatives to the basic paradigms that have been offered to us by physics and mathematics. Behind the well-known affirmation that the whole is more than the sum of its parts lies the idea of an ecosystem that is not simply a juxtaposition of living and inert elements but has emerging properties that are not deduced from just the characteristics of its components. There lies the difficulty. Although many ecologists have a profound conviction that their object of study is not a simple mechanics, they have not been able, till now, to present firm proofs that support this conviction. Perhaps we ought to wait for new developments in our information. Did not Darwin present hypotheses on natural selection that subsequently proved to be entirely relevant, although during his time he knew nothing of the genetic mechanisms that underlie them?

1.3. ECOSYSTEM ECOLOGY AND/OR POPULATION ECOLOGY?

Ecology is often considered a branch of the life sciences, a diverticulum of botany and zoology. This interpretation is not wrong from the historical perspective to the extent that ecology partly draws its origin from the geography of plants from the beginning of the 19th century. But it is reductionist, given that even in the definition of the concept of ecosystem, which is central to ecology, there are explicitly stated interactions between biological systems and physical and chemical systems.

In reality, we generally recognize the existence of two major domains in ecology that are developed in a relatively independent manner, each having its own concepts, theories, and methods: *population ecology* and *ecosystem ecology*. We can in fact talk of two major ecological paradigms (Barbault, 1997) that are characterized by the mechanisms or phenomena studied. To summarize, the paradigm of population ecology concerns the modes of distribution and the dynamics of the abundance of species in time and in space, as well as interactions between species especially in terms of competition for resources. The emphasis is on the living world, while consideration is also given to abiotic factors such as external constraints that control the population dynamics. In the field of systemic ecology, on the other hand, the objects of research are not exclusively biological. We are interested in the cycles of matter and energy that structure the ecosystems: processes and mechanisms of production and transfer of organic matter, decomposition, and mineralization. These biogeochemical cycles are supported by microbial, plant, or animal populations that are grouped on the basis of their role in the regulation of the flows: as producers, decomposers, nitrogen fixers, and so on. Thus, we find ourselves at the interface of life sciences and earth sciences. The systemic approach is often not well connected to the population approach. Often, the biotic compartments are compared to black boxes for the purpose of simplification, but sometimes also out of ignorance.

Each of the two major paradigms is undoubtedly the product of a history but also of a mode of thought. To simplify, one paradigm highlights the reductionist approach and emphasizes interactions between species, while the other is part of a holistic approach and emphasizes abiotic factors as structural elements of ecosystems. The natural extension of the ecosystem paradigm is ecology of the biosphere, as has been developed these past few years. Population ecology tends toward evolutionary sciences because of new research perspectives offered by the tools of molecular biology. New centrifugal forces are at work but, at the same time, in evoking the unity of ecology, there is recurrent support for those who plead for the similarity of the two approaches. For some ecologists, the need to understand the role of biodiversity in the functioning of ecosystems and of the biosphere encourages them to integrate the two paradigms (Pickett et al., 1994; Barbault, 1997). Others are inclined toward a formal recognition of the existence of these two separate paradigms.

During the past few decades, the indication of the role of the living world in the biogeochemical cycles and the composition of the atmosphere have shown that the ecosystem is actually composed of biotic and abiotic elements in interaction. The understanding of the water cycle and nutrient cycle, climatic variations, and processes of erosion and sedimentation is as fundamental in ecosystem ecology as the knowledge of demographic strategies or relations of competition between species. In the editorial of the first issue of the journal *Ecosystems*, Carpenter and Turner (1998) affirm that the science of ecosystems is now a well-established discipline associating fundamental research with the resolution of specific questions and constituting a particularly dynamic sector of contemporary ecology. In the spatial perspective it involves systems with relatively narrow limits, such as watersheds, as well as complex landscapes and the earth itself. In the temporal perspective the science of ecosystems covers scales that range from the second to the millennium.

We can thus seriously question the operational reality of a unified progress for ecology, considering the field that would thereby be covered: from the genetics of species to biogeochemical cycles and climatic changes. Whether it would be desirable on the other hand to encourage a close collaboration between the population and systemic paradigms during multidisciplinary studies is a different kind of problem that allows to each of these paradigms the possibility of developing itself around its basic concepts. In brief, we can consider that the present logic would be to recognize the existence of twin disciplines that have their own dynamics and their own paradigms, even though they share common ground. One argument in favour of this position points out that hydrology, pedology, and geochemistry contribute, just as much as population ecology, to the comprehension of the functioning of ecosystems.

In fact, ecosystem ecology, without repudiating its past, is not yet free of the historic affiliation with life sciences in claiming a dual paternity: the world of the science of living things and the world of earth sciences. Yet, this is what makes it unique and allows it to stand as an entirely separate discipline. It is also this that explains to some extent the difficulty of positioning ecosystem ecology, which is not really recognized by any of the major disciplinary fields.

1.4. ECOSYSTEM ECOLOGY: HOLISM OR REDUCTIONISM?

Ecology, with its search for unifying principles in interactions between living things and their environment, is at the crossroads of many fields of knowledge of nature. It is sometimes presented as an exemplary science of synthesis. Di Castri (1981) illustrates his synthetic conception of ecology by the image of a common trunk towards which many disciplines converge: first of all botany, zoology, climatology, soil science, and physical geography, then biochemistry, microbiology, and advanced mathematics, and finally, sociology, human geography, psychology, and even economics. This idealist and ultimately quite imperialist conception is a more elaborate version of Duvigneaud's (1974) idea that ecology is a vast science of synthesis that calls on the greatest variety of disciplines.

> ### Ecosystem ecology: A science of synthesis?
>
> We often hear that ecosystem ecology is a "metadiscipline", which underlines the fact that it straddles several traditional scientific domains. In these conditions, synthesis is necessary as a way of integrating two or many different scientific domains in order to draw forth a new body of knowledge, a new conceptual progress. Among the examples he proposes, Pickett (1999) cites the notion of engineer organisms in the ecosystems: such organisms are not just a link in the food chain but can have an effect on the availability of resources of other organisms by the modifications they make in the physical system. This concept allows us to make a synthesis between various functions of the organisms in the ecosystems beyond the classic concept of food chain and energy flow.
>
> Yet, beyond declarations of principle, the reductionist approach most often dominates (Pickett, 1999). The difficulties arise partly from cultural questions, especially the primacy of analysis over synthesis. Perhaps we must convince students of the need to integrate disciplines and the advantage of multidisciplinary research for the advancement of knowledge in ecology. Undoubtedly, however, we must first convince the teachers themselves.

Since its formulation as an autonomous discipline by Haeckel, the object of ecology has widened continuously. Limited at first to the study of single ecological habitats, and mostly centred on the associations of organisms in one type of environment, ecology went on to embrace the analysis of ecosystems, taking into account interactions between the *biotope* (conceived as the set of physical and chemical characteristics) and the *biocoenosis* (covering the set of living things). From bacteria to birds, from arctic systems to tropical rain forests, ecology has tackled very different biological models and types of ecosystems.

What is more, the natural sciences have followed a constant course for several decades to identify integrated systems, increasingly integrated. From the field to the watershed, from the ecosystem to the landscape and the ecoregion, we have simply changed the spatial and temporal scales. More recently, ecologists have taken cognition of the global dimension in their research. The biosphere (the planetary ecosystem) is thus made up of interactive subsets, enclosed within each other like Russian dolls. It is customary to think that there is an increasing complexity from the functioning of ecological systems to the functioning of higher levels of integration. First of all, it is essential to take into account the increasing number of elements in interaction.

To argue about the position of ecosystem ecology in relation to other disciplines is a passionate but rather sterile exercise. What is required is *a scientific process that can integrate information from various sources in a dynamic and systemic conception of nature*. This information is provided partly by established disciplines (e.g., hydrology, pedology, geomorphology, biogeochemistry, biology) but generally in a sectorial form. It must be put into perspective in a synthetic process that obliges us, moreover, to collect other information, especially on the dynamics of interactions between the constituents of the system. We must also develop concepts and hypotheses, as well as methods and tools, to understand and explain the functioning of these complex systems. We are now witnessing a scientific progress that has really no equivalent in the academic disciplines presently known. That it relies on other disciplines is a fact that is not specific to ecosystem ecology. In these conditions, we will come to acknowledge that ecosystem ecology is an entirely separate scientific discipline.

1.5. SCIENTIFIC ECOLOGY, POLITICAL ECOLOGY AND ENVIRONMENTAL SCIENCES

Like other disciplines, ecology contributes to a redefinition of relationships between society and nature. Ecological thought indicates the philosophy and reflects the values that the organization of a society is based on (Bourg, 1996) and this ecological reflection establishes a link between ecological science and political action. Thus, there is in theory no tight division between scientific ecology and activist ecology. But there is an obvious risk of confusing the philosophical reflection with the ecological science itself.

Ecologism participates in the diffusion of ecological concepts while searching within ecology for some form of legitimization of its action. Partisans of movements that call themselves ecological, ranging from associations for the conservation of nature to political engagement, have diverse motivations that are social, aesthetic, ethical, or mystical. Some objectively rely on scientific knowledge but others draw only the information necessary for the elaboration of their own argument. Still others are less honest and, under the cover of science, develop an inconsistent ideological discourse, or attribute to ecological science much more than it can objectively do. In other words, the science of ecology can be taken hostage in political and philosophical debates and serve as an alibi in certain power struggles. In the West, science has taken over from religion in claiming the absolute value of knowledge, to represent inevitable progress.

It is in this context that ecology has been regularly instructed to integrate in its progress human activities and their consequences for species and their environment. Undoubtedly, ecology can no longer be a reflection of nature. Nor can it be a reflection of man. A science of systems, addressing itself to species, populations, and communities, it is henceforth constrained to understanding their structure, their organization, and their functioning, constantly referring to man and the history of societies (Lefeuvre, 1989). Till now, it was understood that ecology addresses not only "virgin" systems, but also anthropized systems. Whether man is one driving factor among many is seen from the evidence when we observe European ecological systems that have almost all been modified, even constructed, by human societies. But we demand still more from ecology. Social scientists, for example, reproach ecology for not taking sufficient account of the activities of human societies in its problematic. Others demand that ecologists provide concrete answers to the questions posed by society about the management of environment and resources.

Today, there is a strong tendency among ecologists to justify their activity by the social utility of their studies, and most ecological treatises emphasize the target perspectives of the discipline. Ecology is thus expected to mobilize its expertise to help decision-making on the use and management of resources and land. The concept of goods and services rendered by ecosystems intersects ecology with economy (Daily, 1997), and the concept of ecosystem management proposes a widened framework for the dialogue between scientists, citizens, and managers. The "ecosystem" approach, as it has been developed in the last few years, aims simultaneously to protect the environment, maintain ecosystems in good health, preserve biological diversity, and guarantee sustainable development. Here we find all the ingredients of the present discourse on conservation of species and environments.

A danger for ecology would be to appear as the science that will provide solutions; In fact ecology, no more than other natural sciences, bears social values, while for some people it takes place of philosophy. If one can regret that a society destroys biodiversity, and feel that it is unfair, we cannot claim conversely that the protection of biodiversity deals with social justice among humans. As we have said in the foreword, to compare ecology to environmental sciences is a dangerous drift. To avoid it, we need to distinguish clearly what is spelled out in the ecological process from social needs. *Scientific ecology and ecosystem ecology, in particular, are not sciences of management but sciences of knowledge*. This distinction must be preserved.

1.6. ECOLOGY AND THE MEDIA

Although intelligent ecologists are sometimes irritated by the confusion that arises between science and ideological movements, they nevertheless have an ambiguous attitude toward these movements. They do not hesitate to profit from the perceived social urgency of the ecological crisis to shore up their institutional position (in terms of funding and posts). The belief that the future of the planet depends on a better management of resources and land is recurrent in ecological research and has even become a litany. This attitude is accompanied, or is reinforced, by the use of dramatization. The use of catastrophic scenarios is not unique to ecology. It is a natural part of the process of sensitization initiated, most often, by conservationist movements. Although scientific ecologists are not always the originators of such scenarios, they use them to affirm their legitimacy.

Such dramatization does not necessarily mean that researchers are exaggerating the dangers that threaten the biosphere and making themselves into prophets of the apocalypse. It is true that the transformation of environments or irrational use of resources leads to disturbing situations with respect to the perenniality or renewal of resources. Nevertheless, the immediate consequence of such an approach is the setting up of a confrontation between nature and society. The ecological catastrophe becomes synonymous with social dysfunction and the disturbances of ecosystems are in turn said to be causes of social dysfunction (Fabiani, 1985). In short, ecological danger becomes social danger. Inversely, ecological equilibrium leads to social peace. In extreme situations, we end up, in the name of ecology, concluding the necessity for reducing the human population that encumbers the landscape and destroys natural environments. This is the kind of thinking in which the ecologist takes charge of ecological planning of the biosphere (Lebreton, 1978).

According to Fabiani (1985), dramatic presentation of an ecological research programme and its results allows us to express ecological problems as public problems, and to create recognition for ecology as an important science. But dramatization may have two unexpected consequences: the denunciation of threats that loom over the biosphere tends to be perceived from the outside as a professional ideology, and the social effects of the dramatization may become blunt with use. Then the logic of the dramatization may lead to a prophetic drift. Some ecologists sanction the urgency of ecological problems to treat social questions.

The debates surrounding the hole in the ozone layer, climatic changes, or biodiversity illustrate the difficult and misunderstood liaisons that establish themselves between scientists, the public, and politicians. It is the scientists who, on the basis of their observations, first alerted public opinion and decision-makers about these phenomena. In that they were only doing their jobs. But the observations at our disposal are affected by many uncertainties. We are far from having penetrated all the mysteries or even having modelled precisely the functioning of the gigantic thermal machine that the terrestrial earth constitutes, with its biogeochemical flows and its many retroactive loops. Similarly, in the absence of a proper understanding of the diversity of species, it seems quite difficult to affirm peremptorily that human activities are responsible for unprecedented erosion in biological diversity. However, scientists have been caught up in the twists of media attention and the appeal of dramatic announcements. Various political, ecological, and even religious movements have used the knowledge acquired to demand radical and immediate measures, regardless of the social or economic consequences. Some have tended to force the issue or to presume too quickly that their results have been validated. Oracles crowned with a Nobel prize, or pretending to speak in the name of science, have also been brought to express themselves on subjects they have little command over, because it is far from their speciality.

It is therefore advisable to recall clearly that the role of the science of ecology is not to provide ready-made solutions to decision-makers. It is not for scientists to propose the social and economic choices needed to reduce the greenhouse effect and/or the erosion of biodiversity. It is the policy makers who take decisions on the basis of information available to them, in the understanding that all is not yet known and that, if we wait further, there may be irreversible consequences. In this context, the duty of scientists is to provide the most objective information, in a form that can be understood by various social agents. This transfer of knowledge is part of the vocation of researchers, who must render account to society of the results of research that society has given them the means of producing. In this lies the scientist's full and entire responsibility. And even in this domain, there is much to be done.

Part I
Elaboration of the Scientific Approach in Ecosystem Ecology

Chapter 2

Origin and Evolution of the Ecosystem Concept

A lake forms a little world in itself, a microcosm in which all the elementary forces of life are at work. ... In no part can one more clearly see illustrated what is called the sensitivity of such a complex organization, a sensitivity expressed by the fact that anything that affects a species that belongs to it must have an influence on the other constituents of the whole. This demonstrates the impossibility of completely studying a species without taking into account its relations with others, the necessity of having an understanding of the whole in order to be able to understand satisfactorily the functioning of one part.

S.A. Forbes (1887), *The Lake as a Microcosm*

The formulation of the ecosystem concept by Tansley, in 1935, was an important step in ecology. Years later, with the development of research on matter and energy cycles, ecosystem ecology proved itself as one of the dominant paradigms of ecology and gave rise to the International Biological Programme at the end of the 1960s. Since then, ecosystem ecology has diversified its practices and widened its field of application. It has benefited especially from advances in modelling systems and means of investigation to tackle not only increasingly larger spatial and temporal scales, but also increasingly smaller ones.

This chapter is devoted to a brief summary of the principal currents of scientific thought that led to the elaboration of the ecosystem concept. This aspect is not often addressed in treatises on ecology. Still, with some knowledge of the history and diversity of ideas that have contributed to the elaboration of this concept, we can better understand the present debates surrounding ecosystem ecology. According to a review by F. Golley (1993), the history of the ecosystem concept has referred simultaneously to the influence of scientific personalities, interaction and competition between scientific disciplines, the influence of institutions and scientific societies, the role of cultural paradigms that structure our way of thinking about nature and our scientific endeavours, and other factors. We might add, with Arthur Lovejoy, that the history of ideas shows us how each era tends to exaggerate the shape and finality of its own discoveries and rediscoveries and lets itself be dazzled by those that it cannot clearly discern the limits of, and how each era forgets the fragments of truth contained in the preceding exaggerations against which it revolted (cited by Worster, 1977).

2.1. THE ORIGINAL MAIN STREAMS IN ECOLOGY

The development of the ecology and ecosystem concepts originated in the emergence of the natural sciences in the 18th and 19th centuries. On the one hand, the idea that each species may have needs in terms of the physicochemical environment had its own track. On the other hand, the search for rules to explain the existence of plant associations supported numerous debates from the 19th century onwards. In reality, the question of interactions between the species and their physicochemical environment, which is central to the ecosystem concept, was already largely part of the reflection of botanists and zoologists well before the concept was formalized by Tansley in 1935.

2.1.1. Humboldt's plant geography

Botany in the 18th century was essentially descriptive. It concerned itself in the first place with recognizing and describing species in order to classify them. However, botanical geography, which developed slowly, emphasized the geographic distribution of plants and the causes of this distribution as a function of climate, relief, soils, and other factors. It benefited from discoveries made during many scientific expeditions, which became much more frequent during the late 18th and early 19th centuries and considerably enriched botanical knowledge.

In 1807, after a long expedition in South America, the German geographer A. de Humboldt published an essay on the "geography of plants", a science that considers plants from the perspective of relationships of their local associations in various climates. Without challenging traditional botany, Humboldt laid the foundations of an explanatory system in which the physical parameters (temperature, humidity, nature of soil), themselves determined by the spatial givens (latitude, altitude), determine the characteristics of the vegetation, which in turn influence the distribution of animals and people. Humboldt seems to have been the first to use the notion of *plant association* (Dajoz, 1984). He also demonstrated the utility of recourse to plant geography in the reconstruction of the ancient history of continents, thus establishing relationships with geology. He developed a fertile analogy from the conceptual point of view in suggesting that altitude and latitude have equivalent effects on plants. Finally, he opened up an avenue to animal geography, which grew to its full importance in the mid-19th century with Wallace (1876).

Ideas similar to those of Humboldt were developed by the Swiss botanist A.P. de Candolle in 1820 in his *Dictionnaire des Sciences Naturelles*. Moreover, these two researchers have sometimes together been held responsible for the origin of the concept of plant geography. De Candolle talks of *stations*, that is, environments in which plant species are found, and *habitations*, the regions in which they grow naturally. For example, the station of the *Salicornia* is in the salt marshes, that of the aquatic *Ranunculus* in fresh and stagnant waters; the habitation of these two plants is in Europe. But de Candolle was one of the first to make significant use of the concept of competition to explain the distribution of species. He may have been influenced in this by a meeting with Malthus (Egerton, 1983).

Vegetation and social structures

Although ecology has borrowed part of its vocabulary from economy, another part comes from analogies with human societies. Throughout the 19th and 20th centuries, the terms *community, association, alliance, groups, settlement, society, guild, population, rates of immigration*, and *competition* testify to the constant presence of social metaphors in ecology (Drouin, 1991). Linnaeus, in 1760, made a rather shocking comparison between plants and human society, which is paraphrased below:

God has prescribed to all a precise subordination and, so to say, a regulation. The mosses may be considered similar to very poor peasants who occupy the most infertile land, protect it, and render it softer. Moreover, by their care, they protect other plants so that their roots are not dried out by the heat of the sun or consumed by frost. The grasses in the plant kingdom seem to play the role of farmers. They occupy the greatest part of the land and the more they are trampled underfoot and bruised, the more they work to give strength to their roots. We can compare to the nobility the plants that rejoice in their leaves, shine with splendid flowers, and give to the realm their fragrance, flavour, colour, and shape with an admirable dignity. The trees, comparable to the highest class, dig their roots deep and raise their heads above the other plants. The trees have, as footmen, the mosses and algae which they nourish in a special manner in view of their beauty rather than utility. (Translated from "La police de la nature" in Linnaeus, C., "L'equilibre de la nature". Extracted from a thesis by Henric Christian Daniel Wilcke (1739-1788), a student of Linnaeus.)

Plant geography was practised for some time on the basis of species. However, the work of Humboldt was also the starting point for research on the distribution of plant zones. Taking into account the still embryonic state of plant taxonomy, he described several "growth forms" or "biological forms" that designated the physiognomy of the plant adapted to its environment. For example, there are plants that have the physiognomy of cactus but do not necessarily belong to the family of cacti. Others have the shape of ferns, palms, and so on. The "plant landscape" is the result of all the growth forms of plants that make it up.

In a slightly different approach, A. Grisebach proposed in 1838 the concept of "phytogeographic formation" or "plant formation": a group of plants presenting a defined physiognomic character, such as a grassland or a forest. It is sometimes made up of a single species, sometimes a complex of dominant species belonging to a single family, and sometimes an aggregate of species that, although different, present some common peculiarity in their organization. The alpine prairies, for example, are nearly exclusively composed of perennial grasses. This definition, which also generally poses the principle of discontinuity of vegetation, and thus the possibility of objectively classifying plant groups, is considered the distant but direct ancestor of phytosociology.

4) analyse the problems of plant economy, i.e., the demands of each plant with respect to the environment, as well as the modalities of their resistance with regard to this environment by adaptation of their morphology and their anatomy, which leads to the analysis of types or biological forms of plants (Warming, 1896).

Warming is considered one of the founders of ecology because he gave it a precise and original status in relation to the disciplines from which it originated. In particular, he helped establish a theoretical framework that strongly influenced the developments of geobotany. However, he also opened an avenue to the concept of climax in showing that natural communities do not always remain the same but evolve under the effect of external forces. During the botanical congress held in Brussels in 1910, Warming proposed the term *synecology* for the ecological study of plant groups and *autecology* for research directed toward adaptation of plant species to their environment.

2.1.4. The American school: succession and climax

While the European researchers in a relatively static approach divided and classified the plant cover into identifiable spatial units, the American F. Clements (1916) in the beginning of the 20th century developed the theory of plant successions in the line of the ideas of Warming. He proposed a global system of description and interpretation of the dynamics of the vegetation, which, starting from a bare substrate, goes through a succession of intermediary stages and ends in a ultimate stage or *climax*. The term *climax* had already been used by another American ecologist, H.C. Cowles (1899), who showed that the vegetation modifies the edaphic conditions, which in turn influence the vegetation, and so on. However, Clements made the *climax* the central concept of ecological theory. It was not a matter of classifying plant groups but of understanding the dynamics that leads to the plant formation toward a state of equilibrium, the climax, determined by the regional climate. The course of nature is not erratic but is directed towards a stable state that can be determined precisely by science.

One original aspect of the ideas of Clements, which was the focus of several critics, was that he considered the plant community as a superorganism that is born, lives, and develops until maturity. Such communities are discrete entities presenting structural and functional characteristics that are specific to living things. He wrote:

"The unit of vegetation, the climax formation, is an organic entity. As an organism, the formation arises, grows, matures, and dies. Its response to the habitat is shown in processes or functions and in structures which are the record as well as the result of these functions. Furthermore, each climax formation is able to reproduce itself, repeating with essential fidelity the stages of its development. ... The climax formation is the adult organism, the fully developed community, of which all initial and medial stages are but stages of development. Succession is the process of the reproduction of a formation, and this reproduction process can no more fail to terminate in the adult form of vegetation than it can in the case of the individual plant" (Clements, 1916).

These ideas generate some enthusiasm but also give rise to controversy. American scientists such as Gleason (1926) questioned the metaphor comparing the plant community to an organism and suggested that plant associations are largely the consequence of chance. In his dissertation *The Individualistic Concept of the Plant Association*, published in 1926, he affirmed that plants effectively form associations, but that the associations are accidental groupings. They are too vaguely linked to each other for the whole to be compared to an organized being, so the concept of a succession of stages of development that must lead to a climax is rejected. The question of climax is still the basis of active controversies. According to the thesis of E.P. Odum (1971), the theory of successions has been at least as important for the development of ecology as the laws of Mendel have been for the beginnings of genetics.

2.1.5. Elton's pyramids

Between the two world wars, the study of the dynamics of animal populations developed relatively independently of the study of vegetation, in which specialists were interested mainly in the communities. The publication of C. Elton's work in 1927, *Animal Ecology*, marked an important stage in animal ecology. According to Worster (1977), Elton gave 20th century ecology its most important paradigm: the natural community viewed as a simplified economic system. He emphasized the structure and functioning of communities and introduced several principles describing the way in which, according to him, the economy of nature operates throughout the world.

While adapting the process of Clements with some caution to animal biology, Elton developed the notion of *niche*. In ecological language, the niche designates the result of evolutionary processes of differentiation and specialization. But Elton gave it an essentially functional sense. Niche designates the position of an organism in the community and its relationships to food and its enemies. In highlighting the economy of the community, he reduced the concept of niche in practice to food sources or to a type of food the animal consumes. Elton explains that in each natural community the niches, like the food chains, follow the same fundamental paths, the same basic plans. Each species takes its place in a grand general scheme of life.

On the basis of these principles, Elton summarized the structure of the animal community in the concepts, which later became classic, of the food chain and pyramid of numbers. In each community herbivores, carnivores, and detritivores can be found. Each species, as a function of its food habits, occupies a defined place in the food chain: the producers (all the plants), the first or second order consumers, and so on. In reality, Elton generalized to food chains the terms Thieneman introduced in 1926 (producer, consumer, reducer, and decomposer) to describe the roles played by agents of a specific ecological whole.

The higher one goes up the food chain, the fewer the individuals belonging to a trophic category. Carnivores, for example, are fewer in number than the herbivores they feed on and reproduce slowly to avoid risk of starvation. This reasoning is the basis of the notion of pyramid of numbers and pyramid of biomass: as we go up the degrees of scale, the biomass decreases, as does the number of individuals (see Chapter 11).

2.2. THE PRECURSORS OF THE ECOSYSTEM CONCEPT

Tansley formalized the concept of ecosystems, but he was not the first to put forth the idea that animal and plant communities along with their environment constitute the basic element of ecology. In fact, he profited greatly from discussions that raged in the ecological community. Haldane had already written in 1884 that "the individual organism and its environment had to be seen as a unity".

Mobius (1886, cited in Jax, 1998), who introduced the term *biocoenosis*, which in its original description was made up only of living organisms, later widened this definition:

"Biocenosis ... that is the living community, I call the whole of all effects of the dwelling area, which partly cause the properties and the number of specimens of a species developing there. These effects originate from the chemical and physical characteristics of the medium and also from other animals and plants that inhabit the same area."

Tansley was thus the inheritor of an entire series of studies and reflections. The terms *biome* and *plant group*, for example, can be considered precursors of *ecosystem* (Dury, 1999). The same can be said for the term *organism-environment complex*, which appeared regularly in French studies on ecology from the 1920s till the mid-1930s. In addition, even if he makes no explicit reference in his argument, Tansley was probably aware of the methodological and conceptual developments in aquatic ecology. It is effectively in this field that the notion of ecosystem proved to be operational well before the formulation of the concept.

2.2.1. Forbes and the microcosm

In 1887, the American entomologist S.A. Forbes presented an article titled "The lake as microcosm", in which he developed ideas similar to those of Mobius on the biocoenoses. He undertook in particular a detailed study of the interactions between the physical elements, vegetation, and fauna of a small lake in the American Midwest, arguing that the lake was an entity small enough to be an image of the human spirit (Deleage, 1992). Moreover, the organisms that populate a lake are sufficiently isolated from the terrestrial domain and sufficiently linked to one another to constitute a small world, a microcosm. Forbes emphasized the multiplicity of trophic relations between the species present while studying particularly the food regime of the black bass. He investigated the system of natural interactions that organized this simple combination of plants and animals into a stable and prosperous community.

Forbes describes a lake as "an old and relatively primitive system, isolated from its surroundings. Within it matter circulates, and controls operate to produce an equilibrium comparable with that in a similar area of land. In this microcosm nothing can be fully understood until its relationship to the whole is clearly seen. ... The lake appears as an organic system, a balance between building up and breaking down in which the struggle for existence and natural selection have produced an equilibrium, a 'community of interest', between predator and prey" (Golley, 1993).

In some sense, Forbes's view of the lake as an isolated system in which the cycles of matter maintain an equilibrium between the forces of production and the forces of decomposition has a very modern resonance. We may even suggest that these ideas, presented in 1887, nearly 50 years before Tansley developed the concept of ecosystem, go well beyond Tansley's theoretical developments. Why did Forbes not have a greater impact on the scientific community? Perhaps because he had published his works in an obscure scientific journal in Illinois. Perhaps also because his ideas found no echo in the highly influential botanical community of the time.

The philosophical conclusions that Forbes drew from his studies are of wide import. In comparing the lake and society, he explained that the well-adapted aquatic animal eventually defeats his poorly adapted counterpart as surely as the prosperous businessman eventually wipes out his less fortunate colleagues. From this Forbes concluded two ideas that could explain the order observed. The first is that of a "general community of interests" of different classes of living things in the lake: predators and prey are enemies only in appearance and the system in its entirety can be disturbed by any event affecting any one species. This community of interests functions to ensure the maximum productivity and stability of the whole. The second is that of the beneficial power of natural selection, which imposes an adjustment of rates of destruction and reproduction of each species such that the multiplication of predators does not lead to a disappearance of the prey on which their existence depends (Drouin, 1991).

In Europe, the Swiss scientist F.A. Forel adopted an approach similar to that of Forbes. He laid the foundations of *limnology* by establishing a research project on Lake Leman. In a book that summarized 20 years of work (Forel, 1892-1895), he devoted a great deal of attention not only to physical and chemical factors of the aquatic environment, but also to the analysis of living species and their reciprocal relations. In this monograph we can see the influence of the biocoenosis concept proposed by Mobius. Although Forel describes some food chains well enough, the fundamental character of photosynthesis as a physicochemical process of primary production nevertheless escapes him, so that he ignores an essential trait of the modern theory of ecosystems: primary production necessitates a constant energy input to balance the thermodynamic budget of ecological systems (Acot, 1988).

2.2.2. Thieneman and the *Lebenseinheiten*

In Germany, limnology was developing actively in the beginning of the 20th century. A. Thieneman and his colleagues considered lakes as global systems in which the parts interact to create the lacustrine system. In a work that appeared in 1925, Thieneman presented his overall vision of a lake as a system in which the biological component interacts with its environment. Following Mobius in this, he designated as *biocoenosis* the biological community and as *biotope* the environmental factors associated with the habitat of this biocoenosis:

"The development of life in the water is conceived as a limnological unit of higher order. These units come about through a mutual exchange between biotope and biocoenosis, both standing in functional relationships, one to the other."

Thieneman conceived the biotope and the biocoenosis as a fundamental ecological unit. However, like Clements, he considered that the biotope-biocoenosis entity was so closely overlapped that it could be considered an organism of a higher order (Thieneman and Kieffer, 1916).

"Each lake can be a living unity.... They are a microcosm, and therefore, this is the place to discover a higher structure of limnology, an organism of higher order whose organs interact in close mutual exchange" (Thieneman, 1925).

Thieneman did not use a particular term for these entities, generally calling them *biosystem* or *Lebenseinheiten* (life unit) (Thieneman and Kieffer, 1916). His formulation—environment and biocoenosis as a close, organic unit—was, in fact, a definition of ecosystem before that term existed (Jax, 1998).

2.2.3. Karl Friederichs and the holocoen

In 1927, the German entomologist Karl Friederichs developed the concept of *holocoen* (Friederichs, 1927). For Friederichs, the unit of nature was obvious. The biocoenosis (pond, marsh, beach, etc.) is a unit of life in the sense of Mobius, as a biological system made up of a network of relations and maintained actively by self-regulation. The biocoenosis with its biotope constitutes a unit of life of higher order that Friederichs called *holocoen*.

The term *holocoen* has seen little success, despite the many efforts of Friederichs. The term *ecosystem*, on the other hand, became the dominant concept. For various authors, however, the two terms are equivalent. Nevertheless, Jax (1998) considers that they have opposing origins. Schematically, in the framework of a debate on the academic question of whether the whole does or does not represent the simple sum of the parts, at the time of Friederichs and Thieneman, people widely used the metaphor popularized by Clements comparing the organization of an ecological system to that of an organism. These authors would thus be marked by an organic vision of nature, whereas Tansley proposed the term *ecosystem* in reaction to the organic approach of Clements. Tansley's is a theoretical approach, abstract, constructed by the observer, and applicable to units of different sizes as a function of the questions to be clarified, while the *holocoen* corresponds to a given ontological reality of nature, a conception fundamentally rejected by Tansley.

2.2.4. Sukachev and the biogeocoenosis

The term *biogeocoenosis* was proposed in the 1940s by the Soviet ecologist V.N. Sukachev to designate the interactive complex of organisms and their environment. Sukachev developed a theory of ecosystems similar to Tansley's, but he rejected the term *ecosystem* to emphasize the entity resulting from the interaction between biological and geological complexes. He later defined *biogeocoenosis* as "a combination on a specific area of the earth's surface of homogeneous natural phenomena (atmosphere, mineral strata, vegetable, animal, and microscopic life, soil and water conditions, possessing its own specific type of interaction of these components and a definite type of interchange of their matter and energy among themselves and with other natural phenomena, and representing an internally contradictory dialectical unity being in constant movement and development" (Sukachev and Dylis, 1964).

The concept of biogeocoenosis, like many other ideas originating from Soviet ecology, has not had much impact on the evolution of ecosystem theory. Problems of language as well as difficulties of contact between researchers in the Soviet Union and those in other countries prevented the development of exchanges that would undoubtedly have been fruitful.

2.3. TANSLEY AND THE ROOTS OF THE ECOSYSTEM CONCEPT

After World War I, the ecological geobotany of Warming as well as the plant successions of Clements and the Chicago school laid the bases of synecology, the central concept of which was that of community. In 1916, Clements introduced the concept of biome to designate the geographic entity and unit that constitutes a plant formation and the corresponding animal formation. Gradually, the idea that one could study interactions between species and/or between species and their environment took hold. Nevertheless, many questions persisted, particularly on the nature of relationships between organisms and the mechanisms responsible for the regulation and "equilibrium" of communities. The answer to these questions, as has generally been the case for the development of natural sciences and ecology, had an economic importance and was looked forward to by managers concerned with pest and vermin control (very active at the time in agriculture), exploitation of wild populations, improvement of productivity in agrosystems, and so on.

Moreover, the role of the abiotic environment was variously taken into account till then by naturalists to explain the organization and dynamics of communities, included in Clements' theory of successions. It was certainly acknowledged that living organisms responded to environmental factors in terms of their behaviour and physiological reactions. But the physicochemical environment was often considered a secondary factor, "a stage on which the biota acted a drama", to repeat an expression of Golley (1993). Moreover, there were not many efforts till then to oppose the progress of ecology of communities to that of chemists and geochemists such as Vernadsky (see Chapter 14). Finally, the period between the two world wars saw the emergence of what some called the new ecology, based on thermodynamics and information theory, and modern economy.

The term *ecosystem* was introduced by Tansley in 1935 to name a holistic ecological concept that, in a single system, combined living organisms and their physical environment (the factors of the habitat in the wider sense). In reality, Tansley was looking for a compromise in a controversy that split the specialists of the ecology of plant communities into two camps: those who held the ideas of Clements according to which the plant community is a complex superorganism that is born, develops, matures, and then becomes senescent (Clements, 1916), and those who followed the ideas defended by Gleason, who refuted the reality of plant groups, considering that communities were simply collections of individuals that were together by coincidence, and because they each found, in this place, the possibility of satisfying their needs in terms of habitat.

Tansley did not follow Gleason's ideas but criticized the analogy between community and organism proposed by Clements. Moreover, it became apparent in the late 1930s that Clements had exaggerated the role of climate in the maturation of plant formations and that the system he proposed was much too rigid. Besides, the South African Phillips, one of Clements' supporters and an enthusiastic believer in holism, took the idea to the point of declaring that animal and plant communities were not only similar to an organism, but were a form of organism (Phillips, 1934).

Tansley's reaction to the series of articles by Phillips was rapid and incisive, refuting notably the idea that the community was an organism and denouncing the dogmatic approach put forward by Phillips. While accepting holism and other organic theories as philosophies, he rejected their application to biology. In order to avoid vague analogies with an organism, Tansley thus forged the notion of ecosystem as a new theoretical model of natural organization. He suggested that the notion of biotic community including just living things was artificial, and he proposed that one should speak rather of the *ecological system comprising the set of living organisms and the set of physical factors of the environment, otherwise known as ecosystem*. This idea was not free from the influence of physical science, which talked from the beginning of the 20th century of the fields of energy and systems that allowed us to take natural phenomena into account. Moreover, according to Tansley, the ecosystem was an element in the hierarchy of physical systems, from the universe to the atom. There are large and small ecosystems crossed by energy flows, so that all nature is found in an ordered whole. In the ecosystem, the relationships between the different organisms can be described as material exchanges of energy and chemical substances such as water and the nutrients that are the constituents of nourishment.

As emphasized by Worster (1977), Tansley's ecosystem was admirably convenient to agronomic and industrial concepts of nature as a storehouse of exploitable raw materials. It was a mechanistic concept that was entirely new in the history of the science of ecology, a result of modern physics and not of biology.

Tansley emphasized the fact that ecosystems are not raw givens of nature but the product of a mental creation that gives us the possibility of isolating them from the rest of the universe. It is a whole constructed by the relations that prevail between living species and the physical habitat in which they develop. It is thus an abstract reality formed mainly from elements that are themselves concrete (Dury, 1999). Tansley also considered that humanity is a biotic factor and that human activities are an important factor the analysis of which is drawn from the science of ecology.

Tansley's definition of ecosystem (1935)

"Clements' earlier term '*biome*' for the whole complex of organisms inhabiting a given region is unobjectionable, and for some purposes convenient. But the more fundamental conception is, as it seems to me, the whole system (in the sense of physics), including not only the organism-complex but also the complex of physical factors forming what we call the environment of the biome—the habitat factors in the widest sense. Though the organisms may claim our primary interest, when we are trying to think fundamentally we cannot separate them from their spatial environment, with which they form one physical system.

"It is the systems so formed which, from the point of view of the ecologist, are the basic units of nature on the face of the earth. Our natural human prejudices force us to consider the organisms as the most important parts of these systems, but certainly the inorganic 'factors' are also part—there could be no system without them, and there is constant interchange of the most various kinds within each system, not only between the organisms but between the organic and the inorganic. These *ecosystems*, as we may call them, are of the most various kind and sizes. They form one category of the multitudinous physical systems of the universe, which range from the universe as a whole down to the atom. The whole method of science ... is to isolate systems mentally for the purpose of study so that the series of isolates we make become the actual objects of our study, whether the isolate be a solar system, a planet, a climatic region, a plant or animal community, an individual organism, an organic molecule or an atom. Actually the systems we isolate mentally are not only included as parts of larger ones, but they also overlap, interlock and interact with one another. The isolation is partly artificial, but it is the only possible way we can proceed."

2.4. IMPLEMENTATION OF THE ECOSYSTEM CONCEPT: THE TROPHO-DYNAMICS APPROACH

Ecology as it was understood in the years 1930-1940 inherited two major currents of thought (Worster, 1977).

One, thermodynamics and systems theory developed in physics. In taking a unit of energy, the calorie, as a reference of the activity of biological systems, it is possible to follow a mechanistic approach to the ecosystem.

Two, economic principles to the extent that the concept of energy forged a link with the field of economy. Ecology became the science of natural economy and borrowed its vocabulary from this discipline (productivity, efficiency, yield, and so on) (see Chapter 11). This highly organized economic system was structured, with producers and consumers, networks, retroactive loops, and cycles of transfer of matter and energy. Ultimately, order rules in nature to the greater benefit of humanity. This ecology strongly coloured with bioeconomy is closely linked to the prevailing cultural context. In it we find the influence of the progressivist philosophy of management of natural resources. Because of this, the new ecology provided analytical tools for better exploitation of the earth's resources.

The seminal article by R. Lindeman published posthumously in 1942 marked a major stage in the history of ecology. Lindeman attempted to promote a tropho-dynamic approach, using as a unifying principle the quantification of energy exchanges between all the biotic elements in interaction. Till then energy flows were mostly used to comprehend physiological functioning and metabolism. The ecosystem now appears as an exemplary unity of energy exchanges in nature. But the fundamental notion for Lindeman was the trophic cycle, which links primary producers to consumers and decomposers, thus ensuring the circulation of matter in the ecosystem. Nutrients are recycled within the ecosystem, which ensures a functional autonomy. In this sense, Lindeman joined the biogeochemical approach of Verdnasky. Within about ten years, this paradigm had invaded ecological literature.

As is often the case in ecology, Lindeman's ideas were not entirely original. They had already been advanced by Russian researchers in the early 1930s, but had not received much notice (Golley, 1993). For example, according to Vladimir Stanchinskii, professor at the University of Smolensk, the organisms that make up the communities are constantly changing, recycling matter and energy, and are connected by exchanges of matter and energy (Weiner, 1988). Each species plays a specific role (physicochemical and biochemical) in the community. The dynamic equilibrium of each biocoenosis results from the existence of defined relationships between heterotrophic and autotrophic components, between herbivores and carnivores. The idea of exchanges of matter and energy between the living organism and physical nature was inspired by Vernadsky, but Stanchinskii considered reducing biological phenomena to a common denominator, energy, and expressing their relationships in the form of mathematical equations describing the dynamic equilibrium of the community. From 1931 onward, he made further studies on a simple mathematical model of energy flow in a community comprising primary producers and a heterotroph. The career of Stanchinskii was interrupted for political reasons in 1934 and a book that summarized his tropho-dynamic studies on communities of the steppes was destroyed. That marked the end of research in this field in the USSR.

2.4.1. Matter and energy cycles

The development of ecosystem analysis marks a virtual rupture between ecosystem ecology and other biological disciplines. It allows the integration of biotic and abiotic components by quantifying the processes by the measurement of flow. It involves circulation, transformation, and accumulation of energy and matter across living organisms and their activities.

The publication of *Fundamentals of Ecology* by E.P. Odum is another major stage in the history of ecology. Centred around the theory of Lindeman and driven by the ecosystem principle, this work developed fundamental concepts about energy. The author affirmed:

"Energy is defined as the ability to do work. The behavior of energy is described by the following laws. The first law of thermodynamics states that energy may be transformed from one type into another but is never created or destroyed. ... The second law of thermodynamics may be stated in several ways, including the following: no process involving an energy transformation will spontaneously occur unless there is a degradation of the energy from a concentrated form into a dispersed form."

Odum thus applies the energy theories to food chains, pyramids of biomass, and productivity. The approach to ecosystem functioning by flows of energy thus results in an elaborate theory. This ecosystem theory would be taken up in similar forms by Duvigneaud (1962) and Margalef (1968).

From the 1950s onward, trophic studies became common and the energy approach was formalized by another American ecologist, Howard T. Odum,

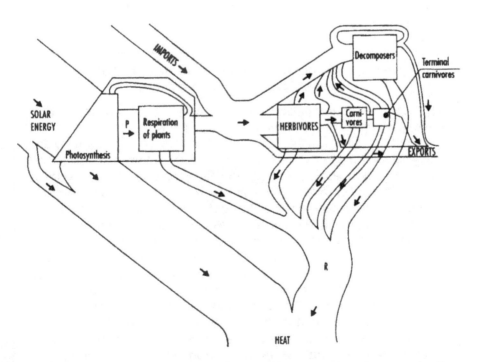

Fig. 2.1. Diagram of energy flows in natural communities proposed by H.T. Odum for Silver Springs, Florida (USA). In this diagram, P symbolizes the raw primary production, R is the respiration of the community. The trophic levels correspond to the square boxes, and the flow is approximately proportionate to the size of the network branches.

brother of E.P. Odum. This author, a specialist in nuclear chemistry, traced the first energy diagrams (Fig. 2.1), which presented a close analogy with diagrams of electric circuits, during a study on the structure and trophic dynamics of Silver Springs, Florida (USA) (Odum, 1957).

With improvement in analytical methods, such as the first absorption spectrophotometers and then autoanalysers, ecologists also developed research on the circulation of minerals in the ecosystems. In the 1960s, some of the research on ecosystems was thus directed toward biogeochemistry. The American ecologists H. Bormann and G. Likens became famous for the studies they made on Hubbard Brook, a small forest watershed in New Hampshire (USA) (Bormann and Likens, 1967 and 1979). The study of this ecosystem as an object clearly demarcated in a hydrological context was focused on the cycle of nutrients. The authors developed a conceptual model in which the ecosystem was an element of the biosphere with which the organic and inorganic elements were in interaction. They recognized four major compartments: the *atmosphere*, the *stock of nutrients* available in the soil, the *quantity of nutrients* available in living or non-living material, and *minerals in the soil and rocks*.

From this information, the authors established a budget of nutrients of the small watershed. A basic result of this work was that the processes were closely coupled to the hydrological cycle. But they also indicated that the forest that occupied a large part of the watershed functioned as a filter or buffer system for many chemical elements, notably the pollutants that enter the system from the atmosphere.

Ecosystems: from operational research in the United States to the expression of German romanticism

Kurt Jax (1998) investigated the reasons why the study of ecosystems developed rapidly in the United Sates after the works of Lindeman and then those of Hutchinson and the Odum brothers, while in Germany and elsewhere in Europe it developed only after the International Biological Programme was set up. Jax suggested that the concept of ecosystem as it was developed by Lindeman was operational and helped launch research on the quantitative functioning of the ecosystem. In contrast, the German researchers, including Thieneman and Friederichs, did not really consider the use of flows to understand the functioning of ecosystems, although this progress was not explicitly excluded from their research. Thieneman, who was in fact interested in the productivity of ecosystems, even doubted that it was possible to quantify the productivity of a lake (Schneller, 1993). The *holocoen* concept of Friederichs and the *Lebenseinheiten* concept of Thieneman had an undeniable heuristic value, but they were not appropriated as a starting point for new scientific approaches or for the establishment of operational research programmes. It was a mode of perception and understanding of nature, in the context of the German philosophy of nature (and of romanticism), rather than an approach to research on an explanation of nature.

2.4.2. Biological production

Forbes had already suggested that the lake system was characterized by the circulation of matter, production, and decomposition of living organisms. A preferred approach to the functioning of the ecosystem was the study of productivity, a very important dimension for ecologists working in targeted domains. Thieneman was one of the first to attempt to define the terms *production* and *productivity*. He distinguished the quantity of organic matter produced per unit of area (called *production* henceforth) from the quantity produced per unit of time and area or volume (called *productivity*). Each of these quantities is a measure of the production but represents a different process.

The study of productivity of ecosystems has encouraged the quantitative description of transfers across food chains, as formalized by Elton in particular. For this, we must determine the number and/or the biomass of organisms present in the environment studied, their food requirements, and their rate of consumption and evaluate their rate of growth, mortality, and production, as well as their metabolism.

This type of approach based on the study of population dynamics requires a significant amount of data and data processing that would have been difficult before the widespread use of computers. Here, as suggested by Elton, the organisms are categorized as producers, consumers (herbivores and carnivores), decomposers, and so on, and the overall exchanges between these major compartments of the ecosystem are calculated.

Biological production: the influence of the economy

The concepts developed about biological production, as well as the terms used (productivity, yield, etc.), clearly evoke those of industrial productivity and economy. This is not surprising: *economy* and *ecology* have had an obvious etymological relationship since Haeckel defined ecology as the economy of nature and since he forged the term *ecology* in reference to the term *economy*. Besides, the development of ecology is nearly contemporaneous with that of economy. Faced with complex systems comprising a large number of elements in interaction, both sciences investigate the emergence of order from disorder.

In the 1930s, there was an extension of the field of economic analysis and an implicit export of its methodology, theories, and tools to the science of ecology. Then, in the 1960s, the use of economic models in the field of theoretical ecology intensified. Ecologists spoke of producer and consumer, production and productivity, and maximizing fitness. The kilocalorie was the monetary unit of ecological systems where the energy yield is established in principle. Ecosystem ecology, developed by the Odum brothers, was particularly marked by such economic references (Vivien, 1997).

2.4.3. The International Biological Programme

In the mid-1960s, the study of ecosystems was popular in ecology. A still poorly organized set of theories was available, notably from the studies of the Odum brothers, Margalef, Slobodkin, and other researchers. All these ecologists, in reality, had tackled the study of ecosystems with different sensitivities, reasoning often by analogy with physical, chemical, and biological systems. Some favoured thermodynamics or the information theory and others the theory of evolution and/or natural history.

In 1965, there was no unified theory of ecosystems, in the absence of a coherent and organized set of information (Golley, 1993). It was thus suggested that systems theory be applied to develop models that would organize the information.

The International Biological Programme (IBP), established from 1967 to 1972 under the aegis of the International Committee of Scientific Unions, aimed to quantify the flows of matter and energy in a large variety of natural ecosystems around the world. The IBP homed in on the biological production of ecosystems as a process necessary to the well-being of humanity. This concept of productivity could interest theoretical ecologists as well as scientists working in the domain of agronomy, forest management, and fishery management. The IBP initiated the elaboration of ecosystem models and the simultaneous development of empirical field studies. Nevertheless, taking into account the limited performance of the information technology available at the time, most of these models remained elegant descriptions of ecosystems with a limited predictive capacity.

The programme was criticized for various reasons, including the fact that human activities were not sufficiently taken into account. In any case, it had significant repercussions and mobilized a large number of ecologists, who accumulated a considerable amount of data, implemented methodologies, and created interesting comparative approaches. Relying on the IBP, but considering man this time as a component of ecosystems, UNESCO proposed the Man and Biosphere programme in 1971, and it took effect from 1974.

2.5. THE ECOSYSTEM AS AN OBJECT OF RESEARCH

In the view of Tansley and other ecologists who followed, the ecosystem was a functional theoretical concept that served to organize thought and research, but not a natural object. However, this concept, abstract in origin, would evolve towards recognition as a concrete object that could be clearly identified in spatial terms.

A lake, a pond, a cultivated field, are entities that are quite well individualized, visually perceptible. The study of the ecosystem as an object would naturally be favoured. The ecosystem would eventually be broken down into sub-systems in which behaviours and linkages could be studied. In theory, a mechanistic representation of its functioning can be proposed on the basis of physical, chemical, and biological principles.

Generally, ecologists have adopted the definition of the ecosystem as an object of research. Bourliere and Lamotte (1978), for example, define the ecosystem as a fragment of the biosphere that can be considered a relatively autonomous entity in relation to the neighbouring ecosystems and in which the structure and functioning can be analysed—this is the essential value of the concept. This concept of ecosystem is useful to eventual objectives of research, to the extent that managers have a good grasp of physical objects such as a watershed, a lake, and a river.

Similarly, according to Likens (1995), an ecosystem is "a spatially explicit unit of the earth that includes all of the organisms, along with all components of the abiotic environment within its boundaries." According to this definition, the ecosystem is a delimited geographic site.

It is interesting to observe that the notion of an ecosystem "without dimension" has remained active among geographers. According to Tricart (1982), the concept of ecosystem in itself is not spatialized. An ecosystem is, above all, a tissue of flows that determine its structure. It is represented by an organigram, not a map.

According to Blanc-Pamard (1982), ecology is the only successful analysis of natural facts but the concept of ecosystem remains very limited. It has no spatial, historical, or dynamic dimension. Moreover, an ecosystem cannot be mapped. It does not take into account the history that records the spatial support, or the geomorphogenesis, or even problems of erosion.

It is possible that the geographers who have supported the concept of geosystem have perceived the ecosystem as a more fluid, even concurrent notion.

The issue of spatialization of the ecosystem has given rise to debate (see, for example, Blandin and Lamotte, 1988), particularly in the domain of operational research, because it is sometimes difficult to delimit the ecosystem. On the one hand, without spatial limits, we cannot identify objects or ecological entities. On the other, the limits of an ecosystem depend largely on the objectives pursued. The watershed, for example, is a spatial unit that is very useful for the study of water and nutrient flows, but it is not the most appropriate unit for the study and management of carnivorous or herbivorous vertebrates.

In reality, the limits of an ecosystem are most often defined operationally as a function of the processes studied or the questions posed. It is natural that the frontiers of an ecosystem would be fluid and ecologists have taken into account what systemicians call the "edge effects". There are transition zones between adjacent ecological systems called *ecotones*. In such zones, it is the interfaces having their own identity that must be studied (Chapter 9).

2.6. IS THE ECOSYSTEM CONCEPT A PRODUCT OF WESTERN SCIENCE?

The scientific concept of ecosystem, as it has emerged since World War II, was in effect 'a machine theory applied to nature" (Golley, 1993). Like physical theories, the dynamics of natural ecosystems is simplified and supposed to be deterministic. The ecosystem is compared to a machine and represented by a computer model. Processes such as the biogeochemical cycles are defined as mechanisms activated by solar energy following, in this, the mechanistic principles of Newton.

Is this concept of ecosystem purely a creation of western science or does it exist in more or less similar forms in different cultural systems? If yes, what can we learn from the experience acquired by other societies?

Colonial origins of ecosystem ecology

According to Grove (1999), the origins of ecosystem ecology can be traced to the 17th century. In various places in the colonial world, the impact of human activities on the environment had been observed for a long time. At Cape Town, at Saint Helene, and in North America, plantations had caused soil erosion and gullying, while deforestation was the cause of floods as well as drying out of rivers and streams. Severe soil erosion due to grazing was observed from the 16th century in Mexico and from the 17th century in the Caribbean Islands and Saint Helene. From the end of the 17th century, the debate on desertification had grown in the context of agricultural development of the islands.

In 1766, the nomination of Pierre Poivre to the post of intendant of the island of Mauritius stimulated French colonial ecology. Poivre was interested in soil conditions, the local climate, and variations in rainfall caused by deforestation. In 1769 and 1776, two laws established a forest reserve on the basis of climatic arguments, and the State simultaneously launched programmes to plant trees in order to prevent erosion and increase rainfall.

The ecological programmes established in Mauritius were the source of later initiatives in the British and French colonies. The King's Hill Forest Act voted into effect on the island of Saint Vincent relied on an ecological hypothesis: a decline in precipitation was due to deforestation. This theory of desertification would generate a coherent intellectual tradition according to which there were climatic changes caused by man.

In reality, it has been discovered that in ancient societies and/or cultures other than European ones there have existed concepts more or less similar to that of the ecosystem (Berkes et al., 1998). Traditional wisdom about ecological systems combined ecology, ethics, and culture in the context of a world view in which humanity was an integral part of nature. Generally, these representations shared two essential characteristics: the entity was often defined by geographic limits, and the biotic components of this entity (plants, animals, and humans) were in interaction. The notion of ecosystems was often based on the watershed to which certain human societies belonged, probably underlining their dependence on local resources. The notion of watershed probably goes far back in human history; it existed among the ancient Greeks, and it was known in Japanese and Turkish societies (Gadgil and Berkes, 1991). The sultan Mehmed

II took measures to conserve watersheds in 1453 after the seizure of Constantinople, prohibiting the felling of trees and overgrazing in the river basin that supplied the city.

Although the concept of ecosystem existed in one form or another in traditional systems of knowledge, the language used is different from scientific language and is often metaphorical. In some cases, the traditional representations of ecosystems also incorporate the spirits of animals or those of ancestors (Berkes et al., 1998). The spiritual dimension of these ecological perceptions has only a limited interest for scientists. But many societies have successfully managed their resources over a long time, thanks to ritual and/or religious representations of resource management. The important point is not the religion itself, but the use of symbols having a strong emotional and cultural power to set up moral codes and modes of management (Berkes et al., 1998). This is to say that movements combining ethical values and beliefs with ecological concepts have a greater chance of succeeding in the operational sense than the science of ecology considered in isolation.

When all is said and done, the spatialized concept of ecosystem and the existence of relationships between the physical and biological components in this ecosystem are notions shared in various forms by many human societies. In their traditional wisdom, humans have almost always been an integral part of the system, and the spiritual dimension is often prominent. The mechanistic vision of the ecosystem, which originated from physics, remains essentially a product of western science. Nevertheless, this observation clarifies the recurring debates between ecologists and specialists in human sciences and social sciences. Social scientists are similar to ecologists in not sufficiently taking humanity into account in their research, or even in deliberately excluding it in studies on non-anthropized systems. In these critics of ecosystem ecology, we can see the impact of the popular perception of an anthropocentric ecosystem. But there may also be a confusion of genres. Ecologists have the right to develop their paradigms around everyday systems in which man is only one agent among many. Ecosystem ecology, just like earth sciences or chemical sciences, has vindicated its status as a cognitive science independent of its possible social implications.

Chapter 3

Approaches and Paradigms of Ecosystem Ecology

The only thing that really interests me is to know whether or not God had a choice in creating the world.

Einstein in Stengers and Schlanger (1991)

Haldane (1963) notes that the process of validating a hypothesis goes through four successive phases:

> *—this is worthless nonsense;*
>
> *—this is an interesting but perverse point of view;*
>
> *—this is true but quite unimportant;*
>
> *—I have always said so.*

Like any natural science, ecosystem ecology aims to describe an aspect of the universe and its functioning. To do this, it must put into operation a set of theories whose validity is judged on their capacity to provide the most faithful representation possible of the real world. In science, nothing has meaning without theory (Wilson, 2000). We look for ecological theories that allow us to make testable and reliable predictions about natural phenomena such as species abundance and distribution, biogeochemical cycles, or global changes.

It is difficult to understand the genesis of concepts and theories in ecosystem ecology without placing them in a more general context. How is scientific knowledge constructed? How do we elaborate explanatory hypotheses and how do we verify them? And more generally, how do we compress the wide range of phenomena observed into a small number of explanatory principles? From the history of science, we can see the advances, experiments, errors, the weight of symbols and intuition, and the influence of ideologies in the elaboration of concepts and scientific theories. History shows us that the reasoning of science is provisional. It evolves and refines itself according to increasing knowledge and technological development. Epistemology, or critical reflection on science, forces us to evaluate the meaning in it, to extract from it the meaning of all human practices. Etymologically, epistemology signifies the theory of science, but it is given a slightly wider meaning of theory of knowledge. We are thus led to question ourselves on the coherence of principles that govern scientific discourse, the adjustment of methods to the object studied, and the basis of interpretations of observations and measurements.

3.1. SCIENCE AND NON-SCIENCE: WHERE IS THE DIVIDING LINE?

According to the general opinion, scientific theories are elaborated with rigour, leaving no room for subjective approaches. But how do we distinguish science from non-science? From facts, science calls on our imagination to organize information and propose theories. The creative spirit, according to Schopenhauer, consists of looking at something everyone has seen and imagining something that no one has yet imagined. However, compared to other activities such as fiction or poetry, science brings to the creative process the rigour of methodology with the possibility of testing and, possibly, refuting its propositions.

A scientific theory is a theory that can be experimentally refuted, according to K. Popper. This refutability is the criterion of demarcation between science and belief or metaphysics. The fate of every theory is to be replaced by another in a long process of progressive increase of knowledge. But all scientific disciplines do not function in the same way. By reference to the physical sciences, for example, some have denounced the lack of rigour in the formulation of hypotheses as well as the poor predictive capacity of theories in ecology (di Castri and Hadley, 1986). Are these criticisms relevant, or do they result from a misunderstanding of the nature of ecological sciences? It is also known that ecologists are not insensitive to philosophical debates that animate society, which may reorient their ideas.

Scientific research is also an art, according to Wilson (2000). It matters little how a discovery is made. Only its truth and its validation count. The ideal scientist thinks like a poet and works like a librarian. And he writes like a journalist. Like an artist before a blank canvas, or a romantic who with closed eyes relives past emotions, he mobilizes his imagination to find subjects as well as conclusions, questions as well as responses (Wilson, 2000).

It is thus particularly important to clarify thoroughly the processes that lead to the elaboration of theories and concepts, and more generally to the construction of scientific knowledge in ecosystem ecology, more so as the conventional representation of science has taken a beating. From accepting the single mechanistic vision of the world dear to physics, and the existence of intrinsic laws of nature, we have begun to understand that the history of the living world is marked by creativity, innovation, and the emergence of new forms and processes, in relation with the capacity to react to changes and uncertainty.

In the absence of a theoretical formalization of their works, ecologists, more than other researchers, risk stumbling over what Bachelard has called epistemological obstacles—and getting stuck in them. Among these obstacles, three are associated with fascination: fascination for the object, for the tool, and for the word or concept.

- In *fascination for the object*, ecologists lose themselves in the increasingly detailed description of an ecological system that is poorly defined or undefined, instead of a precise problem. For example, there is sometimes an avalanche of figures and measurements in quantitative ecology or a terminological inflation in phytosociology.

- In *fascination for the tool*, the mind turns from the object and loses perspective on the problem studied—generally because, here also, the problem has not been formulated or has been incorrectly formulated. In ecology, this perversion is found frequently with respect to mathematical or statistical techniques.
- In *fascination for the concept*, the working hypothesis stands on dogma—and the selective appreciation of facts always allows one to conform to that dogma; the evocative word become a fetish (e.g., theory of catastrophes, adaptive strategy) and the objective gives way to a mania for classification (Houches Seminar, 1981).

The objective of the *scientific method* is to locate the subjective process of elaboration of new ideas within a logical framework of challenging and questioning. An *objective knowledge* must be produced that has been criticized and thrashed out among scientists. Such knowledge is not subjective since it is shared by several persons after a process of validation. It corresponds not only to absolute knowledge but also to the most probable shared knowledge on the basis of the principles that are available.

3.2. A GOAL: THE SEARCH FOR ORDER

According to dogmatic theory, scientific knowledge establishes the truth as an appropriateness of thought to the object: science represents reality as it is. But this theory is also criticized: science is not the knowledge of reality but a set of hypothetical propositions such that the scientific truth must be thought of in terms of probability. In any case, the object of science is to discover the relationships of objects with one another, discover the laws of organization of the world, and explain the phenomena that we observe (Pickett et al., 1994). Science is the organized and systematic enterprise of assembling knowledge of the world and condensing it into laws and principles that can be tested (Wilson, 2000).

We must go back to Plato and the Pythagorans (4th c. BCE) to find the fundamentals of "cosmic religion", which considered the world we live in as a reflection of divine reason (Thuillier, 1991). For the ancient Greeks, the state of the cosmos was the state in which everything was in its place. From an initial magma, chaos, in which absolute disorder reigned, a divine being, the Demiurge, created a world that was mathematically ordered. The Demiurge instituted everywhere the reign of geometric forms and proportions. The work of man was to discover the structures that served as a model for the Demiurge. However, according to Plato, the Demiurge always had some difficulties in imposing order everywhere, so that randomness was not excluded from this world.

Science for a long time was Platonic. To have access to knowledge, it was necessary to discover in this world, ordered and at the same time random, the mathematical forms that God imposed on matter. The task of science was to find the harmonies hidden behind an apparent disorder. It was at the end of the Renaissance that those theories of knowledge were established to which scientists still refer to understand the directions and implications of their work. It was also at the beginning of the 17th century that a grid of mathematical analysis was placed on the world, as well as an explanation of inert or living nature by direct measurement and experimentation in the framework of theories that borrowed their methods from mathematical studies. Theory was established as a medium of knowledge at the cost of the senses and imagination.

Francis Bacon was one of the first to attempt to formulate what became the method of modern science. He claimed that science aims to improve the life of man on earth and methodically combines facts from which it develops theories. This process was subsequently improved by some and argued against by others. But it was during the 20th century that the Austrian philosopher Karl Popper really advanced the debate surrounding the validation of scientific theories by affirming that any theory must be subjected to the proof of refutability.

3.2.1. Understanding the natural world

Generally, science aims to:

— organize knowledge systematically by actively discovering relationships between structures, phenomena, and processes and
— propose explanatory hypotheses on conditions in which certain events are produced, hypotheses that can be tested, i.e., confirmed or disproved (May, 1992).

Very often, the ecological sciences tend to reduce the immense diversity of natural phenomena to a small number of explanatory principles. The generalizations, when they are possible, are nearly always of a probabilistic nature.

Pickett et al. (1994) have said that the comprehension of the natural world is in a way the result of a confrontation between reality and theory, between what researchers observe and what they think and imagine: "understanding is an objectively determined, empirical match between some set of confirmable, observable phenomena in the natural world and a conceptual construct." The result of this confrontation can evolve according to the advancement of knowledge, whether in the domain of facts (e.g., implementation of new methodologies) or conceptual approaches. To establish the dialogue between the facts observed and the concepts, it is necessary to use methods and tools.

• Attempts to *generalize* are part of the dialogue between reality and the conceptual approach. For example, when we study the relations between species richness and the area of ecosystems, we need to assemble a large amount of information on the composition of populations. But any generalization involves a simplification and some form of abstraction in order to identify the essence of the phenomenon studied. Thus, in the relationship between area and species richness, emphasis is laid on the parameters of the equation and the form of regression curves but all the characteristics of the system that are not essential for this generalization are ignored. For example, we are not interested in the nature of the species present, only in their number.

- The *search for causes* consists of determining the processes, mechanisms, interactions, and conditions that are the source of *patterns* or phenomena observed. In ecology, the search for causes may appeal to a large variety of domains, including biochemistry and biogeochemistry, physiology, genetics, behaviour, and evolution. Apart from this, according to systems theory, the cause-effect relations are not just linear, since many types of feedback, positive or negative, govern the dynamics of ecological systems.

- *Validation* is another aspect of the dialogue between concepts and observed facts. It basically involves examination of the validity of a hypothesis, concept, or generalization by asking whether it closely represents the real world. Various means are available to test these conceptual approaches: for example, experimentation, which is the manipulation of a system, comparison and probability studies to indicate similarities between supposedly identical systems, and correlation, which is a statistical relation between the measurements of two properties of an ecological system. If the test is positive, the hypothesis tested is confirmed. If it is negative, it is proved that the hypothesis tested is false or poorly formulated. The vast set of all the hypotheses and theories that can be proved false constitutes scientific knowledge (Peters, 1991).

- Finally, ecological science must be able not only to explain but also to give us the means to predict the most probable events. This capacity of *prediction* distinguishes science, in the minds of some, from other activities such as logic, which allow us to relate causes and effects to identify what is possible, but do not allow us to choose the most likely solutions. What really distinguishes scientific activity is the elaboration of hypotheses and theories that have a predictive capacity and can be tested on the basis of observed facts or experimentation.

3.2.2. Concepts, hypotheses, and theories

When ecologists try to explain nature and its functioning, they propose generalizations often in the form of concepts or hypotheses, which serve especially to isolate a probable event from the set of all possible events.

Concepts are often constructed from many observations and constitute an abstract generalization of regularly observed phenomena (Pickett et al., 1994). Concepts may refer to objects, classes of phenomena, or relationships. Species, for example, is a concept, as is ecosystem, the food web, or competition (Table 3.1). In ecology, most advances seem to result from the introduction of new concepts or improvement of existing concepts rather than from the discovery of new facts, although the two processes are complementary.

The differences between theories and hypotheses are more subtle, since the word *theory* has many meanings. It is difficult to distinguish between a hypothesis and a theory, according to Serres and Farouki (1997), except to say that a theory is generally made up of a set of hypotheses. In practice, a theory corresponds to a wider vision of the world than a hypothesis, which has a more modest import. A hypothesis is a scenario that is proposed from observation to account for a phenomenon. It is adopted provisionally as a directive idea in the search for facts. It is then tested by experimentation to confirm or invalidate it.

Table 3.1. The 15 ecological concepts considered the most important by the members of the British Ecological Society in the late 1980s (Cherett, 1989)

1.	Ecosystem
2.	Succession
3.	Energy flow
4.	Conservation of resources
5.	Competition
6.	Niche
7.	Matter cycles
8.	Community
9.	Biological traits
10.	Fragility of the ecosystem
11.	Food chains
12.	Adaptation
13.	Heterogeneity of the environment
14.	Species diversity
15.	Density-dependent regulation

Definitions suggested by Ford

In an attempt to clarify the terms used, Ford (2000) proposed the following definitions, highlighting the confusion generated by interchangeable use of *hypothesis* and other terms.

• *Axiom*: Undemonstrable proposition that is supposed to be exact on the basis of earlier observations and results. An axiom may indicate that a fact will be found or not, or that a thing influences or does not influence another thing. For example: mature salmon migrate from the ocean toward fresh water to reproduce.

• *Postulate*: Conjecture, new or unexplored idea founded on probabilities but needing to be confirmed or verified. For example: salmon reproduce in the river they were born in.

• *Hypothesis*: This term is used when an affirmation or postulate must be tested specifically. One or several hypotheses may be made to test the postulate.

• *Theory*: Logical construction comprising a set of propositions, some of which are axioms and some of which are postulates. Theories are presented as explanations of an observed phenomenon or a question posed, and they may have a predictive capacity.

The word *theory* seems to have different meanings in biology and in physics. A theory is an affirmation about any natural phenomenon that can be described verbally or mathematically and that can be challenged by reality, which may refute it. In physics, the mathematical language is preferred, and the confrontation between theory and reality is settled by resolving equations. In biology and ecology, theories are expressed most often in verbal form and in less precise terms, and the reconciliation with facts is much more delicate. Darwin's ideas on evolution are called a theory, even though they are only a verbal characterization of some general observations (Bak, 1999). In reality, theory in ecology is a system of thought elaborated from a set of abstract constructions, whether they are facts or concepts, that in some way proposes explanations for observed phenomena. Among the many examples of theory in ecosystem ecology are the theory of niches, the theory of successions, and the island biogeography theory.

3.2.3. Paradigms

A paradigm is a set of hypotheses and theories, as well as the methods and techniques need to apply them, on which all science is based. Paradigms govern the manner in which scientists think and interpret the results of their experiences. This is what Kuhn (1983) has called the *disciplinary matrix*. Scientists within a paradigm have a proven methodological arsenal at their disposal to tackle new problems and try to resolve the theoretical or experimental enigmas in their field, under the aegis of rules adopted within the framework of the paradigm. They are, in a way, the conservators of acquired knowledge. The enigmas they cannot resolve are anomalies. If the anomalies are numerous or serious, they may challenge the very bases of the paradigm. A crisis therefore develops that could lead to the appearance of a new paradigm, which wins the support of an increasing number of scientists till the older paradigm is abandoned. The new paradigm adopted by the scientific community will serve as a reference until it, in turn, faces difficulties that generate a new crisis (Chalmers, 1987). Science is therefore not a gradual and cumulative process of advancement of knowledge. There are long "paradigmatic" periods, with revolutionary personalities from time to time who wake science from its numbness and implant a new awareness (Lazlo, 1999).

Two principal paradigms are usually recognized in ecology: the ecology of populations and communities, and the ecology of ecosystems and landscapes. Each has developed its own concepts, theories, and methods (see Chapter 1). We can also identify other paradigms in ecology, especially those of "equilibrium" and "non-equilibrium" (Pickett et al., 1992). In brief, the paradigm of equilibrium (Simberloff, 1982) emphasizes the equilibrium point of ecosystems and echoes the popular metaphor of the balance of nature (see Chapter 10). The supporters of this paradigm consider the ecosystem a closed, self-regulated system. The notions of stages of succession and climax are associated with this paradigm, which can be called fixed, to some extent. On the other hand, the more recent paradigm of non-equilibrium starts from the observation that natural systems may have several states and that there are various ways of arriving at those states. It takes into account especially the historic dimension to understand the present functioning of ecosystems, which are considered open and heterogeneous. This paradigm is well illustrated by the metaphor of "patch dynamics" and by the increasing influence of landscape ecology (see Chapter 9).

In reality, paradigms of equilibrium and non-equilibrium are not independent of paradigms of population ecology or ecosystem ecology. The paradigm of equilibrium is dominant for scientists who are interested in the dynamics of species and communities over short time scales, emphasizing the interactions between species. That of non-equilibrium is dominant for ecosystem ecologists who take into account large temporal and spatial scales, as well as the role of physical disturbances in the dynamics of ecological systems.

3.2.4. Are there laws in ecology?

Are there general laws in ecology? This rather provocative title of an article by John Lawton (1999) poses a fundamental question for ecologists, who have never found a definitive answer. According to the ideas that prevail at present, ecology is not ready for a deterministic approach of the kind that is successfully used in physics, for example. There are in reality few universal laws in ecology, considering the complexity of ecological systems, which does not mean that there is no order. According to Lawton, ecology has only a few universal laws:

- the first and second law of thermodynamics (Chapter 5);
- the rule of stoichiometry that matter cannot be created or destroyed (which explains the failure of alchemy);
- Darwin's law of natural selection to explain evolution;
- a set of physical principles that control the diffusion and transport of gases and liquids, the mechanical properties of skin and bone, and hydrodynamics and aerodynamics, which alone or in combination define the limits of individual performance of organisms.

To this must be added a trivial but important observation: organisms interact with one another and with their abiotic environment (that is the definition of ecosystem). This simple observation, which makes life very difficult for ecologists, is the basis of the complexity of the living world.

The combination of actions resulting from these laws, rules, and mechanisms leads to the identification of *patterns*, which are in a way phenomena that have an appearance of regularity when observations are made in the natural environment: spatial structure, geographical distribution, etc. One of the most obvious patterns, pointed out by Whittaker (1975), is the existing relationship between the nature of terrestrial biomes and two environmental variables: the average annual rainfall and the average annual temperature. These patterns are simple when we work on a large scale and when we are not looking at details. But although rain and temperature are the major determinants of the existence of terrestrial biomes, their impact is linked to the nature of soils, other climatic details, biogeography, and so on. The fact that the mechanisms and laws we observe vary according to the circumstances and are thus valid only in certain precisely defined circumstances is called contingence. In other words, when we study at smaller scales, it becomes more difficult to identify generalizable patterns such as those indicated by Whittaker. This observation led Lawton to ask what motivates ecologists to still pursue ecological research on a small scale, while the patterns and mechanisms are so contingent that it is very difficult to identify general laws. The development of macro-ecology in the 1990s was intended to give greater importance to just this large-scale approach (Brown, 1995).

3.3. METAPHORS AND ANALOGIES

In situations in which there are few data or experimental demonstrations, ecologists have relied on other sciences or philosophical reflections in an attempt to interpret and organize their observations into a coherent system. Many ideas are born as *metaphors* and *analogies*, drawing lessons from a series of experiences in one context to apply them to other contexts. Ecology thus borrows results and methods from other scientific disciplines, and analogical reasoning has been widely used. Even though they do not have demonstrative value, analogies and metaphors are found upstream and downstream of the administration of the proof: upstream in the invention of the hypothesis and downstream in the didactic communication (Acot, 1988).

An analogy is like an image and establishes parallels between two distinct phenomena of the same kind—one well known and controlled, the other opaque and mysterious—thus building a bridge from the known to the unknown. To compare the properties of continuous electric current to a water current is an analogy that has a didactic purpose (Drouin, 1993).

A metaphor operates between phenomena of different order, comparing, for example, a brain and a computer.

Analogy plays an important role in current life because it is the mode of thought that allows an infant's intelligence to develop progressively, as shown by J. Piaget. It can thus be considered an entirely normal stage in the elaboration of scientific reflection, but its validity can only be assessed by experimental testing of the resulting hypotheses. Well understood, the use of analogies or metaphors is not independent of the philosophical and/or social context in which the author evolves.

One amusing metaphor compares an ecosystem to a machine. Robert Usinger, in 1967, described a river as an assembly line transporting energy and matter to organisms living along the water course, so that the animals can transform the matter and energy. As in any factory, according to Usinger, the productivity of the river is limited by its supply of raw material and its capacity to convert them into finished products. If the abiotic capital becomes scarce, the production of living things will decline. Another example of a metaphor concerns the functional role of biodiversity. To explain that the various species play different roles in the functioning of an ecosystem, the metaphor of an automobile is used (Schulze and Mooney, 1993). In a vehicle, there are some elements indispensable to its functioning (e.g., pistons, carburettor) and some that only serve to make it comfortable but are not essential to the movement of the car (shock absorbers, rear view mirrors). Other elements, such as the horn or spare tyre, are useful in certain circumstances, and some pieces are only decorative. Nevertheless, it is not enough for the parts to be good, they must also be correctly assembled.

The analogical approach that has been most successful in ecology is undoubtedly organicism, which compares the functioning of a social community or ecosystem to that of a living organism. The idea of a correspondence between the organization of the individual as a being whose parts form a distinct unit and that of a multiplicity whose elements form a whole is very old. It has taken various forms throughout history (Acot, 1988). The sociologist H. Spencer (1880), in particular, developed this process in suggesting that there was a real analogy between the biological organism and the social organization. The ecologist Clements, author of the climax concept, was one of the first to suggest, in the early 20th century, that a plant formation is similar to an organism that is born, lives, and dies, and the climax of which is the adult stage. He wrote in 1905 that the plant formation is a complex formation with functions and structures, that it is an organic unit. This type of metaphorical thought process was popular in the time of Clements, who did not otherwise seem to have explicitly supported an extreme position on the subject. In reality, at the beginning of the 20th century organicism was a preferred means of addressing the relations of causality in the plant and/or animal communities undeniably endowed with a relative autonomy. In particular, to compare plant communities to a superorganism was a means of interpreting the existence of properties belonging to the whole (emergent properties) by means of properties belonging to individual structures. This process, which can be compared to a modelling process, is nevertheless ambiguous because it postulates a certain isomorphy between the structures belonging to different fields. Besides, it inherently carries the seeds of ideological perversion (Acot, 1988): although in the mind of its authors the organicist model of ecological communities has always signified that these wholes are living things, some enthusiasts or critics have subjectively interpreted them otherwise.

Excessive analogy: fish communities and social structures

Analogies sometimes seem to get out of hand. Consider this classification of fishes based on a highly peculiar structure of human society (de Boisset, 1947):

- *Knights errant*: salmon, sturgeon, shad, eel, lamprey
- *Lords of the lake*: charr, coregonids
- *Feudal lords*: pike, pikeperch, perch
- *The aristocracy*: trout, brook trout, river charr, gudgeon
- *The bourgeoisie*: carp, tench, burbot, barbel
- *The citizens*: chub, roach, rotangle, bream, dace, carp
- *The little people*: bleak, minnow, spirlin, chub, loach, streber, stickleback, bitterling
- *The wogs*: nase, black bass, sunperch, catfish

The most recent avatar of organicism is comparison of the earth to a living thing. In reality, the idea that the earth is living is an old one. It was a Scottish scholar, James Hutton, who declared in 1785 that the earth was a superorganism and that a study of it should be based on physiology. He compared the cycling of nutrients in the soil and the movement of water from the oceans to the land to the circulation of blood (Hutton, 1788). This idea of the living earth was forgotten in the 19th century, till the term "biosphere" was used by Suess in 1875. More recently, the exploration of space marked the advent of a new vision of the earth often illustrated in the late 1960s by the metaphor of "spaceship earth". James Lovelock was the promoter of the Gaia hypothesis, according to which the earth is a self-regulated superorganism (Lovelock, 1990). He put forth the hypothesis that the earth is living and examined the facts that confirmed or contradicted this hypothesis. In reality, Lovelock defends the idea that the evolution of species and the evolution of their physicochemical environment are closely linked and that the whole constitutes a unique and indivisible system. The earth's richness in oxygen and water, and its relatively stable temperature, are thus directly linked to the presence of life and to its interaction with the terrestrial physicochemical environment.

3.4. ECOSYSTEM ECOLOGY: BETWEEN REDUCTIONISM AND HOLISM

Ecology is often presented as a science of synthesis that puts together results obtained by other scientific disciplines (zoology, botany, pedology, climatology, molecular biology, and so on). Still, the present development of ecology reactivates an old debate between two paradigms: holism and reductionism (Berganid, 1995). Many ecologists studying ecosystems defend the idea that *the whole is more than the sum of its parts* (Ehrenfels, 1890). According to this hypothesis, ecosystems have unique properties (emergent properties) that cannot be reduced to the properties of their components. This approach is also called *holistic* by reference to the concept of holism created by Jan Smuts (1926). Other ecologists consider that it is not true and that the whole is nothing but the sum of its components: therefore, we can independently study each of the components and reconstitute the system once we acquire enough knowledge. These two world views correspond to entirely different research strategies and are at the centre of a fundamental debate on the study of ecosystems (Bergandi and Blandin, 1998).

3.4.1. The reductionist temptation: building cathedrals

Many ecologists are aware of the difficulties they must face to understand the functioning of an ecosystem in its entirety. They know that to achieve this understanding they must mobilize significant manpower and material resources over a relatively long time, which is not always possible. Under these conditions, some will prefer the reductionist route, which consists of reducing the ecosystem into smaller subsets that can be studied more easily. The key to science, according to Wilson (2000), is reductionism, the breaking down of nature into its natural constituents. It is a strategy used to discover openings in complex systems that would otherwise remain impenetrable.

It is complexity that ultimately interests scientists, according to Wilson (2000), not simplicity. Reductionism is the method that allows understanding. The love of complexity without reductionism characterizes art. The love of complexity with reductionism characterizes science.

In epistemological terms, reductionism starts from a simple principle: the behaviour of the entire system can be explained by the properties of its parts. To understand the functioning of an ecosystem, we must try to identify its major components and describe the interactions they have, for example in the form of flows of matter and energy that circulate in the organisms and the environment. The behaviour of these organisms can itself be explained by the set of biological strategies they follow, strategies that are themselves a function of genetic, physiological, and other specificities. This process is that of "building a cathedral" (Peters, 1991). The idea is that one can, at any moment, construct a cathedral for which the plans have been established by theoretical ecologists, using bricks that have been made from observations or experimental results from field ecologists. Reductionism reflects in reality the hope that behind the complexity of nature is hidden a beautiful simplicity, and that the articulation of a small number of fundamental laws could explain all phenomena.

With respect to research on ecosystems, the reductionist approach leaves a wider latitude to the individual thought process or that of small teams than the holistic approach, which requires highly organized research. But it often leads to a loss of perspective of the initial object of the research, and this route has not proved very practical in reality. Most often, we more or less implicitly avoid the ultimate stage, the moment at which the parts must be assembled. The study of elementary processes thus becomes a question in itself, while the holistic approach has served as a pretext to put research in its place.

The reductionist process approaches a mechanistic view of the world (see box) that leads to the deconstruction of ecological systems and to a sum of processes and structures. This vision relies on the application of principles of thermodynamics according to which the biosphere is a machine to use and degrade solar energy. The role of different constituent elements may also be evaluated with respect to this objective.

Although biologists generally do not take recourse to immaterial and supernatural forces to explain the functioning of the living world, they nevertheless reject the naïve mechanistic explanation and emphasize that living things have characteristics that have no equivalent in the world of inanimate objects. If nothing in the processes and activities of living things is in conflict with the laws of physics and chemistry, the explanatory apparatus of the physical sciences is not sufficient to elucidate complex living systems. It is thus impossible to assimilate biology and ecology into the strict domain of physics.

Mechanistic conception of the world and vitalism

The mechanical character of the world view that developed during the Renaissance could be symbolized by the thoughts of Galileo (1564-1642) that nature is a material system in movement subjected to laws. The universe identified with a machine can be broken down into its constituent elements such that its functioning is determined by the action of distinct parts. Descartes (1596-1650), for example, contributed to the expansion of the mechanistic vision of the world by affirming that living things are simple automata, that the human species differs from automata only in that people have souls, and that all science must be based on mathematics. Descartes even felt that organisms are formed by the fortuitous assembly of particles, which implied that nature resulted from the work of blind chance. This attitude was in open contradiction to the notion of natural order that prevailed during his time. Two centuries later, at the congress of German naturalists of 1869, H.L.F. von Helmholtz, one of the leaders of the mechanistic movement, indicated that the ultimate objective of natural sciences is to reduce all the processes of nature to the movements that underlie them as well as to find their motive forces, in other words, to reduce them to the laws of mechanics.

These declarations obviously caused opposing reactions and the emergence of a theological vitalism. The vitalist school was opposed to the mechanists and professed that there were processes in living things that were different from those of physics and chemistry. When Newton proposed his law of gravitation, some biologists invoked the existence of an equally invisible and material force, the *vital force*, to explain the phenomenon of the living thing. Scientists would ultimately affirm that this vital force belonged to a different domain from that of physicochemical laws, thus continuing a tradition initiated by Aristotle and other ancient philosophers. Vitalism had a beneficial effect on the development of physiology in the 18th and 19th centuries. It was popular until the early 20th century, before it fell into disuse, first because it referred to an unknown and probably unknowable factor, and subsequently because certain phenomena that demanded a vitalist explanation were later explained in physicochemical terms (Mayr, 1982).

3.4.2. The holistic mirage

Some ecologists were persuaded that the complexity could not be understood simply by a reductionist approach. The holistic process of thought, unlike the reductionist process, aimed to uncover the rules that contributed to create a functional whole from different parts. The rather mythical expression "the whole is more than the sum of its parts" actually signified that the characteristics of a whole could not be explained simply from the characteristics of each of its parts. The result was that the characteristics of the whole appeared new or emergent. The idea that the characteristics of a whole could not be deduced from those of the parts taken separately is not unique to biological systems. In 1868, T.H. Huxley maintained that the particular properties of water that give it its aqueous character could not be deduced from the known properties of hydrogen and oxygen molecules.

The origin of holism as a system of thought goes far back in time. But it entered biology with the writings of Clements on the "climax", an ulimate phase of the plant community (see Chapter 10). In developing the concept of ecosystem considered as a "base unit of nature", Tansley (1935) attempted to avoid the reductionist and holistic theories that he felt were traps that led to emotional debates between biologists and ecologists. It was the Odum brothers (E.P. Odum, 1953; H.T. Odum, 1957) who contributed to promote a "holistic" conception of the ecosystem, without managing to resolve the methodological problem that accompanied the analysis of "complexity" in ecology or going beyond the historical debate raging between the supporters of reductionism and those of holism.

The holistic process of thought does not imply that research on the different components are useless. From such research we know more about the system and more about its properties. But the study of components alone is insufficient because it does not allow us to reveal the characteristics of the system. A weighty argument of the supporters of holism is that coevolution will allow us to construct systems that function as indivisible units. This process also uses metaphor and analogy in order to develop new hypotheses in a field that is still cloudy.

3.4.3. The current practice of ecology

Ecosystem ecology thus seems to be divided between two trends that coexist, sometimes in a single scientist: one trend that considers the ecosystem as an entity having an autonomous structure and function and one that prefers the analytical approach and studies subsets, hoping thereby to be able to reconstitute the whole at some time.

The first trend, called holistic, affirms that the ecosystem is an entity that has emergent properties. But ecologists are so unable to prove this that holism has no practical repercussions because it is difficult to apply. Emergence is a descriptive notion that seems to resist analysis, particularly in the case of complex systems (Mayr, 1982). In the ecological literature, examples of emergent properties are rare and most often erroneous because the authors have a tendency to confuse emergent properties and collective properties. The International Biological Programme (IBP), which was devised to study productivity of ecosystems from 1964 to 1974, fell within the sphere of influence of the holistic ideas of Lindeman and the Odum brothers on the ecosystem. But its approaches, when looked at in detail, were observed to follow rather the principle "the whole is equal to the sum of the parts": the overall productivity of the ecosystem, for example, was estimated from the sum of productions of each of the trophic levels, these levels themselves being determined from the studies of the dynamics of the constituent populations (Bergandi and Blandin, 1998). In other words, the ecologists conduct their researches as if they have implicitly accepted that the laws of functioning of a given level of integration could be deduced entirely from the knowledge of laws that govern the lower levels of integration. They use in an operational manner approaches based on analysis and the sum of parts. But the constant recourse to the idea of holism clearly expresses their frustration and the fact that the reductionist approach cannot in itself provide an accurate articulation of the reality (Bergandi and Blandin, 1998). According to Bergandi (1995), the holistic discourse in ecology is associated with a defence of the identity of ecologists faced with the risk of seeing the objects of their research taken over by reductionist trends.

However, the holistic process, although difficult to translate into operational terms, does have a didactic value that is demarcated in the usual analytical-additional methods. Moreover, there are vast scientific domains in which the reductionist methods of physics (elaboration of hypotheses followed by reproducible experiments) are not possible. This is particularly the case with disciplines that depend on historical results such as geology, palaeontology, or evolution. The appropriate methods here are of the narrative type, not experimental. What happened can be explained in detail, but what will happen cannot be predicted (Bak, 1999). The science of evolution cannot explain the existence of elephants because life as we observe it today is only one of the numerous possibilities that at the outset were entirely improbable. If a meteor had not hit the earth, the dinosaurs would not have been annihilated, and the course of evolution would probably have been different.

3.5. INDUCTIVE REASONING AND THE HYPOTHETICAL-DEDUCTIVE METHOD

For a long time, at least since the ancient Greeks, people have attempted to explain rationally the phenomena they observed in the living and inanimate world. Aristotle and his disciples were rationalist philosophers. They felt that scientific problems could be resolved by acute reasoning, by deduction resulting from the application of the rules of logic to the analysis of a question. Closer to our time, Descartes also helped promote the idea that we can resolve everything by a simple effort of concentration.

3.5.1. The burden of proof

Since Descartes, the idea that scientists must distinguish their results and theories from absolute proofs has prevailed in the physical and biological sciences (Mayr, 1982). Very often, nevertheless, biologists and ecologists cannot provide such proofs. For example, it is difficult to prove without doubt that natural selection is the directive agent guiding the evolution of organisms. In these circumstances, it is admitted that the valid hypothesis or theory is that which is the most probable on the basis of the available data, or even that which is compatible with the largest number of facts or the most significant facts.

While realizing that is not possible to provide absolute proofs for many scientific results, the philosopher Karl Popper proposed that the burden of proof must devolve on the person who contests a given scientific theory. According to this philosophy, any assertion that cannot be subjected to the test of falsification cannot be accepted in scientific terms. An accepted theory is one that has resisted the largest number of attempts to refute it. New theories thus impose themselves by elimination of concurrent theories.

However, falsification cannot be considered the only process leading to scientific truth, as Popper suggests. The fact that the hypothesis is not contradicted by the experiment is not sufficient to guarantee its validity. For this, it is necessary in principle to prove that all the other possible hypotheses are not relevant, which is difficult. Scientific theories have been rejected at a given time not because they were clearly refuted, but because more simple or more seductive new theories were advanced. Scientists can in reality be satisfied with a theory at a certain stage of their knowledge only until a better theory is proposed. It may also happen that theories that are revolutionary with respect to the ideas of the time will not only be criticized but also be ignored for a short or long period. This was the case, for example, of Wegener's theory of continental drift.

We must also ask, when can we consider that a hypothesis has become a certainty? In practice, by accumulation of experimental confirmation, we end up having a near certainty such that it is less and less probable that the theory or hypothesis will be called into question. Still, that certainty is never definitive (Arsac, 1993).

3.5.2. Inductive reasoning

Induction is a mode of reasoning that progresses from the particular to the general. A scientist using *inductive reasoning* arrives at objective conclusions by recording, measuring, and describing phenomena without previously established hypotheses, without *a priori* conventions (Mayr, 1982). The body of scientific knowledge is constructed by generalization from observations that constitute a firm foundation. Francis Bacon (1561-1626) was the principal advocate of inductive reasoning, which was widely appreciated in the 18th century and the beginning of the 19th century.

This inductive process gives a very positive vision of science. It is also firmly anchored in the minds of many scientists, who feel that there are raw data and objective facts on the one hand and scientific formulations on the other. But it has also been severely criticized (Chalmers, 1987). In particular, the accumulation of observations does not necessarily lead to the formulation of generalizable hypotheses. Besides, observation is not always neutral. It may be biased by scales of observation, sampling methods, and other factors. The theories that might result from them are consequently challenged.

An extension of the inductive process is logical and deductive reasoning. Once in possession of universal laws and theories, a scientist may draw various consequences from them, which will be explanations and predictions (Chalmers,

1987). A logically valid deduction is characterized by the fact that if the premises are true, then the conclusion must necessarily be true. The following example is suggested: water freezes at 0°C (fact established by observation); the lake contains water; if the temperature falls below 0°C, the water in the lake will freeze.

3.5.3. The hypothetical-deductive method

Inductive reasoning has gradually yielded to the hypothetical-deductive method, which consists of first putting forth a hypothesis and then conducting experiments or accumulating observations allowing one to test this hypothesis. According to Popper (1985), imagination comes first in the scientific process: scientists begin by formulating hypotheses, then test them by means of experimentation. The experimentation is intended to obtain a real, measurable observation, precise and inherently impartial, that will be a response to a question that is asked. It is a process of comparing a theoretical discourse with a practical result to confirm it or refute it. It can be said that the experimentation is carried out only because questions were first posed with the theory.

This process is consistent with the conviction that there is no absolute truth and that it is advisable to constantly test the results acquired and the theories established. It also encourages the perfection of new theories, and the carrying out of new experiments. Science progresses by trial and error and only the best-adapted theories survive. When a hypothesis proves to be inadequate, scientists formulate a new hypothesis or an improved version of the preceding theory that can in turn be tested (Mayr, 1982).

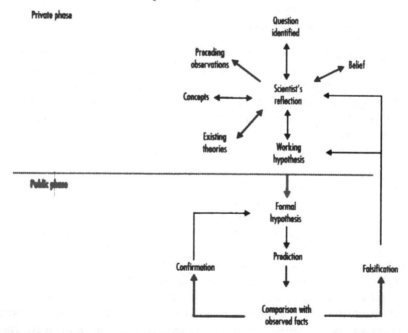

Fig. 3.1. Diagram of the hypothetical-deductive method showing the two phases of elaboration of a theory (modified from Peters, 1991)

"Occam's razor" (after William of Occam) expresses the idea that when two theories are presented with an equal degree of validity, the more simple solution is the more likely one. The principle of Occam's razor thus maintains that when fewer hypotheses are made, the intellectual construction is more solid. This principle does not signify that an explanation of a phenomenon is true because it uses the fewest hypotheses. Nor does it guarantee its scientific verity. But it is one of several means of choosing the causal explanation that will be preferred and subjected to experimentation.

Peters (1991) summarized the hypothetical-deductive method as follows (Fig. 3.1). Science begins with reflection on the part of scientists faced with a problem for which they must find a solution. They begin by confronting various possible solutions with the theory and the facts. This "private" phase supposes that the scientist passes alternatively from a phase of unbridled creation in elaborating new hypotheses to a phase of critical analysis of those hypotheses. At the end of this period, the most promising hypotheses, those most susceptible of being tested with new data, are communicated to the scientific community to be evaluated. During this "public" phase, the hypotheses will be confirmed or refuted. The deductions that can be made from the hypothesis are confronted with observed facts. If the test is positive, it will eventually be repeated. If the observations do not confirm the theory, the theory will be modified or abandoned. In any case, a theory is always a hypothesis susceptible of being subjected to other tests.

3.5.4. Should inductive reasoning be kept out of ecosystem ecology?

The hypothetical-deductive method is the basis of modern scientific research, even though the elaboration of a hypothesis is, in general, preceded by observations and a question (Mayr, 1982). Indeed, the history of science shows that the act of asking a question is always the first step towards a hypothesis or theory. Researchers pose questions, in fact, because they have observed certain facts and have a doubt about how to interpret them (e.g., they do not understand something). If researchers possess important observations at some point and cannot extract a theory from them, it is because they have not asked the right question. Moreover, the theories that constitute our scientific knowledge could themselves be sources of errors and direct observers toward false trails when they try to elaborate a sampling plan relevant to the phenomenon studied (Chalmers, 1987).

In practice, ecologists thus have difficulties in positioning themselves in relation to the hypothetical-deductive process. It is obvious that this process cannot be applied to all situations and that ecosystem ecology, no matter what is said about it, will still largely be based on observation. Brown (1995) defines macroecology as the study of relationships between the organisms and their environment through the search for emerging statistical patterns on abundance, distribution, and diversity of species, for example. In relation to other approaches in ecology, macroecology emphasizes statistical analysis and not the experimental approach, arguing that it is often difficult, if not impossible, to make controlled experiments and duplicate them on large spatio-temporal scales. According to Brown, it is more a question of practicality than of philosophy. Because of this, macroecology may appear more inductive than other approaches. However, it proposes to add to the inductive process, which allows the development of hypotheses, complementary observations that can test these hypotheses.

The sometimes severe rejection of the inductive process by some scientists thus needs to be qualified. In fact, it was observation that revealed the diversity of nature and led to theories of evolution. Observation is the basis of biogeography and many ecological concepts. In their daily work, ecologists often oscillate between periods in which they accumulate observations or conduct purely descriptive and classifying research and periods in which they forge concepts or test hypotheses. Moreover, making observations and organizing them are customary processes of many disciplines that are based on observations: for example botany, zoology, geology, hydrology, geology, meteorology, and paleontology. Allowing for some differences, ecosystem ecology, in its descriptive and comparative methods, is similar to sciences such as astronomy or climatology and, more generally, the earth sciences. It is thus very likely that the two processes coexist and are involved most often in even the most minor of our judgements: Through induction we reach general considerations after having observed the repetition of particular cases. Through deduction we try to interpret particular cases from a general point of view (Besnier, 1996).

Ecosystem ecology and the earth sciences: similar epistemological problems

The epistemological problems that arise in ecosystem ecology are very similar to those found in the earth sciences. The latter focus on the earth, which has evolved continually since its creation. According to Jaupart (2000), it is impossible to reason without considering history or without stumbling over a missing fundamental piece of information. We must always proceed by a series of hypotheses and constantly keep in mind that they will be challenged. The exercise often appears to be that of a tightrope walker. Jaupart particularly emphasizes the following:

• The functioning of a complex system such as our planet involves the combined effects of a large number of phenomena that are difficult to break down into a few essential components. The synthesis of information as well as communication between specialists is difficult.

• There is a tendency to go towards the smallest components, to find a response by refining measurements, as in physics. To understand a geological phenomenon, we must on the contrary go back to the largest scales, which requires us to make many measurements and know the distribution law of values. This operation is rarely mastered for an object of large dimensions.

• A theory of natural environments can rarely be exact in the conventional sense of the term, i.e., in giving precisely measured values. A theory is verified indirectly, and often the number of adjustable parameters is greater than the number of observations. It is consequently difficult to test the theory and to reject or accept it definitively.

• The development of observational tools (satellites, spectrometers, computers) has enabled us to construct a coherent image of the planet. But these tools generate such a wave of measurements that any general theory is incapable of accounting for them in detail. The conventional physical method proceeds by reduction, analysing each phenomenon individually, then combining them two at a time, before attacking the problem of N pairs of phenomena. In the earth sciences, we do not have the certainty that this procedure will bring us to a result close to reality. We can also attempt to simulate the earth by calculation, using computers. Then we must know what to ask the computer, i.e., to define the list of N phenomena that we want to analyse simultaneously. The solutions are probably themselves complicated and cannot be reduced to a few laws by which we can extend a result beyond the conditions in which it was established.

In the earth sciences, as in ecosystem ecology, there is a huge gap between theory and observation. To understand how a heterogeneous environment of large dimensions functions under the simultaneous action of many phenomena is a huge challenge: we must, according to Jaupart (2000), find out a scientific method adapted to complex natural systems.

The procedure proposed by Underwood (1990) is in a way a synthesis of present preoccupations: the model corresponds to a preliminary trial of the information and an interrogation, and the introduction of a null hypothesis in a series of logical sequences is a way to overcome the difficulty of designing experiments (Fig. 3.2).

- *Observations*: Many research programmes in ecology try to identify patterns on the basis of a series of spatial and/or temporal observations. The scientist must explain the phenomenon observed.

- *Models*: Whether it is called a model or a theory, any attempt to explain from observations is acceptable if it allows a realistic interpretation of facts. The researcher thus provides one or several plausible explanations (models) in a mathematical or other form. Many different models can be proposed to explain a single series of observations. The most complex models are not necessarily the most scientifically relevant.

- *Hypotheses* (predictions, deductions): A general explanatory model has to be capable of making predictions whose relevance must be verified. Those deduced from the model are hypotheses, some of which can be tested. If the model is correct, it can be predicted that some observations can be made in certain precise conditions. Thus, there is a tendency to affirm that the model is good if its predictions are validated by observations.

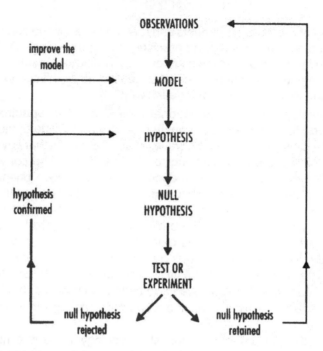

Fig. 3.2. Logical sequence of elements of an experimental procedure that can be refuted and their relationships (modified from Underwood, 1990)

- *Proof or refutation*: After having deduced one or several hypotheses from one or several models to explain a series of observations, we must evaluate their validity. A contradictory case is sufficient in theory to refute a hypothesis.
- *Null hypothesis*: Before the difficult task of carrying out a proof, the method best adapted to ecological problems is to seek to prove that the hypothesis is false and thus to refute it. It must therefore be formulated in the form of a null hypothesis (expressed in a negative form) that we attempt to refute. This null hypothesis must take into account all the possibilities except the one retained by the hypothesis. For example, if a hypothesis predicts that the number of species will increase when we remove the predators from an ecological system, the null hypothesis is that the number of species will not change or decline if we remove the predators. If the null hypothesis is not verified, the first hypothesis will be validated.
- *Test* (experimentation): The objective is to try to refute the predictions of the model proposed and to show that it is false. Thus, the predictions made with the null hypothesis must be confronted with observations made during the experiment or with other series of data (Underwood, 1990).
- *Conclusions* (interpretations): If the null hypothesis is in contradiction with the facts, it is rejected and the model proposed is retained. It could eventually be amended. If the null hypothesis is retained, we must return to observations and develop other hypotheses.

The null hypothesis

The methodology of null models was introduced chiefly by Simberloff, who used Monte Carlo simulations to replace the "control" indispensable to an experimental approach but not feasible in most ecological research. This approach is based on the following principle: before establishing a causal relationship between the distribution of a set of species and one factor or another, we must verify that the distribution is not compatible with the distribution that would be observed in the absence of this factor. This brings us back to comparing the distribution observed to that expected under an alternative "null" hypothesis in which the process does not intervene, and in which the distribution of species is random.

3.5.5. From theory to practice in ecology

All ecologists, or nearly all, recognize the advantage of the experimental approach and the hypothetical-deductive method in elucidating the mechanisms responsible for the structure and organization of ecosystems. However, there is a considerable gap between theory and practice, as shown in a study based on 253 articles published in the journal *Limnology and Oceanography* during the years 1980, 1985, and 1990, which made a rather severe observation on the practices of ecologists working on aquatic systems (Bourget and Fortin, 1995). The study showed that the majority of articles published by limnologists (60%) and oceanographers (70%) were descriptive. Only 30% of the articles used the experimental approach, and among the 27% of articles presenting models, only 3% validate these models using field data. A significant, albeit not quantified, number of articles include statistical analyses without posing hypotheses beforehand, and a large proportion of articles make only a succinct review of the literature, most often to confirm the hypotheses put forth rather than to question them.

Figure 3.3 summarizes the major types of practices developed by the aquatic ecologists in relation to the experimental procedures. These various practices bypass in one way or another the theoretical experimental procedure that is proposed above.

- *Type 1* is typical of the descriptive study in which the experimental procedure is interrupted after construction of the conceptual model. The hypotheses are not tested, and the mechanisms proposed to explain the patterns are purely speculative.

- *Type 2* follows the experimental procedure up to the conceptual model and new observations are made to refine the model, but here also, the hypotheses are not tested.

- *Type 3* corresponds to the case in which the researcher observes a pattern and considers after the fact that the data and the pattern correspond to hypotheses that are not proposed in the literature.

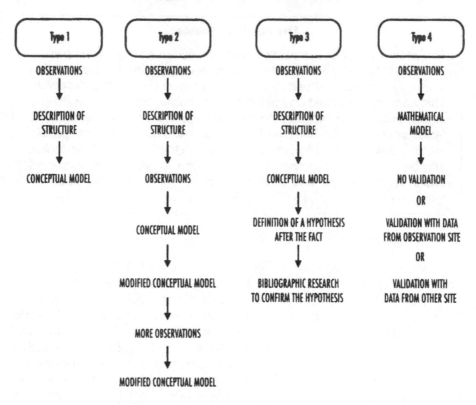

Fig. 3.3. Types of practices of researchers in aquatic ecology (typology based on the analysis of 253 articles published in *Limnology and Oceanography*) (modified from Bourget and Fortin, 1995)

- *Type 4*, which is similar to type 2, puts forth a certain number of hypotheses in the mathematical rather than conceptual form. The models are validated or not validated, but the hypotheses are not tested.

In sum, the scientists attempt to adapt descriptive methods to the study of mechanisms. Instead of testing hypotheses, they refine the descriptive methods as if a better description of patterns could improve the confidence we can have in the mechanisms they propose. In most cases it is not definitively far from the inductive approach.

It is essential, as Bourget and Fortin (1995), emphasize that present practices must be substantially improved. Although it is admitted that it is often difficult to carry out experimental approaches *in situ*, we can nevertheless expect that hypotheses concerning the mechanisms be formulated in a form that can be tested. In other words, a null hypothesis must be constructed and precise biological and statistical criteria must be clearly identified. In the absence of such processes, we are dealing with a vast exercise of collecting information of purely local interest, given that the results cannot be generalized.

As the authors of this study noticed, the abundance of such studies is perhaps the source of the disillusionment of many administrators. Ecologists indicate patterns clearly, but since they rarely test their results, they do not contribute much information on the way in which ecological systems function.

3.5.6. Ecological theories and difficulties encountered in their validation

To develop hypotheses and concepts and then test them to verify their pertinence seems to be a logical scientific process. However, the phase of validation is not always easy to implement because the data are not always available, or because their quality leaves something to be desired. The following example illustrates the difficulty of carrying out the testing phase.

Southwood (1977) is the author of the *template habitat* concept, according to which the characteristics of the habitat forge the biological traits of species occupying it. Over the course of ecological time, certain combinations of biological adaptations are selected, the result of which is an appropriateness between the present biological traits of species and the characteristics of the environment in which they live. This concept, which emphasizes autoecological processes, is similar to the notion of ecological niche, but instead of "*n* dimensions", it considers two: *spatial variability* and *temporal variability*. Besides, it does not stress biological units (organisms, populations) but biological characteristics.

Starting from the hypothesis that there is a relationship between temporal variability of the habitat and the frequency of disturbances, and between spatial heterogeneity of the habitat and the existence of refuge zones that reduce the effects of disturbances, Townsend and Hildrew (1994) were able to predict the existence of different biological traits among the species present in the fluvial habitat: e.g., reproductive strategies, body shape, longevity, and size. They also made predictions as to the species richness and the number of species exploiting certain categories of resources. Their theoretical predictions remain to be validated by field observations.

The predictions were tested on a set of aquatic groups living in the Rhone river (Statzner et al., 1994). Generally, for most of the groups studied, there is a statistically significant relationship between the trends observed in the biological traits and the spatial and temporal variability of the habitat. The central hypothesis of the concept of template habitat is thus confirmed overall by the observations: the habitat acts as a template for biological traits. In details, however, the biological traits observed are highly diverse in each of the taxonomic groups studied and do not often conform to the predictions of Townsend and Hildrew. It is possible, even probable, that the inadequacy between the predictions and the observations is the result of unpredictable combinations or compromises that allow life to survive in the face of variability in the habitat. The species may improve their *fitness* in various ways in their environment, and the combinations of biological traits could be more numerous than predicted. The predictive power of the template habitat is thus not fully satisfying (Statzner, 1997).

In theory, the results obtained by the analysis of data from the Rhone must lead to the rejection of the template habitat hypothesis because it was not validated by the facts. In practice, however, ecologists ask themselves questions about the process they used to test the hypothesis: were the data really reliable, was there a bias in the conduct of the test, were all the relevant data considered? The concept may appear robust and seductive and they do not give it up, even if the proof from tests does not yet appear conclusive.

3.6. DETERMINISTIC OR STOCHASTIC APPROACH?

According to Newton's laws, when we know the state of a system at a given moment, we can deduce its state at any other moment. In other words, variations in the state of a system over time are determined by forces that act on this system at each moment. Newtonian mechanics is thus entirely *deterministic*. Laplace gave a famous formulation of determinism: an intelligence that, for a given instant, will know all the forces that animate nature, and the respective situation of beings that it is composed of (if by the way it is vast enough to submit these data to analysis), and will embrace in the same formula the movements of the largest bodies in the universe and those of the smallest atom. Nothing will be uncertain for this intelligence, and the future, like the past, will be in front of its eyes. All the efforts of the human mind in the search for truth tend to constantly approach the intelligence that we are able to conceive (Laplace, 1814). Laplace created this monster (still called the Laplace demon) only to eliminate it right away by attempting to show, in his works on probability, the need to treat certain processes by statistical means.

The history of determinism is linked to the role of the principle of causality, which itself evolved as a function of tools used by researchers in terms of concepts as well as mathematical instrumentation: factorial analysis, calculation of probabilities, statistics, systems theory, communication theory, etc. In conventional thought, the usual form in which phenomena are explained is linear causality: from cause to effect or from before to after. This mode of explanation is based on a logical chain of causes and effects, spread along the arrow of time. One of the consequences of this deterministic approach to nature is that scientific thought is unrelenting in the search for precision. Physicists have followed this line, probably adapted to their field: the better the measurement, the better the precision. Several philosophers have seen the progress of science in the reduction of imprecision and experimental margins of error. But does not this quest for precision end in the confusion of the measurement for the thing, in the belief that to measure is to control, that to explain is to understand (Moles, 1990)?

The success of the idea has been such that it has been transformed into an ideology: if what is precise is good, consequently what is imprecise is bad. But linear causalty and cause-and-effect relationships no longer appear the only mode of reasoning. In present thought, influenced by the systemic approach, the existence of feedback (see Chapter 5) leads us to talk of circular causalty, which could be part of the self-regulation of ecosystems. Besides, voices have been raised against the dominance of physical and mathematical sciences, requiring that we take into consideration other approaches, some of which describe for example the "science of the imprecise" (Moles, 1990). Indeed, the world in which we live is not a laboratory in which phenomena are purified, isolated, and controlled. When precision, measurement, and conceptualization end up being inadequate, must we therefore turn away from the phenomena studied? Alongside the so-called exact sciences, what place do we give to sciences of the inexact, of the imprecise, of low correlations that approach the knowledge of the real and for which we must construct an epistemology, a metrology, and a methodology? (Moles, 1990).

Ecosystem ecology is especially a field in which it is difficult to apply the paradigms of physics. Several scientists have argued that the deterministic process does not apply to ecology, and some have even challenged the utility of mathematics. To refute determinism is to admit implicitly that ecological phenomena are of a stochastic nature. Thus, generalizations in biology and ecology are very often of a probabilistic nature (Mayr, 1982). Instead of formulating laws, biologists and ecologists have attempted most often to generalize their results by a set of concepts.

The conflict between those who think there is an inherent order in ecosystems and those who defend their stochastic nature is patent: the stochastics, following Gleason and Simberloff, affirm that events are indeterminate and that the formation of communities follows essentially a stochastic and consequently unpredictable scenario. The holists or determinists, on the other hand, following Clements and then Hutchinson and E. Odum, believe in the existence in ecosystems of an order similar to that observed in organisms. Some of them nevertheless consider ecosystems as machines that draw their organization from physical principles rather than compare them to organisms.

This opposition is in reality rather formal because ecosystem ecology is naturally located in this context of the imprecise sciences with which statisticians have associated the concept of statistical error, i.e., imperfection in measurement and in prediction. The acknowledgement of possible errors in observation and in the experiment is inherent to the ecological process because it is often impossible to go through irrefutable demonstrations. The discourse of ecologists is thus essentially a probabilistic discourse. The large part of the public has a tendency to demand absolute truths from researchers, unaware of the basis of the process. Science is thus questioned, sometimes judged, by a lack of reciprocal understanding.

Are "regulatory" mechanisms of ecosystems deterministic or stochastic?

A fundamental question that biologists ask is whether chance is the only master in charge in the long history of evolution or whether there is an "invisible hand", an unidentified force that gives it a direction? Here we find an echo of the economy of nature concept (see Chapter 19). According to Jacques Monod (1970), who was an active supporter of this idea, and many other authors, chance is the only possible answer. In the game of evolution, in which the result is to select an organism that gives the most descendants (principle of fitness), the only true penalty is the possibility of continuing the game in case of success (Levin, 1999). There thus does not seem to be an overall, long-term perspective. Hence the ecologist's dilemma: if evolution is a simple matter of chance, without apparent determinism, it becomes difficult to find an order to nature. Still, some observers have been struck by the fact that the earth functions as an organized system. The study of the functioning of ecosystems and the existence of similar functional processes in varied environments also makes us think that despite everything there is a form of regulation. These observations were the source of the metaphor of a superorganism that was developed at various times by scientists. The last of these was the Gaia hypothesis.

Simply put, regulation implies that there is a control and the identification of an objective. Among the biological systems, we can use the term "regulation" for cells or organisms that are systems pursing one intrinsic aim such as survival and reproduction (deterministic systems). This term becomes ambiguous, however, for biological systems of a higher hierarchical level than the individual. It has never yet been demonstrated that the dynamics of an ecosystem pursues an identifiable objective (Levin, 1999). Ecosystems are not constructs like organisms: they contain many organisms whose interactions are not, in theory, of a deterministic nature but rather stochastic. Nevertheless:

• Some interactions between the components are "quasi-deterministic" in the sense that they correspond to very strong relationships of coevolution (e.g., plants and pollinators).

• The action of an individual often affects the performance of its neighbours. Inversely, its functional role in the ecosystem may change as a function of the activity of its neighbours. This is the principle of retroactive loops or feedback.

• Some species may replace others without the modification of a given functional process.

• Interactions between species in a food web may also contain elements of deterministic control, if only by the sequence of relationships between the eaters and the eaten.

• The functioning of an ecosystem depends very closely on the availability and the spatio-temporal dynamics of resources of water, nutrients, and energy.

Ecologists presently tend to recognize that there are "regulatory mechanisms" at the ecosystem level, by analogy with organisms, but that these mechanisms are generally the result of stochastic events (Schultz, 1995). As we can see, the question remains open and promises lively debates in the years to come.

We conclude this chapter with the statement of Jorge Wagensberg (1997) that the progress of science requires two attitudes: *determinism* and *indeterminism*. Indeterminism is the scientific attitude compatible with the advancement of knowledge about the world. Determinism is the scientific attitude compatible with the description of the world. One corresponds to creative scientists who study a finite number of events and admit in principle the recourse to any theory until they find a model that allows them to account for the different events. The other corresponds to practical scientists who use a finite number of theories because application of science requires a deterministic attitude, until they are confronted with an unpredictable situation.

There is no contradiction between these two attitudes: they may even exist in a single individual at two different periods in his or her study of the complexity of the world.

Chapter 4

Methods of Studying Ecosystems

Contrary to the biology described by Jacques Monod, ecology is not cartesian. It is concerned mostly with relationships, coevolutions, and the multiple relations woven by living organisms, the soils, the chemical elements, the water, and the air that together form the face of the earth. It studies not only the elementary parts, but the totality of nature, the large whole animated by a breath of air, as Alexander de Humboldt wrote in the beginning of his Cosmos, *taking up an expression of Goethe's.*

translated from J. Grinewald (1990)

"Periodically, the ecological literature is swept by claims for new calculating devices, like the logistic, the simulation models of the 1970s, catastrophe theory, fractals, and chaos. These may all be useful tools in certain situations, but initial propagandizing often oversells their importance and utility whereas initial 'applications' seem little better than self-justified abstractions that may only repeat scientific common-places."

Peters (1991)

The study of ecosystems involves four major types of processes.

- *Observation*, which consists of collecting information and then attempting to assemble and integrate them to extract an overall picture of the system, was widely favoured in the early years of ecology. Although its advantages are not to be undervalued, it has shown some limitations, especially with respect to the identification of cause-effect relationships between the constituent elements of ecosystems. The use of mathematical tools to analyse observed data is able to give the appearance of a more rigorous process, but it is always a mode of explaining what was observed (Hairston, 1989).

- *The comparative approach* to structural or functional elements in different types of ecosystems allows us to account for a wide range of geographical and climatic situations. By means of retrospective studies and paleoecology, we can extend the comparisons to other time scales.

- *Experimentation*, which consists of manipulating ecosystems to comprehend their mechanisms, supposes a close interaction between the researcher and the object of study. It makes it possible to work with replications in a process of probabilistic nature.

- *Modelling* and *simulation* provide a simplified and more or less faithful image of the world on which we do not act directly, but the behaviour of which we try to understand.

Considering the complexity of phenomena studied, all these approaches must be implemented in principle to understand the structural and functional characteristics of an ecosystem (Likens, 1985).

4.1. A SCIENCE OF OBSERVATION

Ecosystem ecology is first of all a science of observation. *An ecologist who focuses on theory without contact with the terrain will have only a partial and probably biased vision of the functioning of nature.*

Observation in ecology is based on the acquisition of qualitative and quantitative data, a set of concrete operations that are the basis of any scientific work.

Two major types of approaches were used to study the dynamics of ecosystems.

- The *diachronic* approach consists of observing over time the modifications of communities in a station. It requires periodic observations over a long period.

- The *synchronic* approach is based on analysis of the spatial variation of the structure and composition of communities present at a given time, in a more or less homogeneous space.

One difficulty lies in the collection and representativity of data, and the question of sampling has been the subject of many studies (Frontier, 1983). Not being able to repeat cumbersome and costly research several times, ecologists attempt at the outset to select study sites that are representative of larger areas. Some situations are more favourable than others for observation, which partly explains the fascination that supposedly homogeneous environments have long held for ecologists. In such environments ecologists find conditions more favourable to understanding the often complex phenomena that become practically undecipherable when the environment is too variable. Second, they try to extrapolate the results obtained in one site to other sites of the same nature. This representativity is sometimes inferred on an intuitive basis. There is always a kind of mythology around the concept of "representative sample", which is supposed to represent the "average" environment. These ideas, linked to concepts of environments that are homogeneous and in equilibrium (see Chapter 9), are no longer theoretically relevant when we emphasize the dynamics and spatio-temporal variability.

There are often various possible interpretations of the phenomena observed, so that we are never certain of the validity of conclusions drawn from the observation. However, ecosystem ecology is not different in this respect from other sciences of observation, such as astronomy, geophysics, or meteorology. Besides, observation is the only possible approach for the study of non-reproducible historic processes, such as the establishment of fauna and flora, for which experimentation cannot apply. It is the same for certain ecological processes, which, considering their duration or their unpredictability, cannot be subjected to the proof of experimentation, modelling, and/or statistical verification.

It is true that natural events or those of anthropic origin, often called catastrophes, are unique occasions for ecologists to make observations in near-experimental conditions. There is just one problem: we cannot choose the conditions in which catastrophes occur. It would be impossible to generate an oil spill or create a period of drought on the regional scale.

4.1.1. Sampling

The complexity of any natural system is such that the image that we can draw is most often approximate. This representation, in the general sense of the term, is fashioned by observations that can be realized at different spatial and temporal scales. The manner in which the observations will be collected and analysed is thus not without consequences for the image that is made of the system studied. This is why the question of sampling is a major preoccupation in ecology. It has benefited greatly, over the last decades, from technological advances in the means of observation. The development of statistical methods has allowed the elaboration of modes of management and treatment of data that are increasingly efficient. It is important to recall that until the 1960s ecologists had only a slide rule, and that the first calculators appeared in the 1970s. In such a context, the means of data analysis were necessarily limited although, paradoxically, that did not slow down the progress of ecology.

A sampling plan must follow some fundamental principles (Frontier, 1983).

- It is a question asked of the environment, i.e., we sample according to a certain type of preoccupation. It is thus necessary to define the *objectives*, which implies a choice of variables to study to answer the questions posed. We sometimes talk of *ecological descriptors* to designate the variables to be measured (Legendre and Legendre, 1979).

- The nature of the problem and the way in which it is formulated quite often implies a preconceived idea of the mathematical or statistical treatment that will be used. We can even say that the issue of sampling and that of statistical treatment are so closely related that they must not be treated separately.

- We are sampling not a static object but a system, i.e., a whole that is dynamic in time and in space. It is thus generally necessary to consider several scales of observation simultaneously. Most of the time, however, the means of investigation are limited. In particular, the sampling is subjected to natural constraints (e.g., diversity of habitats, heterogeneity of communities), technical constraints (e.g., reliability of instruments, adequacy of sampling methods to the behaviour of organisms), and human and financial constraints (e.g., frequency and number of samples). The strategy of the researcher is thus to find a compromise between the objectives and the constraints.

In reality, the choice of sampling scale is certainly one of the most difficult stages of a sampling plan, because the choice is fundamental for the observer (see Chapter 8). In physics, a law can be applied at different scales: when it is evident on a small scale, it can safely be extrapolated to a larger scale. This is not the case in ecology. Depending on the questions they want to ask, observers are thus forced to define the amplitude of the field sampled as well as the density of observations needed on the spatial as well as temporal level. Since observations on the small scale are easier to make and generally cheaper than observations on a large scale, the result is that many studies done on the small scale teach us little as to the ecological importance of factors acting on a larger scale. In particular, some phenomena observed on a small scale may not correspond to dominant phenomena responsible for patterns observed on a large scale.

4.1.2. New technologies

Ecosystem ecology, like other sciences, is dependent on technology. The phenomena that it studies require means of observation adapted to questions posed as well as to the different spatial and temporal scales concerned. Over the past few decades, advances in the field of instrumentation have not only allowed us to make innovative research, but also generated an evolution of concepts and the challenging of paradigms.

As Bachelard writes, the expansion of scientific thought in contemporaneous forms is revealed as a solidarity of genius and technology. It is not to be expected in this work that we will treat all of these technological developments, some of which are presented in detail in the work of Legay and Barbault (1995). Here we only give some illustrations.

a) Access to large scales of observation: remote sensing by satellite

Ecologists for a long time have worked on the scale of the land parcel, or even station. The development of means of observation from space have profoundly modified the scale and frequency of measures that can be considered, as well as the modes of working (Blasco, 1995; Bonn, 1996). By means of remote sensing, one of the major advances in ecosystem ecology has been to spatialize occasional measurements without sacrificing their precision.

Observation satellites are particularly useful to indicate changes in the land cover due to deforestation, extension of agricultural activities, sedimentary dynamics in deltaic or coastal environments, or other factors. The most spectacular innovations are undoubtedly those that have allowed the calculation of primary productivity of autotrophic plants, or to measure evapotranspiration by the measurement of surface temperatures.

More progress is awaited with new methods that allow us, for example, to identify forest species or to combine visible data, thermal data, and hyper-frequencies that give us access to parameters of structure and biomass of ecosystems.

Much information is accessible from satellite observation instruments. The following are examples:

- *The physiography of watersheds*, described by numerical models of the terrain, as well as land use and cover. The high-resolution optic satellite instruments of the SPOT series provide information on the land relief and cover at decametric scales. The Vegetation instrument carried on SPOT-4 allows access to optic characteristics of plant cover and land cover, with a spatial resolution of about one kilometre.

- *The quantitative estimation of rainfall*, notably by meteorological radar.

- *The estimation of soil humidity*. The radar instrument of satellite ERS-2 operates at resolutions of a few tens of metres and deduces the surface humidity of soils.

GTOS

The global terrestrial observation system was created in 1996 by five agencies of the United Nations (FAO, UNESCO, OMM, PNUE, ICSU) to offer better access to national and international networks monitoring terrestrial ecosystems. It complements two other global observation systems, GCOS (climate) and GOOS (oceans). The system of information and data management comprises a wide spectrum of databases for *in situ* observations, remote sensing data, and information on land use. It also includes analytical tools and models. The objectives of GTOS chiefly involve the indication of changes in land quality, the loss of biodiversity, climatic changes, and their consequences.

The satellite *Terra*, launched in December 1999, is designed to complement the set of already operational satellites of the Earth Observation System of NASA. Its mission is to collect the information needed to predict climatic evolution of the earth. Its sensors measure, in particular, the radiation balance of the planet, the temperature of the ocean and the soil, the snow cover, the ice cover on the oceans, aerosols, natural fires, deforestation, photosynthetic activity of plants, the productivity of the ocean, and other information. In comparing the deforestation with the rainfall and the local snow cover, for example, we can better understand how the disappearance of trees disturbs the water cycle. The satellite measurements are confirmed by measurements taken with instruments on the soil or on aeroplanes, ships, and buoys.

---

b) Exploration of deep ocean floors: submersibles

The knowledge and observation of great ocean depths has largely benefited from advances in the last two decades in the field of exploration engine technology. While oceanographers for a long time were constrained to operate blindly by dredging and trawling, they can now observe benthic ecosystems directly and at all depths. Modern exploration submarines such as the *Nautilus* allow us to explore large areas, take samples *in situ*, and measure the significant physicochemical parameters. In the late 1970s, they enabled the discovery of flourishing animal communities of a new type near hydrothermal springs (Laubier, 1995). This fauna includes taxa that are not similar to any other form presently known and that belong to genera or orders that were considered forms of the Palaeozoic and Mesozoic ages. Scientists have also discovered the existence of a new mechanism of organic matter synthesis, bacterial chemosynthesis. Thus, thanks to submersibles, we can explore deep ocean environments that were thought to be uninhabited in the 1960s, and discover in them ecosystems that are still unsuspected.

c) Tracers and markers of ecological processes: isotopes

Many elements are composed of several isotopes that differ slightly in their atomic mass and the number of neutrons contained in their nucleus. These isotopes possess nearly identical chemical activities but different physical properties. For example, while standard carbon in the biological tissues is ^{12}C, a fraction of about 1% will be ^{13}C. There are subtle variations in the relative concentration of two elements that depend on the nature of the biochemical reactions involved in the synthesis of carbonate molecules. This ratio is higher in atmospheric CO_2 than in the organic compounds produced by photosynthesis.

The use of stable isotopes of light elements in natural tracking has opened up new research potential in the study of ecological processes (Rundel et al., 1988; Schimel, 1993) (Table 4.1). Better sensitivity of mass spectrometers has allowed scientists, moreover, to measure isotopes that could not earlier be detected, opening up a route to new methods of retrospective analysis. For example, isotopes of C, N, S, H, O, Sr, or Pb have provided long-term data on the quantity of anthropogenic pollutants deposited in natural environments throughout the world.

Table 4.1. Relative abundance of stable isotopes normally used in ecosystem studies (Schimel, 1993)

Carbon	Oxygen	Hydrogen	Nitrogen	Strontium	Potassium	Magnesium
^{12}C 98.89	^{16}O 99.763	H 99.9844	^{14}N 99.64	^{84}Sr 0.56	^{39}K 93.08	^{24}Mg 78.8
^{13}C 1.11	^{17}O 0.0375	D 0.0156	^{15}N 0.36	^{86}Sr 9.86	^{40}K 0.0119	^{25}Mg 10.15
	^{18}O 0.1995			^{87}Sr 7.00	^{41}K 6.91	^{26}Mg 11.06
				^{88}Sr 82.58		

Stable isotopes were first used in the earth sciences for geochemical, hydrological, and sedimentological studies. Applications in ecology came much later because of difficulties of access to isotope mass spectrophotometers. But isotope tracking has proved for some years to be a powerful tool for ecological research. These isotopes are used for the following purposes:

- *To trace metabolism* from the elementary reaction processes (photosynthesis, nitrogen assimilation) to the organisms (determination of alimentary regimes, water use efficiency in plants).

- *To trace processes in the environment,* to indicate the natural denitrification in underground waters, and measure the time of presence of organic matter in soils. The stable isotopes of carbon, ^{12}C and notably ^{13}C, are useful tools in the study of the time organic matter remains in soils (Mariotti, 1995).

- *As paleoclimatic markers* (paleotemperatures). These are powerful tools for the reconstitution of paleoenvironments.

Carbon also exists in unstable radioactive forms such as ^{14}C, the half life of which is 5715 years. In biological material, measurement of the degradation of ^{14}C since its incorporation is a means of estimating age (the carbon dating method).

d) Access to the infinitely small: identification and biochemical characterization of microorganisms

The considerable advances in microbial ecology during the 1990s were made possible by the evolution of technologies of direct counting and the development of immunological methods linked to the use of monoclonal antibodies. The application of techniques of molecular biology to the domain of aquatic ecology, for example, has enabled the precise identification of microorganisms (Lim, 1996; Simonet et al., 1995). These techniques allow us to consider, in the long term, an exhaustive analysis of microbial populations, as well as an approach of qualitative and quantitative modifications of the structure of microbial populations under the effect of biotic and abiotic factors.

We have greatly progressed also in the indication of functional groups thanks to the development of biochemical methods, based on the monitoring of natural organic markers (Scribe and Bourdier, 1995). The basis of this approach is the possibility of describing a multispecies community *in situ* from its biochemical composition. The fatty acids of polar lipids constituting membranes of eubacteria are among the organic compounds that present a taxonomic specificity (Frederickson et al., 1987). Thus, accounting for fatty acids of various origins (bacteria, protozoa, microalgae) in the reserve lipids of pelagic zooplankton could be a new way to tackle transfers of matter within food webs (Desvilettes et al., 1997).

4.2. THE COMPARATIVE APPROACH

Aristotle was one of the first to discover the heuristic value of comparison, and he is recognized as the founder of the comparative method (Mayr, 1982). This is a method of benefiting from the similarities or differences observed between structures or functions of a similar type of biological system (population, community, ecosystem) subjected to a range of varied conditions (different situations of climate, biotope, competition, food resources, and so on). *The principle of the approach is to use, as controlled variables, natural variations of certain environmental factors in space and time to better highlight the effects of these variables on the state of the system studied.* This is equivalent to a "natural experiment" that offers an advantage over static and occasional description in indicating the essential mechanisms of the organization and the dynamics of a type of system studied, not those of this or that local system.

Comparative studies theoretically also consider data observed in time and space as well as experimental results. For phenomena relative to large spatial and temporal scales, the comparative approach is substituted for the experimental process, because the latter thus becomes difficult, even impossible to implement. But only comparable things can be compared. The success of comparative analysis depends, to a large extent, on the preliminary work of identifying structures or processes to be compared and standardizing methodologies. This remark may appear trivial, but very often these conditions are not met in ecosystem research, which leads to the accumulation of unusable results.

The comparative approach is particularly relevant in studies on:

- mechanisms of structuring populations as a function of the level of heterogeneity of the biotope (along ecological successions or gradients, for example);

- mechanism of sharing time or space among species that are similar in phylogenetic or ecological terms, overlapping of niches on the spatial or trophic level;

- the dynamics of populations subjected to factors of different intensity;

- the influence of certain ecological factors such as the predictability of resources or climatic conditions, the intensity of predation, and the abundance of food on demographic strategies used by species.

Ecological gradients are especially interesting areas for examining the relationships between living communities and their abiotic environment. There is an inherent heterogeneity in such gradients. To put it simply, a gradient is a continuum in a biotic or abiotic parameter that is sufficiently structuring for one to be able to observe modifications of the composition and organization of communities along this gradient. Certain spatially continuous gradients generally present a pattern of zonation that is relatively independent of organisms or communities. Mountain slopes, the banks of lakes, or coastal marine environments are examples of gradients in which an environmental factor (here the altitude or depth) plays a major role in the organization of communities. It is a form of organization of nature in which we can look more easily for empirical relations between environmental conditions, the distribution and abundance of species, and the biological traits of these species. In gradients, therefore, we can conveniently put forth and rapidly test hypothesis or conduct experiments in order to improve our predictive capacities concerning the responses of ecosystems to changes in the environment.

4.3. THE EXPERIMENTAL APPROACH

An experiment is any organized procedure of acquisition of information that comprises a confrontation with reality in the perspective of an expressed objective (Legay, 1997). Experimentation is considered the exemplary scientific method by specialists of the physical sciences. It follows rigorous criteria as to the choice of procedures (Hairston, 1989).

Experiments are conducted in order to answer formal questions generally in the form of hypotheses from an observer who ask about the manner in which nature functions. In theory, the experimental process supposes that we can evaluate the consequences, for the ecosystem, of changes in one or several controlled factors, i.e., factors that can be controlled (e.g., temperature, light) and the effects of which can be measured. In reality, the experimental approach does not always lead to indisputable results since it is often not possible to control all the interactions between the elements of an ecosystem.

4.3.1. *In situ* experiments

Over the past few decades, the manipulation of entire ecosystems has been one very useful means of investigation for the study of ecosystem functioning. These true-to-scale experiments have often treated the ecosystem as a black box (Beier and Rasmussen, 1994) but their results have been used to develop and test predictive models, useful in decision making, about the consequences of impacts of anthropic origin on the structure and functioning of ecosystems. We can also benefit from exceptional events. The task of the researcher is thus to infer and reconstruct the conditions in which this "experiment of nature" took place.

a) Manipulation of ecosystems

C. Juday was one of the first North American scientists to use natural lakes for experimental studies. He artificially enriched a small lake in Wisconsin in the 1930s to study the effect of fertilizers on the production of plankton and the growth of fish (Juday and Schloemer, 1938). Subsequently, students of Juday conducted several true-to-scale experiments, of which the one on lakes Peter and Paul was the subject of several publications. These two lakes, of nearly equal size and connected with each other, were separated by an earthen barrage. Lake Peter was treated with lime, while Lake Paul, not treated, was used as a reference system (Johnson and Hasler, 1954). This experiment was a strong inspiration for the setting up of two experiments on large ecosystems in the United States: the Hubbard Brook ecosystem in New Hampshire (Likens and Bormann, 1972) and the zone of experimental lakes in Ontario, Canada (Johnson and Vallentyne, 1971).

The zone of experimental lakes northwest of Ontario comprised several hundreds of lakes, of which 47 were selected as experimental sites. Chemical products were introduced in some of them in order to modify the pH of the water, by the addition of lime or acid. Following these disturbances, Schindler et al. (1985) observed that certain fundamental characteristics of the ecosystem, such as productivity, were relatively strong and changed little under the effect of the treatment. On the other hand, the nature of the dominant species was modified: rare species became common, and species that were earlier common became rare. These experiments, which are still being carried out, indicate the existence of many biological redundancies in these lakes (Schindler, 1990, 1998).

The Hubbard Brook ecosystem consisted of a series of small watersheds that were clearly separated and quite homogeneous in terms of geology, soils, climate, and vegetation. These conditions were ideal for comparative experiments on the ecology of forest landscapes by manipulation of these small basins, using some of them as a system of reference (Likens, 1985).

The effects of deforestation were studied, as well as the effects of forest practices on the temperature of rivers or the consequences of acidification. The research showed that acidification alters the biological and biogeochemical structure by reducing the biological diversity, increasing predator density, and reducing the complexity of food chains. The long series of ecological and biogeochemical data collected on these experimental sites since 1955 constitutes a unique basis from which it was possible to elaborate and test hypotheses on the functioning of ecosystems.

In Europe, the projects Nitrex (nitrogen saturation experiments) and Exman (experimental manipulation of forest ecosystem in Europe) were set up by many North European countries to study the impact of an increasing input of atmospheric pollutants on forest ecosystems, and the possibility of remedying the acidification of soils by the addition of buffer substances or the reduction of pollutants (Rasmussen and Wright, 1998).

These projects had a double objective: to study certain aspects of functioning of ecosystems and to provide the information needed to elaborate national and European policies on the air quality (Beier and Rasmussen, 1994). For example, it was indicated that the suppression of sulphides in emissions led to a rapid reduction (in a few months to a year) of the concentration of reduced sulphates in soils. On the contrary, the reduction in aluminium concentration had an impact only several years after the reduction of acid inputs. The negative effects of acidification could be prevented or reduced by the addition of lime or fertilizers. The major results of these projects served to elaborate a protocol signed in 1994 on the reduction of sulphide emissions in Europe (Rasmussen and Wright, 1998).

Iron content in oceans

An experiment in enriching the iron content of surface layers of the ocean was carried out in the eastern equatorial Pacific (Frost, 1996). It was designed to verify the hypothesis that despite high nitrogen and phosphorus concentrations in sea water, the phytoplankton is not abundant in this region of the ocean because of low levels of iron, an essential element in the synthesis of chlorophyll. Preliminary experiments showed that the addition of iron in bottles filled with Pacific water encouraged phytoplankton development. But the results of experiments realized in microcosm could not be extrapolated to the natural environment without verification.

Experiments of *in situ* fertilization of the ocean by soluble iron were thus carried out in 1993 and 1995. Areas of 64 to 72 km² were treated and monitored during their displacement in the ocean, which was 10 to 100 km a day.

The effect of iron fertilization was immediate and spectacular: the growth rates of phytoplankton doubled and it became 20 times as abundant, reaching levels observed in blooms of coastal environments. Large planktonic algae (diatoms) became dominant. In parallel, the concentration of nitrates reduced by half. This experiment thus met its objective, which was to demonstrate that the iron deficiency was responsible for the lack of phytoplankton in the equatorial Pacific. Yet, one week after the last iron input, the phytoplankton of the treated zones returned to a state similar to that of the untreated zones, as if the iron introduced had not effectively cycled in the environment. It thus seems that episodic inputs of iron produce short-term growth of phytoplankton, but the iron subsequently disappears from the system.

Although *in situ* experiments are particularly useful to study the responses of an ecosystem as a whole, they present some bias. In particular, in field conditions it is not possible to control most of the variables.

The focus of interest is usually one variable (more rarely a small number of variables) that is manipulated as a function of an existing plan, while the other variables fluctuate independent of the experiment. Thus, we make the hypothesis that these variables affect the experimental treatments equally, which we must recognize is not always verified. We must add that these experiments are recorded over time and are quite costly. We must multiply them *in situ* to progress in our understanding of the functioning of ecosystems.

b) Introduction or elimination of species

The introduction or on the contrary the selective elimination of species in an ecosystem constitutes another form of true-to-scale experiment. A good example is the consequence of the introduction of *Oreochromis alcalicus grahami*, a fish species endemic to Lake Magadi (Kenya), into Lake Nakuru (Kenya) to control mosquito larvae. In this saline lake, which had no fish, the cichlid introduced in the late 1950s developed rapidly by feeding on abundant populations of the cyanobacterium *Spirulina platensis*. The most marked effect was the development of a very large population of ichthyophagous birds only a few years after the introduction. The avifauna, which was essentially composed of pink flamingos before the introduction of the tilapia, was greatly diversified and ultimately exceeded 50 species of water birds.

4.3.2. Controlled experimental systems

If we are capable of understanding the functioning of an ecosystem, we must be capable also of reconstituting them artificially or, at the very least, of reconstituting miniature ecosystems that maintain themselves without human intervention. This is a challenge that some ecologists have taken up. *Ecospheres*, for example, are aquariums in which small organisms live without exogenous inputs. Clair Folsome, a professor at the University of Hawaii, conducted an interesting experiment in the 1960s: he placed a type of alga and a crayfish in a closed container filled with water. This ensemble, which successfully survived in total autonomy, revealed the possibility of cycling 100% of the wastes in a closed environment: the algae fed the crayfish, whose wastes fed the algae.

a) Microcosms and mesocosms

Microcosms and mesocosms are artificial ecosystems that simulate the functioning of natural ecosystems. According to Bloesch et al. (1988), microcosms are experimental structures containing the important components and created to study the principal processes that are at work in an ecosystem. The prefixes micro-, meso-, and macro- refer to the size of experimental ecosystems. Such systems have a degree of organization intermediate between the simplicity of laboratory assays and the complexity of natural environments. They thus allow us to study the effects of certain disturbances in simplified and controlled environmental conditions. They are designed to better understand the behaviour, progress, and effects of pollutants within aquatic ecosystems. In mesocosms, the logistic constraints inherent to field studies are eliminated. We can thus multiply the experiments and create replications, which makes it possible to apply statistical techniques.

In ecology, mesocosms are a means of testing ecological theories and better understanding natural processes by simplifying the complexity of the natural environment. They have long been used by limnologists, toxicologists, and microbiologists. Their use by terrestrial ecologists is relatively more recent.

The research carried out in microcosms has played an important role in the comprehension of population ecology, especially with respect to models of competition and predation (Drake et al., 1996). The most recent studies focus on the ecology of ecosystems and communities in the domains of climatic changes, biodiversity, restoration of habitats, and so on. The common aspect of these studies is that they allow us to manipulate one factor of the environment (e.g., CO_2, fluctuation of the water level, diversity) in order to evaluate the role it plays in the structuring of communities or ecosystems (Fraser and Keddy, 1997).

b) Ecotron

The Ecotron (Lawton, 1996) is an experimental structure constructed by the Imperial College of London. It is made up of 16 containers in which natural environments are simulated by the control of factors such as light, rainfall, humidity, and temperature. Ecologists can thus reconstruct small terrestrial ecosystems and monitor their evolution over several months, in a controlled manner. The miniature ecosystems may include up to 30 species of plants and metazoa, constituting four trophic levels (plants, herbivores, parasitoids, and detritivores) in interaction for several generations. There is a possibility of conducting replications and thus of drawing statistical conclusions.

One of the experiments carried out in the Ecotron consisted of directly testing the role of species richness in the functioning of ecosystems (Naem et al., 1994). Ecosystems were reconstituted at three trophic levels, the richest communities of which contained 31 species and the poorest only 9 species. The communities having the lowest species richness were impoverished forms, in terms of species composition, of those more rich in species. It was shown in this way that most of the processes measured varied significantly with the species richness, but not systematically in the same sense. Nevertheless, the consumption of CO_2 and the primary productivity decreased when the species richness reduced.

Biosphere 2

Biosphere 2 is an experiment financed by an American millionaire with a double objective: to understand the functioning of the giant organism earth and to conceive a structure that could be transported into space with a view to colonizing it. An entirely autonomous miniature world was created in the Arizona desert. Within this airtight structure, the major regions observed on the earth's surface were recreated: an ocean, a desert, marshes, a savannah, and a rainforest, as well as an agricultural and livestock area. The 4000 species of plants and animals populating Biosphere 2 were carefully selected in order to obtain the most perfect possible replica of the earth. Even though Biosphere 2 was not considered a system identical to the biosphere, it offered a unique occasion to test the hypothesis that the biosphere regulates its environment by the bias of retroactive processes that microbial, plant, and animal communities effect. On 26 September 1991, four men and four women, called bionauts, entered a giant greenhouse of 200,000 m^3 spreading over 1.28 ha and designed to last 100 years. These bionauts lived two years in the greenhouse producing their own food, without inputs from outside, and recycling the water and air. Still, the oxygen level in the air fell to 14% and oxygen had to be injected to keep it at more than 16%. The experiment was thus partly a failure.

---

c) Representativity of results obtained in controlled environments

The use of microcosms has been the subject of many debates. Some ecologists are sceptical as to the generalization possible from results obtained in an artificial environment to natural ecosystems. Considering their simplicity, microcosms have only distant analogies with complex natural systems. Among the arguments advanced are the following:

- Communities recreated in laboratories are highly artificial and too simple to teach us anything about the real world. For example, experiments carried out on plants isolated in a microcosm show that the carbon dioxide level in the air increases with photosynthetic activity and a better use of water. These results are relevant at the scale of observation. On the scale of the ecosystem, however, plants do not respond homogeneously to the increase in CO_2 level. Some will benefit more than others, which modifies the interactions between species and the structure of communities.

- Species brought together in experiments of the Ecotron type are artificial assemblages that do not share a common history. In other words, the communities are not in equilibrium.

- Ecosystems controlled without exchange with the external world do not allow emigration or immigration, whereas any real ecosystem is open to its surroundings. In particular, mesocosms of an aquatic environment lack water-sediment exchanges and/or exchanges between the pelagic environment and the coastal environment (Schindler, 1998).

- The scales used are inappropriate, often too small: some processes act on larger spatial or temporal scales, or some organisms are too large in relation to the mesocosms. The temporal scales are often too small to evaluate with precision the response of organisms that respond slowly.

Schindler (1998), drawing conclusions from several years of research on the Canadian lakes, showed that the results of experiments in microcosm and mesocosm were different from those obtained by manipulation of actual lakes. Application of the results obtained in controlled environments to lake management would thus lead to erroneous decision-making. Although certain biases such as the rate of renewal of water or water-sediment interactions can be corrected, many other biases cannot. Thus, it is difficult to introduce terminal predators, which are rare but important, or highly mobile species. The conclusion of Schindler is unambiguous: experiments that do not address the scales relevant to ecosystems are inappropriate if we are trying to predict the responses of lacustrine systems in their entirety.

Clearly, the dialogue is difficult. Some ecologists, such as Carpenter (1996), do not hesitate to say that the major advantage of microcosms is to allow scientists to publish results in a short time, which is indispensable to a successful career. Certainly the experiments in controlled systems of the Ecotron type can never replace observations in the natural environment or analyses of long series of data. They are also not highly relevant in tackling questions that cannot be addressed in a natural environment or that require expensive operations. It is obvious that we need a large variety of approaches to understand the functioning of nature. Each has its advantages and disadvantages. The fundamental issue is to maintain a critical perspective about the results obtained.

4.4. MODELS AND SIMULATION

Systems theory plays a central role in ecological theory and practice involving the ecosystems and the biosphere. The creation of models and simulation is based on many methods that are the most widely used by the systemic approach.

When there are only a small number of variables, ecologists use simple analogical models established from a previous analysis, seeking to combine the principal elements of a system to put forth hypothesis on the behaviour of the whole. This exercise becomes impracticable when the real world observed is too complex and many variables are in play: e.g., diversity of components, complexity of interactions, non-linearity of phenomena, multiplicity of cause-effect relations, retroactive loops. Modelling is meant to tackle these problems: a model is elaborated to try to take the complexity into account while reducing the number of parameters in question. The idea one has of the real is represented as well as possible, then the representation of this idea (the model) is confronted with data from observation. The model summarizes and explains; it includes hypotheses on the mechanisms that can explain the results observed. Modelling thus becomes an essential tool of theoretical analysis of ecosystems.

Models and simulation

Analysis of system, modelling, and simulation are the three fundamental stages in the study of dynamic behaviour of complex systems.

• *Analysis of the system* consists of defining the limits of the system to be modelled, identifying the important elements as well as the types of linkage and interaction between these elements, and organizing them in a hierarchy. In particular, we must identify the flow variables, the state variables, the delays, and the retroactive loops or feedback as well as their influence on the behaviour of different subsets identified within the system.

• *Modelling* is the set of activities intended to represent a real object by one or a few models, either in a schematic and didactic form in order to render this object intelligible to investigators, or in a mathematical form. Using a model, we can mentally express our perception of the "anatomy" and "physiology" of the object of study. According to Jorgensen (1997), a model can be considered primarily a means of synthesizing elements of knowledge involving a system. The quality of the model thus depends on the quality of knowledge about the elements of the system and the available data.

• *Simulation* pertains to the behaviour of a complex system over time. It allows us to study the evolution of a system by varying one or several factors and combining "calculated" values with "observed" values. The results of the simulation must not be confused with reality. The simulation does not give an exact solution to a problem posed but can draw general trends of behaviour from a system, the directions in which it will probably evolve. It allows us to verify hypotheses without having to make experiments that may destroy the system studied. In this sense, simulation is sometimes considered a tool to help in decision-making when a choice needs to be made among possible scenarios.

4.4.1. The origin of systemic models

Model and *modelling* are words that have pervaded the scientific vocabulary for about 20 years. It was between the world wars that L. von Bertalanffy developed systemic ecology, which is part of the general theory of systems. This science of systems, still nascent, aims to uncover the rules by which parts are assembled to make functional wholes. It is interested in the possibilities of reproducing this behaviour in other systems. The lack of technical means (computers) limited the operational development of models for some years. It was around the 1960s that a real epistemological breakthrough occurred: the mathematical representation that would come to be called "model" was no longer seen as the most reliable formalization possible of a reality, but considered a simplified and idealized image, even a caricature of this same reality (Pavé, 1997). The model served to represent the world as we perceive it, as we believe it to be, and not as it is. This being so, there are a wide variety of practices, approaches, tools of modelling, and types of models. The demand is great, while the competencies are still limited in the field of ecology.

Mathematical modelling of the structure and functioning of ecosystems has greatly influenced the scientists who defined the objectives of the International Biological Programme between 1965 and 1972. One of the objectives of the IBP was to validate this process. Still, the results were partly deceptive and the objective of modelling ecosystems did not give the results expected because the question finally proved much more complex than imagined at first. In particular, it was not possible, considering the knowledge and statistical capacity available in the early 1970s, to develop models that would be sufficiently precise descriptions of ecosystems or to use them reliably for a predictive purpose. Although scientists continued to use models, modelling lost its appeal as a method of structuring research on ecosystems. On the contrary, the reductionist approach took over and models became tools to organize data and support reflection. The biocentric tendency was found to be confirmed and ecologists contributed their effort to the trophodynamic functioning of ecosystems and biocoenoses, much more than to studying the system itself. Since that time, progress in statistics rather than in modelling has given a new momentum to modelling, which is now perceived to be essential to comprehension of the functioning of ecosystems (Blasco, 1997; Blasco and Weill, 1999; Coquillard and Hill, 1997).

4.4.2. The use of models

The objective of modelling has been to imagine how nature functions and to formulate a mathematical model of the process. The advantage of models is that it obliges the users to order their observations and put forth specific hypotheses. However, the manner in which models are constructed often allows the possibility that hypotheses will be incorporated that are not always verified. Moreover, a frequent bias is the explanation of what is already known. Modelling thus has limits and its use is not without risk, even if the models are closely fitted within a statistical software. In any case, models are incomparably more efficient tools than the empirical approaches. Modelling actively intervenes in the three major functions of scientific research: identification of questions; problematic and acquisition of knowledge; and definition of actions and study of their consequences (Pavé, 1994).

In terms of use, we can distinguish three classes of models:

- Models as *cognitive tools* that allow the formalization of knowledge and hypotheses in a scientific process.

- Models as *normative tools*, which must represent reality as well as possible through strict validation. Such models also contain an explanatory proposition to be tested.

- Models for *decision-making*, which provide various predictions according to the scenarios chosen. These are the most in demand, but also the least developed.

Several stages are necessary before a model can be retained and used (Schmidt-Laine and Pavé, 2002) (Fig. 4.1):

- It is constructed as a function of what is to be represented as well as the use to which it will be put.

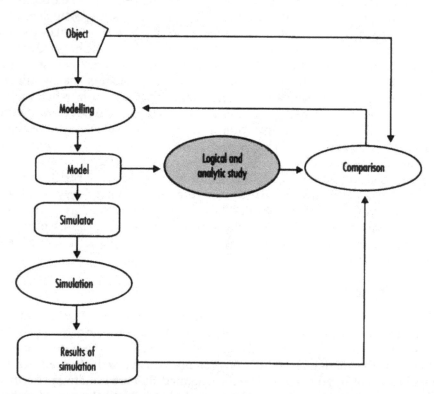

Fig. 4.1. General methodological process of modelling and simulation. Two stages can be distinguished: a model may be suitable or unsuitable, and the simulator may give correct or incorrect results. Therefore, adequate verification is necessary (Schmidt-Laine and Pavé).

- It must be conformed to what is expected and must have a qualitative behaviour consistent with what is observed. If the model is faulty, this indicates an error in the formulation, which brings us to question the phenomenon modelled, to acquire more knowledge of the phenomenon, and to elaborate a new model integrating that knowledge.

- Throughout the study and use of the model, we use simulators that must correctly express the properties of the model, qualitatively and quantitatively.

- A model must be practically usable.

If the models are mainly elaborated to represent a process or phenomenon studied in a disciplinary framework, the combination or association of models, even the elaboration of new models combining the various aspects and factors of evolution of a "complex" system, becomes a means of integrating the various components of these systems (Schmidt-Laine and Pavé, 2002).

Part II
Structure and Organization of Ecosystems

Chapter 5

The System Concept and Attempts to Apply Physical Principles to Ecosystem Ecology

La nature est un temple où de vivants piliers
Laissent parfois sortir de confuses paroles.
L'homme y passe à travers des forêts de symboles
Qui l'observent avec des regards familiers.

Charles Beaudelaire, *Les Fleurs de Mal* (1857, Spleen et idéal)

"What thermodynamics does do is settle the age-old question of whether we eat to live, or live to eat. According to thermodynamics, we live to eat."

J. Roughgarden (1998)

"Nature can produce complex structures even in simple situations, and can obey simple laws even in complex situations."

Goldenfeld and Kadanoff (1999)

The "Newton syndrome" has characterized the evolution of science for the past three centuries. This founder of modern physics showed that by using a small number of theoretical concepts, such as the universal law of gravitation, it was possible to interpret the movement of planets, a phenomenon that intrigued human beings for thousands of years. This process has sometimes given the impression that science can always allow us to predict, with great precision, the evolution of natural systems from the knowledge of their present state. Since Newton's time, scientists have attempted to persuade themselves that science can describe a world whose complexity is only apparent using relatively simple parameters the dynamics of which is determinist and predictable (Solbrig and Nicolis, 1991). The formulation of the second law of thermodynamics in the 19th century further reinforced the conviction that systems evolve spontaneously towards a universal state: thermodynamic equilibrium. The result was that in the beginning of the 20th century, physicists were nearly unanimous in acknowledging that the laws of the universe were deterministic and reversible.

Nevertheless, by the end of the 20th century, the Newtonian paradigm was boldly challenged. It was thought, on the contrary, that the fundamental processes that were at work in nature were irreversible and stochastic. The laws of physics, reversible and deterministic, that were expressed by simple mathematical equations were powerless to explain the functioning of complex entities such as ecosystems. Indeed, as soon as we leave the domain of physics, we observe an unbelievable complexity and the limit between the simple and the complex, between order and disorder, is not as distinct as it earlier seemed (Goldenfeld and Kadanoff, 1999). In many cases, a slight modification of experimental conditions suffices to make new forms of complexity appear spontaneously: it is the deterministic chaos that has indicated the determinant role of stochastic effects on the evolution of systems. As pointed out by Nicolis and Prigogine (1989), today, no matter which way we turn, we find only evolution, diversification, and instability. We live in a pluralistic world in which we find phenomena that are deterministic as well as stochastic, reversible as well as irreversible.

From this rather unsettling observation, it remains only for ecologists to try to adapt to their fields of research the concepts and methods developed in other scientific fields to answer the fundamental question of the organization of nature. This is a matter of knowing whether certain laws stated for the physical world can be transposed into the living world and, if so, under what conditions to take into account the specificity of life. The following is a summary of the present thinking in this field, keeping in mind that it remains highly controversial:

— it is possible to apply systems theory to ecological systems;

— ecological systems are complex;

— they are structures in a communication network (cybernetic cohesion);

— they respond to laws of thermodynamics (thermodynamics of dissipative systems);

— they are organized hierarchically (hierarchy theory);

— they are dynamic in time and in space (non-linear dynamic systems theory);

— they can adapt and evolve (adaptive systems theory);

— they are self-organized and self-regulated (self-organized criticality theory).

These various approaches have provided useful information in answering the essential question of the organization of ecological systems, but that information remains fragmentary in the absence of a general ecological systems theory. However, we are making progress in this direction, even if the operational implementation of some principles is difficult.

5.1. A FUNDAMENTAL NOTION: THE SYSTEM APPROACH

The study of interactions between species and their biophysical environment involves an attempt to identify the causes of observed phenomena. The notion of *causality* is moreover implicit in most ecological thinking. However, in seeking to identify the "causes" to understand their "effects", we explicitly introduce the notion of a system in which causes and effects are closely linked. From this reflection, various scientists endeavoured to build a general systems theory during the 1950s from cybernetics and information theory. A "second systemics" appeared in the 1970s and 1980s, integrating two other essential concepts: that of communication and that of autonomy or self-organization (Durand, 1996).

5.1.1. The macroscope

Ecology is a field in which application of the *system approach* is preferred. This approach was symbolized by the concept of *macroscope* in the works of Joel de Rosnay (1975). The author indicated thus that we must consider this approach not as a *science*, a *theory*, or a *discipline*, but as a "new methodology" allowing us to collect and organize data with a view to greater efficiency of action. The concept of system is in fact an intellectual construction designed to understand and explain the functioning of a part of the universe that we are studying. It is one way of considering complex phenomena. *The system approach attempts to look for invariables, i.e., general principles, structural and functional, that are common to various systems.* These principles can be used to organize data into models in order to facilitate communication and to use the data as a basis for reflection and action.

Unlike the analytical approach, the system approach encompasses all the components of the system studied and, in a dynamic perspective, takes into account all their interactions. For example, the paradigm according to which abiotic factors control biotic elements in an ecosystem has long prevailed in ecology. According to this paradigm, the climate and soil control the distribution of the vegetation and its biological efficiency. This concept has gradually been overtaken by the idea that physicochemical systems are also controlled by living organisms. This is particularly the case with the composition of the atmosphere (see Chapter 16). Similarly, the alteration of rocks is as much the result of the activity of microorganisms as of physicochemical phenomena. The various elements of the ecosystem are thus subject to a double effect of action and retroaction.

5.1.2. Structured and organized systems

An essential characteristic of systems is *organization*. The word is difficult to define. It is a property that is independent of the components of a system. It is a network of relationships between components or individuals resulting in a whole that has qualities that each of the components taken individually does not have. Behind the notion of organization there is the idea of some optimization of the agency of components of a system.

Organization has a structural aspect and a functional aspect. Its principal structural traits are the following:

— The *elements* of the system that can be identified, numbered, and classified: biological species, individuals, developmental stages, as well as physical components (e.g., temperature, rate of flow) and chemical components (chemicals) of an ecosystem.

— The *reservoirs* in which energy, matter, and information may be stored. The existence of reservoirs is essential to the proper functioning of a system because it allows the system to adapt its functioning.

— A *communication network* that allows exchange of information, matter, and energy between the elements and the reservoirs. The food web is an example of a communication network, very popular in ecological literature. Other types of networks are those that allow species to communicate among themselves (e.g., sounds, pheromones, vision).

— A *frontier* that separates the system studied from its environment and is more less permeable.

The functional traits that characterize a system are the following:

— The *inward and outward flows* of the system that represent the relationships of the system with its environment. These relationships are more or less numerous and intense according to whether the system is more or less open to the outside world.

— *Flows* of various kinds within the system: of information, energy, or elements that circulate among the reservoirs. These flows circulate in the communication networks and the rates of various flows are modulated and controlled by "gates" or *regulatory systems*.

— Information loops called *feedback* that play a determinant role in the functioning of systems and combine the effect of reservoirs and flows.

— *Delayed responses* allowing adjustments over time that are necessary for an efficient operation of the system.

5.2. COMPLEXITY

Complexity is another difficult concept to define precisely, since it belongs to our everyday vocabulary. We have come to use the word to signal our perplexity in the face of a situation that is difficult to analyse. The ecological literature, however, points out repeatedly that biological and ecological systems are complex.

According to Serres and Farouki (1997), a material structure is called complex if, simultaneously:

— it is composed of many elements belonging to multiple categories;

— these elements are grouped in subsets organized in successive hierarchic levels;

— these elements and the subsets are linked by numerous interactions and are of multiple types.

We can also distinguish several forms of complexity for ecosystems (Jorgensen, 1997):

- *The number of organisms and species on earth is very large and they are all different.* The magnitude of the number of species is about 10^7, while that of individuals is about 10^{20}. This large number of species give rise to a vast number of interrelationships. Thus, the complexity arises from the multiplicity of intra- and interspecific relationships between individuals, interactions between food webs, and interactions between organisms and their abiotic environment in time and in space. This observation has initiated a debate on the relationship between the stability of ecosystems and the biological diversity that has emerged for the present into a disappointing conclusion: there is no simple relationship between the two.

- A complex system is made up of a *wide variety of components or elements having specialized functions.* These elements are themselves interconnected by linkages whose nature and intensity may evolve over time. Such *relationships* are called *non-linear* because they may depend on other variables. Natural processes such as the propagation of waves, turbulence of fluids, and transport of sediments are often non-linear (Werner, 1999).

- There is a great deal of *feedback* in ecosystems. In particular it allows organisms to respond and adapt to changes in the conditions of their environment. An important characteristic of complex systems is the idea of *delays* that result from different rates of circulation of information and flows, and different durations of storage in the reservoirs. These delays play an important role in the phenomena of amplification and inhibition, which are characteristic of complex systems.

- The components of the ecosystem (its structure) and the processes are *organized in a hierarchical manner*, from genes to communities. Each level constitutes a unit influenced by higher and lower levels of the hierarchy.

- There is a high *spatial and temporal heterogeneity* because ecosystems are dynamic systems in which all the biotic and abiotic components modify themselves constantly. This high heterogeneity explains, in large part, the diversity of species. But it is also an additional difficulty when a system has to be modelled.

- Ecosystems and their biological components *"evolve" over the long term towards a greater complexity.* All species are faced with the question of how to survive in a changing environment. Mutations and natural selection are the origin of evolution and the appearance of new species better adapted to environmental conditions that are known to have changed constantly in the past under the influence of climatic factors (see Chapter 17).

- The direction and amplitude of changes that affect an ecosystem depend on pre-existing conditions. It is increasingly thought that *the history of ecosystems is an essential factor of their dynamics*. The importance of initial conditions has been highlighted with chaos theory.

- Finally, complex systems are those in which the whole is greater than the sum of its parts, not in a metaphysical sense, but in an important, pragmatic sense: given the properties of parts and the laws of their interaction, it is not easy to infer the properties of the whole (Simon, 1962).

An example of a complex system: the climate

The climatic system is an example of a complex system with multiple interactions. The rain that falls on the ground allows the growth of plants, whose transpiration (physiological process) contributes with evaporation (physical process) to return the water in the form of vapour into the atmosphere. This water vapour forms clouds that reflect part of the solar radiation towards the atmosphere, thus reducing the input of energy on the earth's surface and, consequently, the primary production. But water vapour also absorbs the infrared radiation coming from the earth, thus contributing to a greenhouse effect that modifies the thermal regime of the planet. Besides this, living things emit a certain number of so-called greenhouse gases that contribute to the warming of the planet. The temperature differences on the earth's surface create a heterogeneity of atmospheric pressure that generates winds and tornadoes. The winds create turbulence that favours the evapotranspiration of the soil surface and also contribute to the oceanic circulation, which in turn influences the global temperature. In this cycle, therefore, there is a constant involvement of the physics of clouds, the physics of states of surface, atmospheric radiation, the dynamics of oceans, and biological processes such as primary production, all in interaction (Rind, 1999).

More pragmatically, Pave (1994) distinguishes the following:

— A *structural complexity* that corresponds to sets of many elements in interaction. This is a topological notion that may result from the complexity of relationships observed between constitutive elements. It corresponds for the natural systems to a spatial structure that is more or less complex.

— The *complexity of behaviour* linked to the dynamics of a system that generates complex trajectories of variables of the state of this system. This spatio-temporal dynamics can modify the structure (topological or spatial) of the system itself.

In conclusion, the interactions between the constituents of an ecosystem are not only numerous but also diffuse over time and space, which makes it very difficult to identify the cause-and-effect relationships between them. The description of dynamic behaviour of ecosystems requires us to use non-linear functions, delayed response, and so on, to the extent that the pertinence of quantitative predictions cannot be evaluated with conventional statistical methods (Maurer, 1998). We are still far from having overcome this methodological constraint, which remains the cornerstone for the development of ecosystem ecology.

Biocomplexity

Biocomplexity is a fashionable term in the United States that can be defined as *the interplay between life and its environment*. Biological complexity results in fact from functional interactions between biological entities at all levels of organization, and the biological, chemical, physical, and human environment at all the levels of aggregation. It thus finds its origin in the multitude of physical, chemical, biological, and social interactions that act on biological organisms or are modified by these same organisms, including man. Among the characteristics of biocomplexity can be mentioned non-linear and chaotic behaviour, interactions at different spatio-temporal scales, unpredictability, and the need to study the system as a whole and not in parts.

Given that all biological systems, from the molecule to the ecosystem, are intrinsically complex, it has been difficult to understand their role and effects on the systems they are a part of. In this context, biocomplexity is thus presented as a sytemic and multi-disciplinary approach (involving biologists, geologists, chemists, physicists, sociologists, economists, statisticians, and so on). These collaborations make use of new technologies with a view to understanding the functioning and complexity of ecosystems. Research on biocomplexity has focussed, for example, on the following questions:

— How do systems with living things respond and adapt to disturbances?
— Are the adaptations and changes predictable?
— To what extent will climatic changes modify the distribution of species?
— Can we predict the combined effects of climatic and socio-economic changes?
— To what extent does diversity (of species, genetic, cultural) act on the stability of systems?

In fact, the theme of biocomplexity serves as an umbrella over major themes identified by the National Science Foundation (NSF), namely global changes, biodiversity, and the dynamics of ecosystems, or even the human dimension of the environment.

5.3. INFORMATION THEORY AND CYBERNETICS

The theoretical advances in cybernetics, establishing physical and mathematical concepts, offer tools to ecologists in understanding the mechanisms at work in ecosystems. Cybernetics is the study of communication processes and information control. According to Engelberg and Boyarsky (1979), the very essence of cybernetic systems lies in the existence of a communication network that ensures a connection between the various parts of the system to make it an integrated whole. The functions of this network are to pilot and regulate the system and to control the flows of matter in space.

Cybernetics, or the theory of command and communication in animals and machines, was invented by Norbert Wiener in 1948. The term *cybernetics* comes from a Greek word, *kubernesis*, which signifies the act of steering a ship and, in the figurative sense, the act of directing or governing (Durand, 1996). Although the definition of cybernetics given by Wiener is very wide, since it concerns machines and animals, its practical content is more limited and essentially treats the command of machines. However, according to Patten and Odum (1981), ecosystems are of cybernetic nature: the interaction between cycles of matter and flows of energy, under the control of an information network, generates a self-organization. In a cybernetic system, it is the transmissions of information that ensure most of the interactions and couplings of constituents among themselves and with the environment. These transmissions contribute to the construction of the whole by ensuring its cohesion.

Information theory developed in parallel with cybernetics, of which it at one time seemed to be only one aspect. But this new concept extended its scope considerably. Information theory has been developed further by telephone engineers of the Bell System laboratories, who were interested in channel networks in which messages flow. This theory is used to estimate the diversity of channels of the network, or the information measure of Shannon-Wiener, named for its authors. The engineers also sought to establish a relationship between the diversity of channels and the capacity of a channel to transport information. In communication theory, the quantity of information I carried by a signal is defined as a function of the inverse of the probability, inherently p, of the realization of the event: $I = f(1/p)$. In other words, when an event occurs it carries as much more information as it was more improbable beforehand (Frontier and Pichod-Viale, 1998).

It is from cybernetics and information theory that the general system theory gradually emerged in the 1950s. This theory would subsequently integrate the concept of self-organization as highlighted by Nicolis and Prigogine (1977) with dissipative systems.

The Shannon diversity index

One way of characterizing the species diversity in an ecosystem is to quantify the information that it carries, in the sense of information theory. In particular, the Shannon diversity index is used to measure diversity in biological communities by combining in a global index the estimation of the species richness and that of the greater or lesser "regularity" of the distribution of individuals between the species. There is a vast body of ecological literature on the art and manner of applying this formula. But the results are generally deceptive because we often end with trivial observations that could be made independently of all these mathematical treatments. Such a process that greatly modifies the elementary information may be advantageously replaced, whether in chronological series or comparisons of observations, by the study of species richness associated with a factorial analysis that allows a classification based on the nature and relative abundance of species. It is true that factorial analysis appeared chronologically later than the Shannon index.

Till now, ecologists may be thought to have only wasted their time. But some have fallen into a trap in believing that the Shannon information measure that describes the stability of the phenomenon in telephone networks can also be used to describe the stability of biological systems (Colinvaux, 1982). In fact, systems that function through a network of intersecting circuits representing possible choices for the circulation of information or energy are more stable if they have more intersections. Therefore, the error lies in having assigned to plants and animals the same role in a food web as the intersections of a telephone network. In nature, on the contrary, organisms block the flow by storing the nutrients rather than facilitating the flow of nutrients. The analogy between systems pushed too far results in inconsistencies that suppose that living organisms behave otherwise than they are known to in reality.

5.3.1. Feedback

In a simple system of cause and effect, inputs (energy, matter, information) coming from the environment provoke a response from the system that in turn transmits energy, matter, and information to the environment (Fig. 5.1a). In more complex systems, an input may be determined, at least partly, by an output (Fig. 5.1b). This is known as *feedback*.

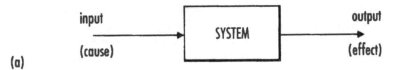

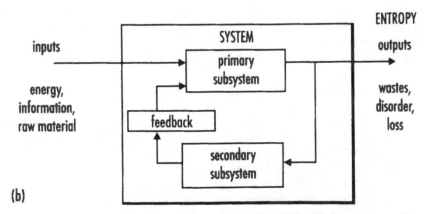

Fig. 5.1. (a) A basic input-output system. (b) An input-output system with feedback in which control mechanisms are diffused as in an ecosystem (Patten and Odum, 1981).

Cybernetic systems are retroactive systems in which the information network is closed. The action of any element in the system ultimately exerts an influence on every other element. Because of the existence of feedback, the results of an action are returned to the input end of the system in the form of information. *The idea of feedback thus creates a rupture with the theory of linear causality: the cause acts on the effects and the effects on the cause.* This mode of regulation makes the system autonomous.

A feedback is called *positive* if a modification leads to an action that goes in the same direction as the principal action, amplifying it in some way. The effects are thus cumulative. For example, in species competing for the same resource, a reduction in the density of one species (A) benefits the other species, whose density increases, which contributes to weaken species A further. The idea of positive feedback has been extended to organisms that benefit from one another through interactions of the symbiotic or mutualistic type. Whereas the role of competition has been the object of a great deal of attention in the study of communities, the role of positive interactions has probably been underestimated because ecologists feel they play only a minor role in the structuration of natural communities (Stone and Weisburg, 1992). It is thus a domain that is not well explored.

If, on the other hand, the new conditions have the result of triggering processes that go in the reverse direction of the earlier phenomena, they constitute a *negative feedback*. A feedback is negative if a change leads to the appearance of forces that oppose or slow down that change, which results in a regulation of the phenomenon (Frontier and Pichod-Viale, 1998). For example, an increase in phytoplankton biomass leads to the exhaustion of nutrient salts, which will slow down the increase, and to a rise in the population of zooplankton that feeds on the phytoplankton, which will also slow down the increase of the phytoplankton biomass. Prey-predator relationships thus have a regulatory effect on each of the elements of the chain, leading to a cascade of negative feedback. The effect of this negative feedback is to stabilize the cybernetic system, to regulate it, and to direct its functioning. When the system is disturbed, the feedback reduces the effect of the disturbance by regulating certain variables, which consequently maintains the equilibrium of the system. This is called resistance, as opposed to resilience in the distinction made by ecologists between these two forms of stability (see Chapter 10).

5.3.2. Information networks in ecosystems

The major function of an ecosystem is to ensure the perenniality of life. It must particularly clear the residues of preceding biological cycles and convert them into a form that allows them to be reused. Consequently, the basic functional network of an ecosystem is that which involves energy flows (trophic relationships) and cycling of matter (processes of mineralization). According to Patten and Odum (1981), however, there is also an information network that is superimposed on the basic network and helps to regulate it. Without the information network, these authors say, nature would be chaotic and disordered.

The invisible threads of nature that constitute this network (Odum, 1971) are all the factors, processes, and interactions involved in the control of flows of matter and energy. These are, for example, all the modes of use of the physical environment (air, water, soil, sediment) as well as the numerous visual, mechanical, auditory, olfactory, gustatory, and other signals that help regulate the flows of substances and energy in the biosphere.

It should be mentioned at this point that another property of information systems is *amplification*: small causes may give rise to significant effects. Many biochemical agents are active at low levels, even traces. The behaviour of organisms, body colour, pheromones, and many other signals may provoke direct or indirect responses in various components of the system. It seems that the amplification of signals is far from being an exception in nature. Ecologists normally take into account these elements in an individual and sectorial way, rather than seek to understand what role these signals collectively play at the level of the system as a whole. The explication of a vast communication network that structures the ecosystem will probably be a focus of research in the next few decades.

5.3.3. Communication networks in ecosystems

Communication can be defined simply as the transmission between a transmitter and a receiver. The transmitter must have a specific mechanism to produce stimuli and the receiver must have specialized receptors that allow it to capture these signals and respond to them specifically: messages of recognition between individuals of a single species or different species, messages of aggression or cooperation, submission, fear, or anger, messages of seduction in relation to reproduction, or messages contributing to structure social life. The living world is now understood to be strongly structured by a system of communication that has long been unknown. The evidence of these numerous communications between species (Darnet and Tordjman, 1992) supports the hypothesis of Patten and Odum (1981) about the cybernetic nature of ecosystems. The role of these communication networks between and within species in the dynamics of ecosystems needs to be seriously investigated. The few examples below are simply meant to illustrate the recent advances in this field.

a) Communication within and between species in aquatic systems

In aquatic systems, there is a wide variety of communication signals.

- *Visual signals* are used in water that is clear enough so that they can be perceived. In many fish species, mating displays use visual means, including movement, colour, and the position of partners.

- *Chemical signals* require a large variety of substances but are relatively slow and non-directional. Communication by chemical mediators is thought to be the first means of exchanges between species and individuals (Saglio, 1992).

- *Electrical signals* are particularly developed in certain species of fish, especially those belonging to the family Mormyridae, which use a rich repertoire of signals (Kramer, 1990). Such signals allow these species living in generally turbid water to maintain elaborate social relationships.

- *Sound signals* and *vibrations* are known to be used in vertebrates and invertebrates. Sound is propagated more rapidly in water (1500 m/s) than in the air, so it constitutes a good means of communication at long distance when the communicator is not visible. In reality, aquatic systems appear to be more noisy than has been believed till now. Many species of fish seem to have the means of producing sounds, as well as other signals, especially by the vibrations of the wall of their swim bladder.

During the past few years, much progress has been made in the understanding of information transfer between organisms through chemical substances. Interspecific communication by chemical substances apparently controls the spatial distribution of many planktonic species. It must be the origin of certain patterns of spatial distribution that were earlier thought to be the consequence of abiotic factors (Larsson and Dodson, 1993). There is a daily pattern of vertical migration in phytoplanktonic and zooplanktonic species that corresponds to a phenomenon of avoidance of predators (Jones, 1993; Lampert, 1993). These reactions may considerably modify the functioning of aquatic ecosystems by the transfer of nutrients between the surface and the deeper waters. At least four major types of signals are used in the aquatic environment:

- *Alarm substances* produced by tissues of injured fish alert the congeners and provoke reactions of fear (Carr, 1988; Smith, 1992) and similar phenomena have been observed in invertebrates (Hazlett, 1990). Benthic organisms may change microhabitats in rivers when they perceive chemical signals emitted by predators (Carr, 1988).

- *Repellents* may also be used by prey to keep predators away. Many algae are toxic for zooplankton but the latter appear to recognize toxic algae and avoid feeding on them (Larsson and Dodson, 1993). It has also been shown that the presence of cyanobacteria may modify the rate of filtration and the selection of prey in planktonic crustaceans (De Mott et al., 1991). These responses may modify the efficiency of energy transfer between the primary producers and the consumers. The browsing of phytoplankton is a question not only of quantity but also of quality: zooplankton feed more or less successfully on the phytoplanktonic species that are present.

- *Kairomones* produced by predators cause reactive modifications in the behaviour, morphology, and biological characteristics of prey species (Dodson et al., 1994). For example, the substances emitted by *Daphnia* lead to morphological reactions in green algae of the genus *Scenedesmus*, with the formation of colonies of 4 to 8 cells and the appearance of longer and more rigid thorns (Hessen and van Donk, 1993). It was subsequently shown that the presence of rotifers and copepods could also provoke the formation of colonies in *Scenedesmus*. In several species of *Daphnia*, the presence of predators such as larvae of *Chaoborus* or fish has been shown to cause the appearance of a casque and cephalic spurs (Tollrian, 1994). Besides, many studies have proven an influence of substances emitted by predators on the size at maturity and the fertility of *Daphnia* (Lampert, 1993).

- Some *substances* select the habitat: many species of aquatic invertebrates are sessile in the adult stage but have a planktonic stage that allows the dispersal of populations. It has been shown that larvae of coelentera, bryozoa, annelids, echinoderms, and other animals use chemical signals to choose habitats suitable for adults (Carr, 1988).

b) Communication in terrestrial plants

As with animals, individual plants may be interconnected and communicate in various ways:

— By their root systems.

— By mediating organisms such as mycorrhizal fungi that can establish linkages between plants, allowing them to communicate with one another. The contacts are not limited to conspecific relationships but may also include plants of different species. In both cases, the transport of substances (minerals, carbohydrates) has been proved.

— The air or the water may also ensure the transfer of chemical elements from one plant to another. A particular case is the role of volatile substances such as ethylene (Mattoo and Suttle, 1991), which induces a series of physiological changes in plants.

Relationships between plants have mostly been studied in the context of nutrition and competition. Recent studies have also demonstrated that communications linked to volatile substances of plant origin could bring non-infested plants to trigger defence mechanisms when they are near infested plants (Bruin et al., 1999). It remains for us to understand in detail why and how the volatile substances are emitted by plants that are under stress and how healthy plants use this information. But the phenomenon seems well established. It may be thought that the plants are able to anticipate certain risks by means of these signals, which inform the receiver of an impending threat. In the case of plant-predator interactions, for example, the receptor plants may devote a part of their energy to defence mechanisms such as increase in the level of toxins, inhibitors of digestion, repellents, or indirect defences.

5.3.4. Cybernetic cohesion of ecosystems

In light of new research, the idea of Patten and Odum that there exists a communication network superimposed over the network of matter and energy flow seems increasingly probable. Although many discoveries are still to be made, the results obtained so far seem to prove that communities are not random collections of species that are independent of one another. There are strong and diverse interactions between species, whether they belong to the same taxonomic groups or different groups. These interactions are expressed in the existence of a wide range of signals that maintain the cohesion of plant communities and/ or modulate the behaviour and biology of species. There are in addition many symbiotic and mutualistic relationships that contribute to give the whole a greater cohesion.

All these interactions apparently exist in all environments and for all groups. We can speak of a *cybernetic cohesion*, which originates from the network of invisible connections between the biotic components of an ecosystem. This field is still to be explored but is bound to become the focus of active research in the coming decades. Scientists are already investigating the role that synthetic chemical substances or their by-products may play in the cybernetic cohesion of ecosystems. Can artificial compounds be made that mimic natural substances or affect the behaviour of species? This question is not without a basis.

5.4. THE INPUTS OF THERMODYNAMICS

In the search for universal laws of nature, some ecologists have explored the possible relationships between thermodynamics and the maintenance of biological order (Jorgensen, 1997; Patten et al., 1977; Straskraba et al., 1999). Thermodynamics emphasizes the fundamental duality between the processes that are the source of order within a system and the tendency to disorder (increasing entropy). Two principles form the basis of classical thermodynamics. They regulate all the physicochemical transformations that occur within physical systems and that are essentially isolated or closed systems.

• The *first principle* or *law of conservation of energy* states that energy can be transmitted from one system to another in various forms, but it cannot be destroyed or created. The result is that the quantity of energy present in the universe is constant.

• The *second principle*, in the original version, describes the *irreversible evolution of an isolated system toward a final state of thermodynamic equilibrium with an increase in entropy*, which is a measure of the disorder of matter and energy in the system. From the observation that heat never moves spontaneously from a cold body towards a hot body, this second principle affirms also that the transfer of energy is irreversible in natural processes. During the course of processes involved in the transformation of energy, energy of good quality degrades into energy of poorer quality.

However, the classic thermodynamics of physicists is a science of equilibrium, of final states that do not undergo reversible changes, in which the factor of time never explicitly appears. This is not the case with the world in which we live. The fundamental question is thus whether these laws are really applicable to ecological systems.

5.4.1. Basic principles of thermodynamics

a) Open and closed systems

An *isolated system* is defined as one in which interactions with the environment are such that there is no exchange of matter or energy with the external world. Such a system is an abstraction of physicists that has served to establish the laws of physicochemistry but is hardly found in ecological systems. A *closed system*, on the other hand, exchanges energy but not matter with the external world, and an *open system* exchanges energy and matter with its environment. Thus, the earth as a whole (the biosphere) is a closed system that receives the energy of solar radiation and emits radiation into space, without having exchanges of matter with space (or having negligible exchanges)(Fig. 5.2). On the contrary, many ecological systems are open systems that exchange matter and energy with other systems.

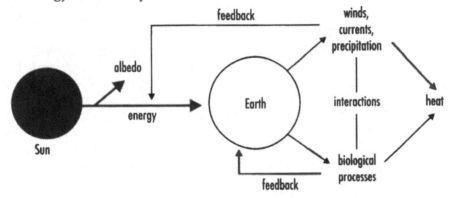

Fig. 5.2. Simplified diagram of thermodynamic functioning of the biosphere. Solar energy is transformed in the biosphere and some heat is radiated into space.

b) Conservation of energy

The first principle of thermodynamics, also known as the principle of Carnot, who stated it in 1824, affirms that the total quantity of energy of the universe remains constant. This principle of the conservation of energy tells us that *there is no destruction or creation of energy but only the potential for transformation of one form of energy into another*. It also establishes the equivalence of different forms of energy (e.g., radiation, mechanical, electrical, thermal, chemical) and the laws that regulate these transformations. The change of energy in a system (ΔE) is equal to the work accomplished by the system (ΔW) plus the energy released (ΔQ) in or by the system: $\Delta E = \Delta W + \Delta Q$. For example, the transformation of solar energy into chemical energy by plants conforms to the first principle of thermodynamics: the solar energy assimilated by photosynthesis is equal to the sum of the chemical energy contained in elaborated organic matter plus the caloric energy dispensed by respiration. The energy balances established on the level of food chains also are based on this principle of conservation and transformation of energy.

c) Thermodynamics of equilibrium and entropy

In 1848, James Joule established experimentally the principle of reciprocal conversion of heat into work and vice versa. But the study of exchanges of energy in thermal machines also proved the existence of a *hierarchy between the different forms of energy*. Mechanical, chemical, or electrical energy can be fully transformed into heat, but the inverse transformation (from heat into mechanical work) cannot be done fully without external input and without a necessary loss of energy in the form of irrecoverable heat. *Any transformation of energy supposes thus that high quality energy will be degraded irreversibly into energy of lower quality*. Energy is not destroyed but becomes unavailable to accomplish the mechanical task. High energy photons of the yellow band of the solar spectrum are thus transformed into infrared photons of lower energy: 20 infrared photons are produced for each photon of degraded solar energy. The second principle introduced, in this manner, a fundamental asymmetry, absent from the first principle, between the noble forms of energy and heat, which is the most degraded form of it (Boutot, 1993).

The second principle of thermodynamics or the principle of degradation of energy introduces a new function of the state of the system, *entropy* (from the Greek *entropê*, which signifies change), a measure of the quality of energy that is traditionally interpreted as a measure of disorder. It indicates that an isolated system, in the absence of external forces, evolves spontaneously towards a state of thermodynamic equilibrium, so that entropy increases in a monotonous and irrevocable manner until a maximal value. This occurs by dissipation of energy and matter, processes during which there is degradation of energy. The law of increasing entropy is thus a law of progressive disorganization: at thermodynamic equilibrium, the system by definition no longer evolves and is theoretically found to be in a state of complete degradation characterized by a uniform value of all its thermodynamic parameters (e.g., temperature, pressure).

These two principles of thermodynamics call into question the theories of Newton that systems are closed, deterministic, reversible, and universal. In particular, the principle of the reversibility of phenomena is called into question by the second principle, which can also be formulated as follows: time is irreversible, and as a result all processes move in a single direction that cannot be reversed because that would violate the second law of thermodynamics (Jorgensen, 1997).

A third law of thermodynamics, less known than the first two, stipulates that the production of entropy is a function of the temperature. Disorder does not exist at absolute zero (–273.15°C) but appears as the temperature rises.

5.4.2. Thermodynamics of dissipative systems

It was in the 19th century that the laws of thermodynamics and those of biological evolution were formulated. But while thermodynamics talks of the tendency to disorder in the evolution of physical systems, the idea of evolution in biology is associated with increasing organization, the formation of structures that are increasingly complex. This multiplication of highly ordered systems seems incompatible with the second principle of thermodynamics that any macroscopic system evolves towards a state of equilibrium marked by an increase in entropy, with its characteristic degradation of order. What then are the mechanisms that sustain the organization of living things?

For an ecological system to maintain its internal organization and structures (spatial and temporal organizations), it must necessarily conserve a level of low entropy. In reality, as Prigogine (1968) has remarked, living systems cannot be compared to isolated thermodynamic systems but only to open or closed systems in interaction with their external environment. An open system, so far as the external reservoirs of energy and matter are sufficiently large to remain constant, may tend towards a stationary regime other than that of thermodynamic equilibrium. This is a stationary state of non-equilibrium. Exchanges with the external environment thus allow it to overcome the law of increasing entropy imposed by the second principle of thermodynamics on isolated systems. In the language of thermodynamics, these systems crossed by an uninterrupted flow of energy are called *dissipative systems*. Dissipative structures (i.e., those in which energy is not conserved) are associated with an entirely different principle of order, which is an order by fluctuation. They are dependent on energy flow and disappear when the energy flow itself disappears.

The ecological systems that make up the biosphere exist and maintain themselves by means of energy flows that allow the circulation of matter and information.

It is by the exchange of energy and matter across the limits of the system that a stationary state far from thermodynamic equilibrium can be attained and maintained. These flows that structure ecological systems characterize what is usually called their functioning. Without this continual input of energy they would rapidly degrade.

The ecodeme

Johnson (1981) postulated that the ecodeme, which he defines as a group of similar individuals interacting with one another within the limits imposed by the frontiers of their ecological niches, presents the characteristics of a functional and organized dissipative structure that may actively acquire energy and conserve it. The ecodeme corresponds in a way to a population (collection of individuals of a single species) that consumes energy, accumulating part of it in the form of biomass and using another part to maintain the biomass. An example of an ecodeme is that of fish populations composed essentially of large individuals of relatively uniform size in which juveniles are virtually absent, as in the virgin Arctic lakes (Johnson, 1994). This structure, characterized by a bell frequency of size, seems stable over time and may be compared to the climax of plant ecologists, i.e., a stable state that may vary according to the place, as a function of the species present and the local environment. This structure also corresponds to the principle of least dissipation of energy (or least entropy) for a given energy input. Indeed, the aged individuals have a reduced metabolism, and the renewal rate (P/B) is low. The Arctic lakes do not constitute an isolated case and this type of system has been observed in different parts of the world.

Theoretically, an ecosystem is made up of a set of ecodemes that interact in different ways. The tendency of an ecodeme to approach a state of least entropy is thus counterbalanced by the overall system, which tends on the contrary towards a state of greater entropy. Johnson (1994) ultimately developed a theory that the ecosystem is the result of two opposing but nearly equivalent forces: one tends to increase the energy in the system and defer its transfer (case of ecodemes), while the other tends to accelerate it. Thus, the ecosystem is at the intersection of two opposing tendencies. On the one hand, on the scale of ecological time, which is that of succession, it tends towards a stable state, climax. The dominant tendency is towards homogeneity. On the other hand, on the scale of evolution, which is that of diversification, with an increase in the number of species and the complexity of ways in which energy travels, the tendency is to increase the flow of energy across the system.

Change of entropy in a dissipative system is presented in the form of a balance between an energy flow of low entropy coming from external sources and crossing the system, and the production of entropy within the system itself (Nicolis and Prigogine, 1977). These are "input-output" systems (Patten et al., 1997) in which the anabolic and catabolic processes oppose each other. The former tend to construct an organized structure far from thermodynamic equilibrium, while the latter draw the system towards thermodynamic equilibrium. The subsequent chain reaction summarizes the preceding considerations:

— The *source* of energy of low entropy is solar radiation (high energy photons).

— *Anabolism* (charge phase) is the incorporation of this high quality energy into biochemical structures, which keeps the system far from thermodynamic equilibrium.

— *Catabolism* (discharge phase) corresponds to the deterioration of the structure with the release of accumulated chemical energy and its transformation into work and heat over the course of a large number of geochemical, biochemical, and biogeochemical transformations.

— The *sinks* correspond to the dissipation of degraded energy in the form of heat into space (low energy photons).

There is thus a continuum that Patten et al. (1997) call *the cascade of energy in the food web*. The various stages of the continuum are organized around a general process, a charge-discharge cycle of energy that is a fundamental attribute of the energy functioning of ecosystems. Each cell or organism has its own charge-discharge cycle. In the charge phase of ecosystems, organic and inorganic matter of the biosphere are synthesized several times over in the form of energy-rich tissue in successive cycles of primary and secondary production. The biomass is then transformed into work and heat in the course of the discharge phase.

Thus, organisms, ecosystems, and biosphere possess in common an essential thermodynamic characteristic, to create and maintain a high level of organization (or conditions of low entropy) by means of a continuous degradation of energy of great utility (light and nutrients) into energy of low utility (heat), which is dissipated towards the exterior of the system (Jorgensen, 1997). For this, the system must be connected to a source of energy of low entropy and to sinks of energy of high entropy, i.e., with respect to ecosystems, the incident solar radiations and the long wave radiations that are lost in the cosmos. It is the catabolism of the ecosystem (respiration) that constantly produces a disorder (heat) that paradoxically makes possible the maintenance of order in the ecosystem.

We must nevertheless keep in mind that the existence of an energy source is not enough. The most important factor is the existence of a system capable of transforming energy into work, in such a way as to construct and maintain complex living systems, from its building blocks or molecules. Thus, the earth is an open thermodynamic system subjected to a high gradient of energy imposed by the sun. The earth system is forced to reduce this gradient by all the physical and chemical means available. We can even say that life on earth is a means, a sophisticated means, of ensuring the dissipation of energy from the sun. In fact, life is not an isolated phenomenon but corresponds to the emergence of a category of processes whose objective is the dissipation of thermodynamic gradients (Jorgensen, 1997). According to this hypothesis, growth, development, and evolution are inscribed in the perspective of best dissipating energy. The system that dissipates most efficiently carries along the others (Schneider and Kay, 1994).

Living organisms and ecosystems present many characteristic traits of dissipative structures. They have a tendency to approach a state of least production of entropy in the course of their development, given that their rate of metabolism diminishes with time, while they accumulate energy in the biomass.

At maturity, an organism enters a stable state during which a disturbance may provoke an increase in metabolism, but the metabolism returns more or less quickly to its initial level when the disturbance ceases. When the disturbance persists within limits compatible with the survival of the biological system, another stable state with production of minimal entropy can be established in these new conditions. However, at least two characteristics of dissipative structures are missing in biological or ecological systems: the first is continuity, since their life span is limited to that of their life cycle, and the second is that they can only adapt to changes in the environment within the limit of their adaptive capacities, which are the result of hereditary constraints in the case of organisms.

Energy

Energy is defined as the capacity to achieve a task, on the basis of the physical principle that work requires energy. It can be expressed in different ways: e.g., mechanical, kinetic, calorific, chemical, electrical,

electromagnetic, atomic. All forms of energy can be transformed into heat, so that it is relatively easy to express them in thermal equivalents.
The basic unit is the calorie, which corresponds to the quantity of caloric energy need to bring 1 cc of distilled water from 14.5°C to 15.5°C at atmospheric pressure. The joule is equivalent to 4187 calories.

Emergy

Emergy is a concept introduced by H.T. Odum (1983) to designate the energy needed for the elaboration of organisms at different trophic levels. Emergy (expressed in joules) can be defined as the quantity of solar energy needed directly or indirectly to generate a flow, a product, or a stock. To evaluate it, the different sources of energy that have participated in the elaboration of the product or flow must be identified and transformed into solar energy equivalents. Some scientists describe emergy as the "memory of the energy" that is degraded by a process of transformation. The more the elaboration of a product requires work, the more there is transformed energy, and the more the product contains emergy. It can be said that emergy is the work provided by the biosphere to maintain the system far from equilibrium (Bastianodi and Marchettini, 1997).

Exergy

A dissipative structure such as the ecosystem consumes energy and a flow of exergy is needed for the system to function. Exergy measures the quantity of free energy of biomass incorporated in the structure (Jorgensen, 1996). The more living matter there is in the system, the more the system contains exergy. If the system is in thermodynamic equilibrium with its environment, exergy will be equal to zero, but ecosystems try to evolve towards higher levels of exergy to stay far from this equilibrium. The systems that are able to acquire the most exergy, taking into account the existing conditions, have a comparative advantage and will be more apt to resist adverse conditions of the environment. According to Jorgensen (1997), the use of the notion of exergy seems preferable to that of entropy to describe the irreversibility of processes observed because it uses the same units as energy (it is also a form of energy), while the definition of entropy is more difficult to associate with concepts normally used to describe reality.

5.4.3. A fourth law of thermodynamics?

The concept of limiting factor is contained in the principle of conservation of matter: the quantity of elements on the earth is constant. In these conditions, ecological systems cannot develop by fabricating more organic matter using more inorganic matter, but can only make the best possible use of the available inorganic elements. This is possible by means of better organization of the structure: diversification of ecological niches and growth in the number of life forms in order to better respond to the variability of constraints. This evolution of the structure of the ecosystem allows the fixing and storage of more exergy (Jorgensen, 1997).

However, a system cannot capture more exergy than that furnished by the solar system. The storage of exergy will in turn become a limiting factor when a sufficiently organized system is constituted. In this context, the additional constraints on the ecosystem are to better use the exergy that is entering: a smaller part of this exergy must be used for the maintenance of the system, and a larger part for the development of the structure, in such a way as to increase the exergy stored in the ecosystem. In theory, systems are organized with time to use most efficiently the incoming light energy stored in the form of chemical energy in biomass. They will develop mechanisms that allow them to stabilize their internal chemical processes and maintain their functioning in the eventuality of changes in the environment. What is more, the ecosystem will be capable of evolving and adapting its structures in order to improve its energy utilization (Kay, 1984).

On this basis, Jorgensen proposes a fourth law of thermodynamics (or the law of thermodynamics of ecology): any system that receives an energy flow of high quality (energy of low entropy or exergy) will have the propensity to exploit this energy to maintain itself far from thermodynamic equilibrium. If the system has several alternatives to use the energy flow it will also have a propensity to select the organization that can provide it with the maximum exergy possible in the existing conditions. This fourth law is an expansion of Darwin's theory of natural selection: the individuals most apt to respond to changes in the environment by rapid adaptations that allow them to survive are favoured. The most apt to survive are organisms having characteristics that allow them to maintain and even increase their biomass in the existing conditions, i.e., those that contribute best to the accumulation of free energy (or exergy) in the system to which they belong, because of their biological characteristics.

In this context, ecosystems have had the opportunity for four billion years to make trial-and-error experiments, testing and selecting the best scenarios to keep them far from thermodynamic equilibrium. Life has thus chosen various solutions, some of them highly sophisticated, in the endeavour to accumulate the greatest possible exergy in highly varied conditions, from the ocean depths to mountain peaks. This diversity of conditions and solutions that have been brought about by evolution explains the great complexity of the biosphere as we know it, as well as the existence of large variety of strategies of reproduction, growth, and survival. Finally, the long period of selective pressure over the course of evolution has been the origin of the development of a large number of symbiotic and/or mutualistic phenomena, as well as many phenomena of feedback, which explain the emergence of the Gaia effect (Jorgensen, 1997). This brings us to the logic of adaptive systems.

Chapter 6

Abiotic Factors and Structure of Ecosystems

"We may conclude then that in every respect the valley rules the stream. Its rock determines the availability of ions, its soils, its clay, even its slope. The soil and climate determine the vegetation, and the vegetation rules the supply of organic matter. The organic matter reacts with the soil to control the decay of the litter, and hence lies right at the root of the food cycle. One could go on and on, building up an edifice of complexity, all linked and cross-linked in the manner beloved by drawers of food nexes. ... We must, in fact, not divorce the stream from its valley in our thoughts at any time. If we do we lose touch with reality."

Hynes (1975)

An "ecological factor" is any element that may act on living things during any part of their development. Conventionally, a distinction is made between abiotic and biotic factors. Abiotic factors are all the physical or chemical factors that characterize the environment being studied. Biotic factors, also called density-dependent factors, correspond to all the interactions that occur between individuals (e.g., predation, competition, mutualism). The respective role of biotic and abiotic factors in the composition and structure of plant and animal communities has been an inexhaustible source of often sterile debate in ecosystem ecology. With the consideration of interlocking spatial and temporal scales and the emergence of a dynamic perspective of ecosystems, most of this "biotic vs. abiotic" debate is no longer relevant. In some cases, and at some scales, abiotic factors play a dominant role. In other cases, interactions between species do. It is also known that the biotic-abiotic dichotomy is not as rigid as it was once believed to be and that the living world acts in turn on the abiotic world, helping to modify it and make it evolve. Still, the physicochemical context is a constraining framework on the structure and dynamics of living communities. A good understanding of this system of constraints is a prerequisite to any ecosystem study.

6.1. ABIOTIC FACTORS: A NETWORK OF CONSTRAINTS

Many treatises on ecology devote many pages to questions about the structuration and dynamics of populations and communities but give only a marginal importance to the major abiotic factors that control all the physical and biological functioning of ecosystems at various scales of time and space.

Sometimes even the physicochemical processes are presented as simple factors of disturbance (e.g., fire, storms). In reality, the authors more or less consider that these subjects belong to other disciplines: geography, climatology, geomorphology, and hydrology, to name only a few. Still, in a hierarchical approach (see chapter 7) (Fig. 6.1), it has long been known that abiotic factors determine the framework of constraints within which the qualitative and quantitative distribution of various parameters such as temperature, humidity, and soil composition control the presence of species and the structure of plant and animal communities. These same factors also control the evolution of spatial heterogeneity and temporal variability. To put it in an extreme way, albeit one that has some foundation, the soil, flora, and fauna are functions of the climate and geomorphology (Bailey, 1996).

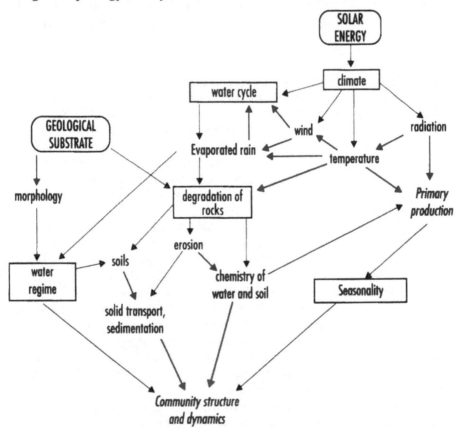

Fig. 6.1. Characteristics of the abiotic environment that control the dynamics of the biological compartment in a hierarchical structure

The dynamics of an ecosystem, as a function of its geographic location, is controlled by several interacting abiotic parameters derived from the climate and from the geological context. A knowledge of elements of this "bottom up" hierarchy is essential if we wish to understand how an ecosystem functions. Nevertheless, it must be observed that very

*often, by ignorance or because of the absence of specialized knowledge, ecosystem studies
miss out on these questions.*

6.2. GEOLOGICAL PROCESSES

Geomorphology is the study of the configuration of the earth's surface, and
geochemistry is the study of the distribution and flows of chemicals. These are
disciplines that are practised respectively by geographers and geologists to
understand the dynamics of the lithosphere. Their contribution to ecosystem
ecology has long been marginal. Fortunately, it tends to reinforce itself to the
extent that multidisciplinary studies have shown their advantages of a better
consideration of the physicochemical environment to understand and predict
the evolution of biological systems (see Amoros and Petts, 1993, for example).

6.2.1. Geomorphology

On short temporal scales, most geological processes that influence the functioning
of ecosystems are generally catastrophic. They are rare but may cause significant
modifications in certain regions. These processes include tremors, exceptional
floods, mudslides, and intrusion of salt water, as well as their corollaries such as
erosion (of soils or of coasts) and sedimentation. Some volcanic eruptions block
up rivers and thereby initiate the formation of lake systems. The eruption of
Mount Pinatubo in the Philippines released into the atmosphere 25 million to 30
million tonnes of SO_2, modifying the radiation balance of the earth and causing a
reduction in the global temperature by 0.6°C (Kerr, 1993).

Over longer time scales, geological variables (lithology, geological structure,
disturbances) and their interactions with hydrological, atmospheric, and
biospheric processes acting at various spatial and temporal scales have modelled
and continue to model the relief and landscapes. The action of the climate on
the relief is manifested in the differential erosion of geological structures and
the accumulation of eroded elements in certain sectors; in this way the relief is
modelled. In its elementary aspect, the relief or orography summarizes a set of
topographic surfaces called slopes that are characterized by their numerical
slope and can be expressed mathematically. The scientific description of relief
aims to characterize, inventory, and classify systematically the forms of
landscapes and their associations.

Geomorphology is the science of description and explanation of the various
ways in which the terrestrial, continental, and ocean-bed reliefs evolve. There
are various types of morphogenic systems on the earth's surface (e.g., glacier,
aeolian, or fluvial) that correspond to characteristics of erosion, deposit, and/
or sedimentation. The fluvial system has been the focus of much interest because
watersheds constitute a dynamic system of erosion, transport, and sedimentation
that actively models the earth's surface. It was after World War II that climatic
geomorphology, the study of relief in its relationships with the climate, developed,
with the use of aerial photography and satellite imagery (Coque, 1993).

The age of lakes and the nature of lake ecosystems

The ecological functioning of a lake system differs greatly according to the age and geological origin of the lake. In lakes considered ancient and of tectonic origin, such as Lake Tanganyika or Lake Baikal, which have existed for several millions of years, the coevolution of species and environments is the source of a great diversification of species, most of which are endemic. This diversification of species is associated with the diversification of food regimes, so that all the resources of the environment are exploited (Lévêque, 1997). In contrast, many of the lakes of Europe and North America are "young" lakes that resulted from the retreat of the icecap about 15,000 years ago. Their fauna of recent recolonization is poor and practically no species are endemic. In these conditions, we can hardly speak of a coevolution of species and environments, and the ichthyological community is a random collection of species that result from recolonization opportunities in a recent past. The ecological functioning in terms of the food web, all things considered, is very different from that of the ancient lakes.

6.2.2. The cycle of erosion, transportation and sedimentation

The breaking up of rocks by climatic impacts and biological agents initiates the contact of the lithosphere with the atmosphere and the biosphere. This process is of utmost importance because it ensures the transformation of rocks and minerals located on the earth's surface into mobile products. The process of alteration comprises physical degradation followed by chemical degradation (hydrolysis, hydration, oxidation, solution) as well as a series of organic processes involving plants and animals. This alteration of surface rock has two essential results—the formation of soils and the shaping of the landscape— both of which are dependent on the nature of the underlying rock.

Mechanical erosion tends to reduce the thickness of soils and rocks by mechanically grating the particles, which are then carried in suspension by the flow of rivers or blown away by the wind. On the other hand, chemical erosion tends to deepen soils at the expense of rocks. The primary minerals of the rocks are dissolved or hydrolysed, releasing elements in solution that are leached by draining waters and emptied into rivers or ground water. Mechanical and chemical erosion are thus two concurrent mechanisms, but chemical erosion often precedes mechanical erosion and the two forms of erosion are ultimately responsible for the peneplanation of continents. Continental erosion thus results from interactions between the atmosphere, biosphere, hydrosphere, and lithosphere.

The evolution in the morphology of watersheds over time is the result of processes of erosion, transport, and deposit. The classical cycle of erosion, transport, and sedimentation involves most river systems, i.e., most of the emerged surface of the earth. *Water is a significant agent of erosion of rocks and land. It constantly shapes the earth's surface, attacking the relief in two ways: mechanical breaking up and chemical dissolution of rocks.* Water erodes and dissolves rock, transports the broken elements, and then deposits them when its kinetic energy diminishes, sometimes very far from the initial point. Water courses have a high erosive capacity and rapidly flowing rivers play a significant role in landscape evolution. They can hollow out deep valleys, sometimes canyons bordered with vertical cliffs, such as the gorges of Verdon, in which the cliffs are 700 m high.

The mechanical effects vary as a function of flows and the current speed. The steepness of the slope determines the power with which running waters break up and transport rocky material. The fine elements are transported in suspension in water, while coarse elements are rolled or carried by successive leaps, or dragged along the bottom. The result is that the size of transported material and the mode of transport of sediments evolves perceptibly from upstream to downstream because of changes in the conditions of flow (turbulence, current speed) and modifications of the slope and the nature of the rocks. The torrents have the greatest capacity for eroding and transporting. In the lower course, where the current is weak, the transport of fine material in suspension is dominant.

The load of suspended matter transported by rivers depends on the soils of the watershed and the vegetation that covers them. Each year, the Rhone receives on average nearly 320 t of matter for every km^2 of its river basin and carries more than 30 million t of matter to the Mediterranean Sea. In comparison, the Ganga and the Brahmaputra, which descend from the Himalaya, erodes 1500 t/km^2 of watershed every year, while the Amazon erodes only 70 t/km^2 and the Danube around 35 t/km^2. On the global scale, the total mechanical erosion is estimated at around 15×10^9 t/year.

6.2.3. Soils

Soils result from the interaction of surface layers of the lithosphere with the atmosphere and the biosphere. Pedology (from *pedon*, soil) developed in Russia at the end of the 19th century with the works of Dokoutchaiev (1846-1903).

a) Pedogenesis

The genesis of soils (or pedogenesis) from a mother rock involves two different processes.

- *Alteration of the mother rock* under the effect of water, temperature, and living organisms. The disintegration and decomposition of the mother rock results in fragmentation and transformation of these initial components into more simple minerals. For example, a granitic rock yields arenite, sandy formations composed of a combination of quartz grains (pure silica) that are virtually unchangeable and colloidal elements such as clays (hydrated aluminium silicates) derived from the hydrolysis of initial mica and feldspar. In addition there are mobile mineral ions, in solution in the soil water or fixed on the colloidal elements, such as Ca^{++}, Mg^{++}, K^+, and many trace elements.

- *Contribution of organic matter* by living things. Vegetation gradually colonizes the disintegrating mother rock and produces an increasing quantity of organic matter in the form of aerial or underground debris. The mass of litter provided is variable, from about 4 t dry matter/ha/year in the temperate forest to 15 t in an equatorial forest. The fauna that develops after the vegetation is established is also the source of organic matter input.

These two processes continue more or less intensively over time. As burrowing organisms progressively mix the minerals derived from the alteration of the mother rock with the organic matter, an organomineral complex is formed that is an essential characteristic of the soil. In addition to these processes, there is migration of certain constituent elements (soluble or colloidal) under the effect of water circulation in the soil.

b) Physical and chemical characteristics of soils

The constitution of a soil depends on the size of its components, which determine its texture, and on the way in which they are arranged, which characterizes its structure. The *texture* is indicated by granulometric analysis, which attempts to classify the mineral particles of soils according to their diameter. The *structure* of soils depends on the state of the particles it is made up of. When colloidal, organic, or mineral particles are stuck together, they form aggregates with larger particles. In other cases, the particles are dispersed and the soil elements remain independent of one another. In the first case, the structure is *aggregate*, with different types according to their size and form, while in the second case the structure is *particulate*.

The soil properties are linked to the texture and structure. The texture and structure determine the soil's *porosity*, which is in a way the proportion of volume of lacunae in relation to the total volume. Porosity allows the circulation of gases and water, and more generally the soil aeration required for the activity of aerobic microorganisms.

Another property of soils is their capacity for water retention, which is related to the structure. The retention of capillary water constitutes in effect a capital water reserve for plants during a dry period. A soil that is clayey or rich in humus retains water more effectively than soil of sandy texture.

c) Properties of soils and distribution of vegetation

The geochemical properties of rocks determine the geochemical characteristics of soils and the quantities of minerals available to plants and animals. There are many examples in which these properties control the presence or absence of plants. This is the origin of the landscape mosaic. Geological maps thus contain a large amount of information of potential value to ecologists. In reality, geologists have long used the nature of the vegetation as an aid in geological mapping. Incidentally, major elements for the nutrition of living organisms (Al, Ca, Fe, K, Mg, Na, Si) are also the major elements used to classify the minerals. The difference is that living organisms use these elements as components of a basic structure of carbon, hydrogen, nitrogen, and oxygen, while the rocks use the same elements as components of a basic structure of silica and oxygen (Dobrovolsky, 1994).

Although the composition of the vegetation depends on the nature of soils, it also acts in turn on the soil characteristics. Each type of plant formation intervenes by its specific microclimate (conditions of humidity, temperature, and so on) and its structure. For example, depending on the species, the development of deep roots will favour or hamper water circulation and aeration. The chemical characteristics of the plant litter directly influence the processes of organic matter transformation. These processes are generally more intense when the plant debris is richer in nitrogen, calcium, and soluble organic matter. This is true of deciduous species such as ash, elm, or oak as well as legumes. Inversely, coniferous forests and moors with heather provide litter that is nitrogen-poor and highly acidic.

6.3. THE MACROCLIMATE

The weather is a passing, transitory state of the atmosphere that varies from one moment to the next throughout the day, and from one day to the next. These states are characterized by physical magnitudes, essentially temperature, precipitation, wind speed and direction, atmospheric pressure, and the humidity of the air (Foucault, 1993). Climate is the succession of states of the atmosphere in a given place, close to the ground. It is like an uninterrupted film in which each image consists of the weather at a given moment.

To characterize the climate of a region of the earth we must, to put it very simply, determine for each season the conditions of temperature and rainfall, as well as the variabilities of these parameters, i.e., the climatic regime. If these climatic components change, many other abiotic or biotic components will also change in reaction. *Therefore, the climate is the principal factor controlling the distribution and dynamics of ecosystems.*

6.3.1. Sunlight

Many forms of energy are indispensable to the functioning of ecosystems. For example, it is known that photosynthesis is not the sole metabolic process for the production of living matter from minerals and that there are species that elaborate their proteins by chemosynthesis. But the earth's principal source of energy is solar radiation. All this energy is created by nuclear fusion. Each second, 700 million t of hydrogen fuses to yield 695 million t of helium and 5 million t of electromagnetic energy that irradiates the solar area in space. While the energy emitted by the sun is around 63,000,000 W/m^2, the quantity of energy intercepted at the limits of the terrestrial atmosphere is only 1370 W/m^2.

Solar energy serves to heat the surface of soil and water and makes photosynthesis possible. The wavelengths of solar rays range from 0.2 to 4 μm, i.e., from ultraviolet to near infrared, including the visible radiation between 0.30 and 0.67 μm, the most useful for photosynthesis, which represents around 40% of the total energy (Fig. 6.2). The ultraviolet is almost totally absorbed by the ozone layer located in the troposphere around 25 km altitude, thus protecting living things from lethal radiation levels. The ultraviolet radiation is also absorbed by a thin film of liquid water. The infrared is absorbed by the air, soil,

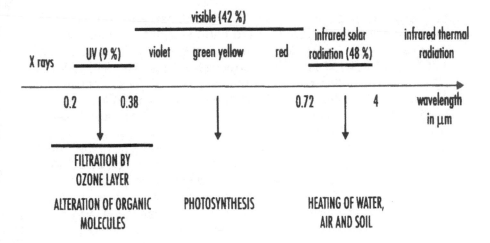

Fig. 6.2. Composition in wavelengths of solar radiation and ecological effects of different wavelengths (modified according to Frontier and Pichod-Viale, 1998)

and water. On average, the quantity of incident energy is 350 W/m²/day, or 7.2 × 10⁶ calories/m²/day. Only about half of this energy reaches the soil because it is partly reflected by the atmosphere (see Chapter 15).

One of the factors of climate variability on earth is the inclination of the sun's rays as a function of the latitude. The Greek term *klima* signifies inclination (of the sun in relation to the zenith). While these rays are perpendicular to the earth's surface at the equator, they spread over an increasingly greater area as we go towards the poles because of the curvature of the earth. The result is a reduction in the energy that reaches the soil's surface, from the equator toward the poles, which explains the existence of climatic zones whose latitudinal limits are more or less parallel to those of the equator. In the intertropical regions in which the sun is close to the zenith, the temperature is high and varies little throughout the year. The average annual temperature is generally higher than 20°C for low altitude zones. In the polar regions, on the contrary, where the solar rays are very oblique, the temperature does not exceed 10°C at any time. In a rather artificial way, we thus distinguish climates that correspond to these regions, which roughly form belts parallel to the equator.

Auxiliary energy

Part of the incident solar energy is stored by living things by means of photosynthesis and chemosynthesis. It goes through the biomass across the food chains. But another part of the solar energy is also involved in the functioning of ecosystems without entering the biological pathway. This is *auxiliary energy* (a concept proposed by the Spanish ecologist Ramon Margalef). This energy, which is purely physical, is used for example to move fluids (e.g., air, water) or transport elements from one place to another (Frontier and Pichod-Viale, 1998).

On average during the year, around 3.1 W/m²/day in relation to 350 W/m²/day of solar energy received by the earth serves to generate winds. Because of the friction resulting from wind, they are transformed into heat and return to space in the form of terrestrial radiation (Hayden, 1998). A significant part of this energy is dissipated by the friction caused by vegetation in terrestrial ecosystems and by the waves in the sea. For example, the wind speed is 12 m/s in the oceanic domain close to Great Britain, while on the continents it does not exceed 6 m/s.

Living things are themselves a source of auxiliary energy to the extent that their active displacements can ensure the migration of biogenic elements between different elements of the ecosystems. This is true of vertical migrations of certain aquatic organisms in the lake or marine ecosystems. It is also true for birds that consume marine organisms and deposit guano on the terrestrial environment. The burrowing organisms ensure the mixing of soils by bioturbation. Finally, human beings introduce auxiliary energy by fertilization or irrigation.

6.3.2. Temperature

Temperature is a parameter that is conventionally measured in the shade. Therefore, we measure not incident radiation but the average temperature of an environment protected from radiation.

Temperature affects the rapidity of biological processes. The Van't Hoff law stipulates that the speed of chemical reactions is approximately doubled for an increase of 10°C in the ambient temperature (but this law has many exceptions). For any organism, there is generally an optimal temperature at which its biological activity is maximal and a maximal temperature beyond which the organism dies. Heat generally causes an irreversible degradation of constitutive proteins while cold causes death by crystallization of the water contained in tissues and cells.

In terrestrial environments, there may locally be significant variations in the temperature of the air depending on the more or less direct exposure of the physical environment or organisms to solar radiation. The difference between the highest and lowest temperatures of the day (daily amplitude) is an important ecological factor. In aquatic environments, the temperature is buffered but there may be a vertical stratification that results partly from the action of wind.

6.3.3. Rainfall and humidity

In addition to the thermal factors that structure the distribution of climates on the earth's surface, there is a distribution of precipitation that is partly conditioned by the circulation of air masses and that sometimes considerably modifies the preceding distribution (see Chapter 15). The precipitation is the result of cooling of the humid air causing the condensation of water vapour. Pluviometry is the measurement of precipitation. The average annual rainfall (quantity of water received from precipitation) as well as the precipitation regime (seasonal distribution of precipitation, regular or sporadic nature of precipitation) varies according to the longitudinal position of the observation points.

The humidity of the air is defined as the quantity of water vapour contained in a certain volume of air, expressed in grams per cubic metre. The air can contain only a limited quantity of water vapour, a quantity for which it is called saturated. This quantity is a function of the temperature. Often, the humidity of the air is expressed as a percentage of the humidity at saturation: the term *relative humidity* is therefore used.

6.3.4. Climate and world structure: the biomes

The climate, directly or indirectly, plays a primordial role in the modalities of erosion and the determination of aspects of the relief. But the combination of precipitation and temperature allows the globe to be divided into major morphoclimatic domains. At a very macroscopic scale, we can identify four ecoclimatic zones that are distributed more or less zonally around the globe: the hot and humid tropical zone, the humid temperate zone, the polar zone, and the arid zone (Bailey, 1996). At a smaller scale, different regions of the globe in which the climatic conditions are identical are observed to be occupied by comparable ecosystems. The vegetation presents the advantage of being a reliable indicator to express at relatively large spatial scales the play of various factors such as geomorphogenesis and the climate. The limits of major plant formations thus outline remarkable discontinuities of the natural environment and even carry information on the probable climate in zones for which good climatic data are not available. It is the biomes that are regional macrosystems, homogeneous from the climatic point of view (temperature and precipitation).

The isotherm of 10°C of the hottest month of the year, which corresponds approximately to the northern limit of the great conifer forests of the northern hemisphere, separates the high-latitude cold environments from middle-latitude temperate environments (Coque, 1993). Similarly, the isotherm of 20°C of the coldest month of the year quite clearly delimits the environments characterized by constant heat. Within this vast domain, the hydric regimes differentiate the great rainforest, which is always hot and humid, from savannahs characterized by a dry season.

The number of biomes identified depends on the resolution desired, and 10 to 100 biomes are distinguished, depending on the authors (Table 6.1). In most cases, the physiognomy of the vegetation serves as the basis for the delimitation of the biomes: forests (24% of the area), savannahs (15%), prairies and tundras (15%), etc. To this we must add the cultivated areas, which represent more than 10% of the area of emerged land, as well as the deserts and frozen wastes (30%) (Saugier, 1996).

Table 6.1. The major terrestrial biomes (Dajoz, 1996)

Biomes of cold temperate regions	Biomes of humid tropical regions	Formations of arid and semi-arid regions	Mountains
• Conifer forests of boreal or taiga regions	• Equatorial tropical forests of evergreens and pines	• Tropical steppes	• Mountains of temperate Europe
• Conifer forests of the Pacific coast of North America	• Tropical forests with seasonal rhythms	• Continental steppes with a temperate climate	• Mountains of North America
• Deciduous forests of the temperate regions	• Laurel forests	• Desert	• Mountains of tropical regions
• Evergreen forests of Mediterranean regions	• Savannahs	• Tundra	
• Natural herbaceous formations: grasslands and steppes			

6.4. WIND

Winds originate from the differences in air pressure at the earth's surface. This pressure is a function of the temperature. If the air heats up, its density reduces and it has a tendency to rise. Colder air masses thus replace the hot air, and these displacements cause winds. Winds blow all over the earth's surface, but their intensity is variable.

The wind plays an important role in the water cycle: it increases evaporation, which consumes energy, and thus has a considerable cooling power. It is an ecological factor that is often underestimated in the study of functioning of ecosystems (Ennos, 1997).

6.4.1. Exceptional events: tornadoes and hurricanes

Chronic winds must be distinguished from exceptional events such as tornadoes and hurricanes. Tornadoes are among the most violent and unpredictable meteorological events. Some winds reach 400 km/h. In forests, the consequences of violent tornadoes are known: by causing tree fall, they contribute to the restarting of the processes of succession, thus helping to maintain a high species richness by the re-establishment of pioneer species.

Hurricanes (or typhoons) develop mostly in the oceanic tropical regions. In France, however, a storm in November 1982 ravaged nearly 10 million m³ of woods in France, affecting 68% of the forest communities (Touzet, 1983). Subsequently there was the storm of December 1999. Because of a difference between tree species with respect to resistance to wind (conifers are more fragile than deciduous species), such an event modifies the composition of forests. Also, many trees and branches fall. This type of event does not seem exceptional, although its frequency is low. Winds greater than 100 km/h have occurred at least once in 15 years in some part of the French territory, and an exceptional storm once in 30 years (Touzet, 1983). They maintain a state of heterogeneity that has important consequences for the dynamics of forest ecosystems.

6.4.2. Wind as an agent of erosion and transport

Part of the wind energy serves to transport particles that accompany erosion or deposit, depending on the case. Thus, wind transports sand and dust over sometimes very long distances. It is not rare in some years to find deposits of Saharan dust on cars in Paris. This transported dust results from the erosion of soil by the wind. Observers have also recorded particles from Africa transported as far as Brazil and Florida.

Saharan dust

Each year, wind transports some hundreds of thousands of tonnes of dust and sand from the northern Sahara to Europe, especially France and Germany, but sometimes up to the Scandinavian countries. The input is not regular but the result of a few violent episodes: dust storms. For example, on March 8, 1991, Tunisia was struck by violent winds. On the next day, this storm caused the appearance of coloured snow on the Pyrenees and red rain in central France. More than 150,000 t of dust was thus deposited over 150,000 km². On March 10, around 50,000 t of desert dust coloured the snow over a vast region close to the polar circle located between Sweden and Finland. The material deposited was composed of minute mineral particles and pollen from alder, hazel, and sagebrush that had travelled over 5000 km.

Wind transports more than just mineral particles. It is an important vector of dispersal of animals and plants. In June 1998, significant quantities of pollen from pine (*Pinus banksiana*) and spruce (*Picea glauca*) were observed north of Canada, in the Arctic zone. The pollen (30-55 µm) had been transported over more than 3000 km by the wind (Campbell et al., 1999). Also, insects were found in the Antarctic that were transported over thousands of kilometres. Such "aerial" plankton is formed of animals (mostly Arthropods) that are carried by air currents. This transport over long distances is an effective mode of colonization for some species.

Such aerial transport has other ecological consequences. In particular, it carries nutrients (iron, phosphorus) that favour the growth of algae in marine waters that are normally poor in these substances.

Atmospheric transport of minerals and anthropogenic elements

Wind transports natural or anthropogenic elements over long distances. The ice of Greenland has preserved traces of the history of copper production from antiquity (Hong et al., 1996). The analysis of glaciers has also shown that lead contamination dates from at least three millennia ago (Hong et al., 1994).

Detailed analyses in the pond of the Gruere in the Swiss Jura (Shotyk et al., 1998) have indicated that dust due to soil erosion has been the most significant source of lead in the peat for around 3000 years. About 8500 years B.P., the dust came mostly from Scandinavia, but after 7700 years B.P., it was dust of Saharan origin that was the major source. After 3000 years B.P., the anthropogenic sources dominated the atmospheric emissions of lead, with an important increase at the time of the industrial revolution and a very clear peak after the introduction of leaded fuels.

6.4.3. Wind and the structure of aquatic systems

In aquatic systems, the force of the wind and the distance it travels (or fetch) determines the size of waves and the maximal depth at which the action of the wind can be felt. The water that is stirred becomes homogeneous over the depth of the turnover layer, which is as thick as the wind is strong. This process ensures a good oxygenation over the entire thickness of the turnover layer. In deep lakes, the homogenization of the surface layer by the wind affects the stratification observed in the summer, with a thermocline separating the epilimnion from the hypolimnion.

During studies on a shallow tropical lake, Lake Chad, it was calculated that a fetch of 10 km with a wind of 4.5 m/s (or 3 km for 9 m/s) created a turbulence of 4 m depth in the water column. Given that the average depth of Lake Chad was around 4 m between 1965 and 1970, and that the wind speed was greater than 5 m/s on average for 6 h/day, the turbulence affected most of the free zones. There was thus no vertical stratification of the water mass (Lemoalle, 1979).

Another aspect of the effect of wind in littoral zones and in shallow lakes is the resuspension of sediments when the depth of the turnover is greater than the actual depth of the water layer. The consequences of this turnover are many. There is an increase in turbidity and reduction of light penetration, which limits the growth of macrophytes and the thickness of the euphotic layer. The destabilization of the bottom related to the action of the wind is unfriendly to the development of benthic fauna. Finally, the resuspension of the sediment may lead to anoxia of the water mass because of redox phenomena, which could lead to massive animal mortality.

In vast extents of water, when a strong and regular wind has blown for two or three days, a particular turnover regime of the surface layer appears, called the *Langmuir circulation* (Fig. 6.3). Cells of more or less permanent convection appear, generating successions of zones of sinking of the surface water (or convergences) and rising of the water towards the surface (or divergences). This circulation is often visualized by bands parallel to the wind of floating bodies in the zones of convergence. The wind causes a circulation of water, dissolved elements, and living and non-living particles, and a vertical structuration of the water layer affected by the phenomenon.

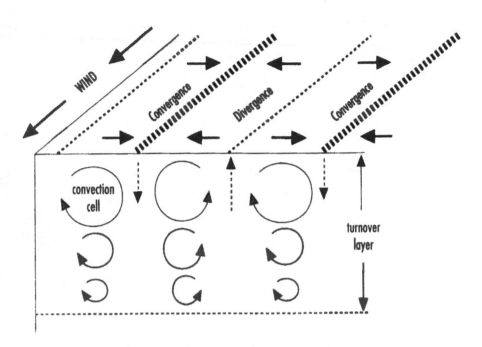

Fig. 6.3. Langmuir circulation (modified from Frontier and Pichod-Viale, 1998)

Another effect of the wind in aquatic systems is its contribution to thermal stratification and the appearance of the thermocline (Lévêque, 1997). Since the density of water differs according to the temperature (it is greatest at 4°C), a water column has a natural tendency to stratify in the absence of turbulence according to a vertical temperature gradient. But the penetration of solar radiation in the surface layers (or epilimnion) causes a heating of the water that is attenuated with depth in the absence of turbulence (Fig. 6.4a). This situation is not stable because the action of the wind on the surface leads to a thermal homogenization of the surface layer (Fig. 6.4b). When the turbulence of the

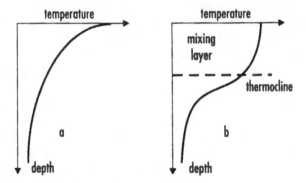

Fig. 6.4. Formation and disappearance of a thermocline. (a) Theoretical profile of temperature of a perfectly stagnant lake. (b) Formation of a turnover layer and a thermocline due to the action of wind on the surface.

water becomes insufficient to ensure turnover, i.e., when the depth is too great for the action of the wind to be felt, the turnover of the water no longer takes place and the temperature decreases rapidly with the depth. *Thermocline* is the zone in which the greatest temperature gradient is observed. In the temperate lakes, the gradient could be from 10 to 15°C in a few metres during the summer. This thermocline is of major importance for the functioning of aquatic systems because it constitutes a barrier to vertical transfer of nutrients between the epilimnion and the hypolimnion.

6.4.4. Winds, upwellings and coastal fisheries

A coastal upwelling is a rise along the continental shelf of deep, cold waters that are rich in nutrients that allow the development of a considerable biological production (Herbland et al., 1983). It is a physical process resulting from the action of the wind that sets in motion the surface layers of the water along the coasts and moves them towards the open sea, leading in reaction to a vertical flux ascending along the continental shelf to compensate for the disequilibrium at the coast. The result is the formation of a strong temperature gradient from the coast towards the open sea. On the Atlantic coast of Africa, the trade winds, which have a direction that is approximately parallel to the coast, are the origin of constant or seasonal upwellings that develop along the coasts of the Gulf of Guinea up to Mauritania (Roy, 1991).

The physical and chemical components of upwellings are found to be highly variable from one region to the next and generate particular structures and dynamics (Cury and Roy, 1991). The coastal pelagic resources such as sardine, anchovy, and *Sardinella* colonize mainly the upwelling zones, which are the most productive zones of the ocean. While they represent only a tiny part of the ocean surface, close to 40% of the fish captures are realized in these regions.

Fishery specialists have observed the very high variability of fish populations exploited in coastal upwellings: to the plethoric periods succeed periods in which the resource becomes rare and disappears. This uncertainty focuses attention on the mechanisms responsible for the variability observed and thus the nature of relations between the dynamics of stocks and their environment. Some of those mechanisms have been understood from research conducted to better understand the interactions between the variations of the resource and the fluctuations of coastal upwellings along the Atlantic coasts (Cury and Roy, 1991) and on the coasts of Peru, California, Chile, West Africa, and Southern Africa. The essential result is the indication of the existence of an "optimal environmental window" (Fig. 6.5) that favours the development of larvae and juveniles of pelagic fishes (Cury and Roy, 1989). There is in effect a relationship between successful reproduction and the intensity of the wind that generates upwelling: when the wind intensity is close to 5 to 6 m/s, it is optimal. Below this value, the upwelling is not sufficiently established. The development and survival of pelagic fishes are limited by a lack of food. Above this limit, the sea begins to foam and the turbulence breaks up and destroys the swarms of larvae and plankton. Thus it is clear that environmental fluctuations have a major

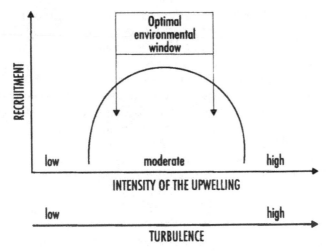

Fig. 6.5. Concept of environmental window (modified from Cury and Roy, 1989). In the left part of the curve, the available nutrients increase with the intensity of the upwelling, while in the right part the greatest turbulence breaks up the larval swarms.

impact on the survival of larvae and determine in large part the level of future catches.

6.5. FIRE

Fire is often perceived as a destructive element that has eliminated entire ecosystems. This is partly true. *But fire is not a human invention. It is a natural component of many ecosystems.* It is caused especially by lightning and volcanic eruptions. In the Yellowstone region, lightning caused at least 369 fires between 1972 and 1987 (Jones and Caloner, 1991). Natural fires occur more frequently in dry and hot regions with inflammable vegetation. But temporary episodes of dryness may increase the probability of natural fires in more humid regions.

Fires have significant impacts on the ecology of ecosystems, wild fauna, air pollution, and economic development. But the nature of impacts varies according to the type of plant cover. Simply put, in forests dependent on fire, like many boreal or dry tropical forests, fire is an event that is welcome and necessary for the proper functioning of the ecosystem. It periodically ravages the taigas of the Old and New World. It leaves vast extents of bare soil that are gradually colonized by seeds from outside. The result is a mosaic structure in which the various elements are made up of trees of different ages (Dajoz, 1996). On the other hand, there are forests that never or very rarely burn spontaneously, such as tropical rainforests that have seen little human intervention, in which fires could modify the ecosystems in the long term. At present in the large rainforests, the rates of humidity remain always high, so that fires cannot start. Still, many traces of paleo-fires are found in the Guyana forest from 10,000 years B.P. This phenomenon is apparently possible only in case of great anomalies of the rainfall regime leading to marked droughts.

Fire is also an intrinsic part of the ecosystem of savannahs, which cover nearly a fifth of the earth's land mass. Savannahs have many characteristics that favour fire: alternation of dry and humid climatic episodes, dry storms and lightning, dead wood and grass that serves as fuel, and a particular kind of plant cover. They certainly burned well before the presence of humans, even though humans increased the frequency and extent of fires, whether to maintain pastures, clear land, eliminate parasites, hunt, or some other reason (Menaut, 1993). Generally, the effects of fire are expressed in the transformation of vegetation of the savannah. Late fires (at the end of the dry season) are highly destructive and only resistant species survive. On the other hand, early fires (at the end of the rainy season) slow down the regeneration of the plant cover but do not affect mature individuals. It is shown, from experiments conducted over 40 years at Kokondekro (Côte d'Ivoire), that the density of woody species has decreased from an initial density of 3300/ha to respectively 11,000, 9800, and 250 per hectare on a protected parcel, under early fires and late fires (Menaut, 1993). In experimental systems, the major effect of protection against fires is to increase the density of woody species at the expense of the herbaceous stratum over the long term, leading over the very long term to establishment of a forest system.

Using fire systematically to obtain cleared areas, the Neolithic cattleherd-cultivators and their descendants caused profound changes in the Mediterranean ecosystems, mainly of three kinds:

— forest substitutions: replacement of deciduous oak groves by sclerophyllous oak groves in the Mediterranean region;

— formation and extension of ecosystems composed of small ligneous species; and

— expansion of all the ecosystems with a basis of heliophilous and nitrophilous herbaceous species in the areas that have some contact with humans (Pons, 1995).

But fire may have beneficial effects on ecosystems, in particular:

— It opens up the canopy by causing tree fall, which allows pioneer species to emerge. In certain ecosystems, forest fires also reduce the total area of the forest, allowing the development of prairies and steppes.

— It favours the germination and dispersal of seeds, given that some seeds germinate only when there has been a fire on the ground.

— Pests are destroyed when the fire is intense enough to kill pests and pathogens while not destroying the trees.

— Nutrients are spread in soils, particularly in conditions in which the decomposition is slow. In the Mediterranean region, fire is thought to be important for the uptake of nutrients from sclerophyllous plants (plants with tough leaves) that otherwise would have decomposed slowly.

The result of these effects in forests dependent on fire is to increase the variety of the forest with respect to its composition, size, and age.

In 1988, a fire of unusual magnitude devastated about a third of the Yellowstone National Park in the United States. Some ecologists did not hesitate to talk of ecological disaster that would modify the landscape of this region as never before. Ten years later, the assessment is much more moderate. The shrubby and herbaceous vegetation has re-established itself and the present landscape is as heterogeneous and diversified as before the fire. In the rivers, the invertebrate communities have been modified, but the fish communities were reconstituted rapidly. In other words, all things being equal, natural biological systems are rapidly regenerated. At the time, only a few ecologists resisted the scenario of catastrophe and predicted that the ecosystems would reconstitute rapidly. They turned out to be right (Barkin, 1999).

Unfortunately, many fires burn in the wrong place, in the wrong season, and with too much intensity. Fires are deliberately started to clear soils, hunt, harvest timber, or simply commit vandalism. Most accidental fires are due to loss of control of deliberately started fires. They result in damage of many kinds:

— Alterations in the vegetation and elimination of poorly adapted and less competitive species. At the worst, a forest type of vegetation may disappear and be replaced by another type.

— The loss of wild fauna.

— Air pollution causing smoke and acid rain and contributing to an increase in greenhouse gases. For example, six months of smog resulting from forest fires in Indonesia in 1997 were equivalent to a year of emissions from all sources in Western Europe.

Fire and Mediterranean ecosystems

In the Mediterranean region, fires were often used until World War II to rejuvenate pasture land. The primitive Mediterranean vegetation was almost entirely destroyed by fire. The present landscape is for the most part inherited from the impact of past fires. It is essentially made up of plant groups that have resisted fire or benefited from it. In effect, four groups of vegetation are distinguished:

• Species that are reduced in numbers after a fire (hardy herbaceous species and small woody species that have poor regeneration of stock and/or a low reproductive capacity by seed).

• Species that resist fire. This is true of shrubby and woody plants such as thickets of green oak, scrubs of heath, and so on, whose underground vegetative parts survive the destruction of aerial parts and put out shoots after a fire. The cork oak has a bark that resists fire and protects the living tissue underneath. Many Mediterranean species (lily, iris, amaryllis) have underground bulbs, rhizomes, or tubers. They are not affected by fire and their survival is ensured.

• Species for which fire is more or less necessary for survival. They germinate because the flames burst the fruit and release the seeds. It has been observed that the Aleppo pine releases a large quantity of fertile seeds in the weeks following a fire. After a fire, thousands of small plants of cistus and broom are also observed.

• Plants that, because of their low capacity for interspecific competition, profit from the occurrence of a fire and establish themselves in burned areas in which there is less competitive pressure. These are species that invest greatly in sexual reproduction by seed. They are heliophile plants looking for maximum sunlight, such as pines and shrubs such as cistus.

Since the end of the 19th century, the rural exodus has manifested itself in a gradual modification of forest landscapes linked to natural regeneration. In Provence, France, aggressive species such as Aleppo pine and Scots pine increased their cover respectively from 35,000 and 30,000 ha to 161,000 and 230,000 ha between 1878 and 1989 (Barbero et al., 1989). This expansion of conifers has been accompanied by many shrubby species that are highly combustible, such as cistus, broom, furze, rosemary, and myrtle, which are also fast-growing shrubs that rapidly recolonize burned areas.

Since World War II, the abandonment of vast regions previously devoted to pastoral uses and now invaded by combustible species, and a decrease in the use of groves for firewood, has been accompanied by an increase in risk of fire.

6.6. WATER

Water is an element that acts at various levels on ecosystems:

— It is a limiting factor of primary productivity of terrestrial ecosystems because all the vegetation and the living world in general require large quantities of water to grow and survive.

— It is an agent of erosion and a vector for transport of particles and dissolved elements (see above).

— It is a medium of life for aquatic organisms.

6.6.1. Water and terrestrial ecosystems

The water cycle (see Chapter 16), which involves only a small part of the volume of water present on earth, provokes the constant exchanges between surface water and atmospheric water by the combined action of precipitation and evaporation. The objective of hydrological assessments is to quantify the volumes of water present in the different compartments of the water cycle at different spatial and temporal scales. The general equation of the balance is simple: $P = E + Q$, where P corresponds to precipitation, E to evaporation and transpiration, and Q to water flow. The quantification of elements of the budget, which is mostly done in the framework of a watershed, is more difficult. In reality, the budget must take into account the water reserves of soil as well as plants.

The hydrological year is punctuated by the relative values of rainfall and the climatic demand, i.e., the quantity of evaporation that the atmosphere can absorb (potential evapotranspiration or PET). Conceptually, the following are distinguished (Cosandey and Robinson, 2000):

— A *hydrological winter*, a period during which the rainfall is sufficient to satisfy the evaporation demand. This is the period of reconstitution of reserves: the water reserve of soils is reconstituted.

— A *hydrological summer*, a period during which the evaporation is greater than the input from rainfall. The water reserves in soils and plants are gradually exhausted.

As in the case of wind, there are exceptional events, such as droughts and floods. Major droughts may cause massive tree mortalities: the drought of 1976 caused the death of 100,000 m³ of firs in the province of Alsace in France (Bouvarel, 1984). Exceptional floods regularly clear the way for chronic floods, but they cause more economic loss than ecological damage.

6.6.2. Fluvial systems

The *hydrological regime* is the principal driving factor of functioning of continental aquatic ecosystems. Through a statistical approach it describes the variations in time of physical phenomena such as the flow of rivers (volume of water, expressed in m³/s, that passes during a given time across the section of the river). The *average annual flow* of a river is the annual average of instantaneous or daily flows and the *annual flow volume* is the quantity of water that has flowed annually in the river being studied.

The flow is strongly influenced by meteorological phenomena and the relief. It is highly variable but difficult to predict in the medium and long term, since it depends on climatic variations. The *hydrograph* is the curve that describes the instantaneous variations of the flow measured at the outlet of a watershed. After a shower, the flow of a water course increases rapidly, reaches a maximum, then decreases. This is called *high water*. However, this term is also used in a slightly different sense to designate simply the period of high waters, over a short or long period, following exceptionally heavy showers. Inversely, *low water* is the lowest annual level reached by a water course. Low and high water, which actually define the extreme forms of the flow, are characterized by their frequency, their duration, and exceptionally high or low values of volume flows of water.

The seasonal evolution of flow of a river is closely dependent on the local and regional climatic conditions. One or more periods of seasonal high waters are observed when the flows are most abundant, and one or more periods of low waters are observed. These seasonal fluctuations of flow reflect the seasonality of precipitation.

Four major types of regimes are found in Europe:

- The *pluvio-oceanic* or *Atlantic* regime characterized by a maximal flow at the end of autumn or in winter and relatively moderate seasonal variations. This is true of rivers such as the Seine, which are essentially fed by rain water.

- The *mountain nival* regime for rivers, such as the upper course of the Rhine, that are mostly fed by snow melt. The period of high waters occurs at the end of spring and the beginning of summer.

- The *glacier* regime characterizing rivers strongly influenced by the melting of glaciers, such as the upper course of the Rhone. The flow is maximum in July-August and minimal in winter.

- The *plain nival* or *continental* regime for the continental European rivers, in which the greatest flows are observed during the spring thaw.

The current is a key parameter of the functioning of river hydrosystems in which the dynamics is controlled by hydraulic variables (speed, flow, shearing, etc.) and geomorphological variables (width, slope, depth). The hydraulic parameters explain the large-scale morphology of water courses to the extent that the channel is constantly adjusted to tend towards an equilibrium between solid flow and liquid flow and to thus optimize the dissipation of energy (Bravard and Petit, 1997). Any constraint imposed by an event on the hydromorphological variables will cause a reaction that may counteract the disequilibrium that is caused.

The current and other hydraulic variables are among the major factors structuring biological communities in the water courses, as well as the distribution of organisms on the habitat and micro-habitat scale. By its physical effect on organisms, the current favours or limits the presence of certain species. Statzner et al. (1988) and Gore (1994) have shown the significance of the erosive force and the Froude number in predicting the composition of communities of benthic invertebrates and fishes. The organisms, because of their morphology and their metabolism, survive within a particular range of hydrological conditions (Statzner et al., 1988).

6.7. ABIOTIC FACTORS STRUCTURE ECOSYSTEMS

The various abiotic factors act synergically on the structure and general functioning of ecosystems. Some examples are given below.

6.7.1. Lakes

The flows and exchanges of energy and matter in a lake system are numerous and may take various routes. Figure 6.6 represents the major climatic and geological factors that control the functioning of the system as well as the structure and composition of biological communities.

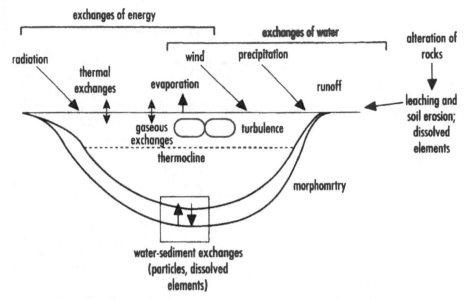

Fig. 6.6. Major climatic and geological factors controlling the lake environment (modified from Talling, 1992)

The geological nature of the lake basin determines the morphometry of the lake, especially its depth and the shape of the basin. It is the context that will determine a large part of the hydrological functioning of the system. The climate is indicated by the exchanges of energy and water with the lake environment. The energy exchanges involve solar energy, which affects the lake temperature and its primary production, as well as auxiliary energy such as wind, which by stirring up the surface water layer and creating turbulences causes a thermal stratification and gives rise to a thermocline. Part of the energy is used for the evaporation of water, the cycle of which involves rains as well as input from runoff in the watershed and, eventually, outputs by means of effluents. The runoff also brings in minerals, which are partly cycled in living organisms and partly stored for a long or short time in the sediments.

6.7.2. Watersheds

Watersheds are made up of aquatic and terrestrial environments in interaction. The longitudinal dimension of water courses, in which water resulting from precipitation flows in a cascade by gravity, greatly determines the functioning of the entire aquatic system.

In reality, water moves between and within the storage zones on the surface, in the soil, and in the underground environment (Fig. 6.7). Precipitation is mostly intercepted by the vegetation and a certain volume is lost by evaporation. Then part of the water running over the soil infiltrates it or is also lost by evaporation. A fraction of the water contained in the soil recharges the ground water. Ultimately, the flow of the water course will integrate the various steps of

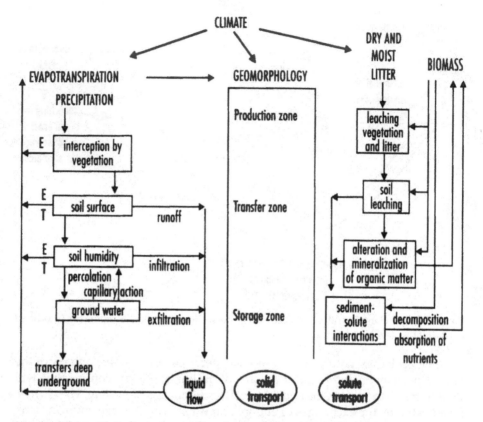

Fig. 6.7. Major abiotic factors controlling the functioning of fluvial systems (modified from Amoros and Petts, 1993)

this cascade: surface runoff, infiltration of the soil, inputs from ground water, etc. In particular, a significant part of the low water flows comes from ground water.

During movements within storage zones, the quality of water is modified by its contact with the abiotic and biotic components of the ecosystem. The dynamic processes involving transfers of water tend to determine the complex process of spatial and temporal variations of the water chemistry.

The surface flow, associated with erosion and transport of sediments, structures the fluvial landscape. Along the course of a river, a marked evolution of geomorphological characteristics is observed, with which are associated characteristic biological communities. In the zone furthest upstream, for example, water courses are narrow and often have a high slope. Despite the low flow, erosion is intensive because of a high current, and the beds are stony. This part, the upper course, is sometimes called the *production zone* or *erosion zone* because several such small streams feed the fluvial system with water and sediment.

Further downstream, the slope and the speed decline, while the flow and the width increase. The beds are made up of pebbles and gravel and, as the slope diminishes, there is more sedimentation than erosion. This part of the fluvial system, the middle course, is also called the *transfer zone* of water and matter towards the alluvial plain.

Finally, in the plains, where the slope is low, the current slows considerably and a large part of the transported material is deposited to form wide alluvial plains, or deltas. This is the *storage zone* or lower course in which there is accumulation of alluvium resulting from the demolition and mechanical and chemical transformations of surface rock.

Chapter 7

Hierarchies, Levels of Organization and Typology of Ecological Systems

The whole is more than the sum of its parts.

(Smuts, 1926)

There are many ways to study ecosystems. For the ecology of communities, for example, it is the population networks, interacting in a physicochemical environment, that determine the framework of possible situations. The functional approach on the other hand mainly emphasizes energy transfers and the dynamics of the nutrients in the ecosystem, favouring biogeochemistry as a point of entry. *The organization of an ecosystem thus appears to be controlled by two sets of constraints: structural constraints that act on organisms and functional constraints that act on processes.* This representation of the ecosystem as a double hierarchy is convenient from the conceptual point of view but raises questions about the links between these hierarchies, because there is no simple relationship between species and the functions of ecosystems (O'Neill et al., 1986).

The two approaches are not orthogonal, for all that many authors have attempted to quantify the participation of organisms in terms of certain processes. We cannot reduce the functional processes to a simple list of species, because it is clear that:

— Various species can fulfil the same functions or similar functions in an ecosystem. This is the principle of ecological redundancy that leads to the recognition of guilds.

— A single species may fulfil various functions in the course of its biological cycle. This is true of holometabolic insects, in which various stages of development occupy different ecological niches.

— A single species may fulfil different functions, at different periods, and at different places.

It is clear that neither the population approach nor the functional approach can individually provide a theoretical basis for the analysis of ecosystems, each favouring a limited point of view. It is also quite true that we still have no theory that allows us to integrate these two approaches in an operational way by giving them an equivalent weight. To develop an ecosystem theory, we must identify ecological units (Patten, 1978).

7.1 THE SEARCH FOR ORDER IN ECOSYSTEM STRUCTURE

The study of the dynamics of complex systems, as is the case for ecosystems, is based on a large number of paired variables that are constantly evolving and are difficult to integrate. Ecosystems are in fact made up of more or less independent entities but linked among themselves by exchanges of energy, matter, and information. Ecologists have long attempted to ignore or avoid the complexity by working on simple systems. But to the extent that they discovered that nature is structurally heterogeneous, complex, and variable, the hope of the existence of an intrinsic order in nature has evolved into the search for an order that experimental and judicious scientists can prove.

Therefore, ecosystem ecologists generally end up looking for simplifications and attempting to break down their objects of study, which are vast and heterogeneous, into simpler entities that can be characterized by a few physical, chemical, or biological criteria. In this they have followed a process practised by systematicians: to identify the elementary units on the basis of easily identified characteristics, to name them, and to classify them before ordering them within a larger context. According to Benzecri (1973), a typology is a type of classification that allows us to distinguish several categories or types within a heterogeneous set of stations or ecosystems of the same kind. For example, populations are easily identified objects in ecosystems, and the use of populations as a basic unit to study the organization and functioning of ecosystems has produced a consistent theory in the field of population and community dynamics.

7.1.1. The whole is greater than the sum of its parts

On the basis of a general theory of systems, it is possible to break complex systems down into simpler components or subsystems. In theory, for physical systems, almost any property of the system can be deduced from the individual properties of subsystems. The subsystems can thus be studied separately, and at present the whole can be reconstituted by means of modelling, for example. According to this approach, the whole is the sum of its parts.

Researchers have attempted to adopt this approach in studying ecosystems, but ecosystems do not behave like physical systems. Life cannot be explained by the simple juxtaposition of molecules and atoms. *Biological systems are peculiar in that the characteristics of the whole cannot be deduced from those of the parts.* A cell, for example, is an active "machine" that carries out biochemical functions, while the properties of cells are not present at the molecular level. Similarly, an ecosystem has a regulatory effect on the biogeochemical cycles that the constituent species taken individually cannot assume.

In reality, from the molecule to the biosphere, the living world is organized. At each level of integration, properties appear that cannot be analysed only from the mechanisms that have an explanatory value at lower levels of integration. This appearance of new characteristics at the level of the whole that did not exist at the level of the components is called emergence.

> ### Emergence
>
> "Emergence is the theory that the characteristic behavior of the whole could not, even in theory, be deduced from the most complete knowledge of the behavior of its components, taken separately or in other combinations, and of their proportions and arrangements in this whole" (Broad, 1925).

It is also apparent that the constituent elements of ecosystems often behave differently when they are separated from the systems they are part of. The result is that laboratory observations on isolated elements are difficult to extrapolate to the natural environment. Whatever their value, such observations often provide only limited information on the role these elements actually play within the ecosystem (see Chapter 4).

7.1.2. Hierarchies

To affirm that the whole is greater than the sum of its parts is a declaration of principle that nevertheless remains difficult to implement. That is why many authors (Allen and Starr, 1982; O'Neill et al., 1986) have recommended the application of the hierarchy theory to the study of ecological systems, arguing that any object of study in ecology is composed of a set of subsystems and itself constitutes one element of a larger set. *The hierarchical approach tends toward a reductionist approach that attempts to identify the simplest sets to be studied but includes them within a network of constraints and interactions.* The biosphere, for example, can be divided into different ecological systems or biomes. In this context, the tundra or the tropical savannah are relatively independent systems, each of which can be studied without explicit reference to the other.

An important hypothesis of the hierarchy theory is that the units constituting the various hierarchical levels do not have independent dynamics. The processes that have taken place at higher levels exert constraints on the dynamics of processes occurring at lower levels. *The hierarchical organization of ecological systems can effectively be perceived as an inclusion of entities of lower order within larger entities of a higher order, like Russian dolls.* This is a pyramidal form of organization. The higher hierarchical levels in an ecosystem are considered by the external factors of climatic and geomorphological nature that determine and constrain the framework in which the system can evolve. The other levels are subordinate to them.

To speak of the hierarchical system implicitly supposes that all the factors that control the structure and functioning of ecosystems do not have the same influence. Some are in a way more important than others depending on the level of organization one is looking at. Intuitively, we must also acknowledge that the existence of a hierarchical structure stabilizes the system by giving it coherence. In any case, any ecological system seems to have an inherent tendency to break down into independent subsystems that nevertheless are organized in relation to one another (Auger, 1990). This hierarchical organization probably results from differences in the speed of processes.

The hierarchy theory is perhaps not operational but it has great heuristic value and constitutes a highly useful framework for organizing our thoughts. The identification of hierarchical levels facilitates the identification of processes that are important for the functioning of the ecosystem at the selected level of integration.

7.2. ORGANIZATION LEVELS OF BIOLOGICAL SYSTEMS

Instead of hierarchical level, some authors prefer to use the term *organization level* (Pave, 1994). One of the characteristics of the living world and the abiotic world is in fact organization. Atoms are organized into crystals or into molecules, and those molecules are in turn organized into cells that can reproduce. The cells can aggregate and cooperate to make up multicellular organisms that, in turn, are organized into groups, populations, and multispecies communities. If the abiotic environment in which the organisms develop is also considered, we

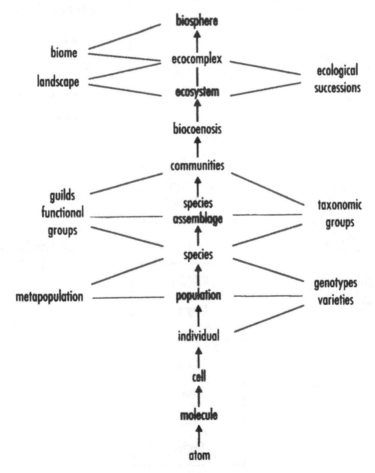

Fig. 7.1. Some typological concepts associated with a scale of organization in ecology

have increasingly complex entities called ecosystems, landscapes, and the biosphere (Fig. 7.1). At a given level of organization, therefore, there is a generic set of elements that have relationships and interactions with one another. These interactions structure all of the elements into groups that constitute the level of organization. Similarly, the elements of an organization level in turn constitute the elementary units that are structured to compose the higher organization level. This scale of organization thus leads to increasingly complex hierarchical levels. At each of these levels, new structures and properties emerge that are the result of interactions between the elements of a lower level.

The existence of several levels of organization is recognized in systems according to the types of entry that are favoured (Fig. 7.2).

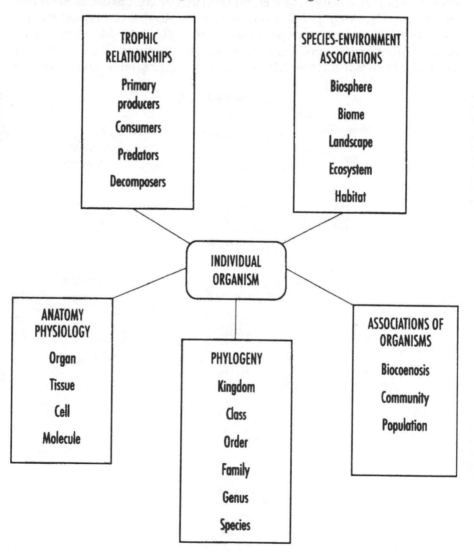

Fig. 7.2. Principal levels of ecological organization (modified from Pickett et al., 1994).

7.2.1. Taxonomic hierarchy: the search for order in species diversity

Taxonomy is the scientific discipline that names and describes living organisms. *Systematics* is the study of the diversity of organisms and relationships between these organisms Its objective is to classify the species and to find out what the phylogenies are, i.e., the degrees of relationship between species.

The classification of the living world is important for the understanding of ecosystems. On the one hand, it allows comparisons to be made on the basis of species or taxa of a higher order, postulating that species belonging to the same taxa share a certain number of biological and ecological characteristics that can be different from those of other taxa. On the other hand, the biodiversity is a structural component of the ecosystem and, according to the phyla represented, certain ecological processes are possible or impossible.

Even though the concept of species is controversial, it is still a reality in biology. The Linnaean typological concept of species based on criteria of resemblance in relation to a "type" has, of course, been replaced by the biological concept of species based on not only resemblance but also interfertility of the individuals constituting a population, the descendents of which are themselves interfertile. Within a single species, subsets can be distinguished, which are called races, strains, varieties, etc. This intraspecific diversity may be apparent in the existence of morphologically well-differentiated forms: the many races of domestic animals are a good example. However, the biological concept of species is often difficult to apply because it is materially impossible to cross most of the wild forms to verify their interfertility. Besides, it is applied rigorously only to species with bisexual reproduction, while many plants and animals have parthenogenetic reproduction. Thus, where possible we continue to use largely the morphological description to identify species. Recently, tools of molecular biology have become very useful in identifying bacteria and viruses that cannot be identified by morphological criteria.

The classification of the living world must be hierarchical because groups are totally included within larger sets, of which they are only a part. Taxonomy, initially based during the Renaissance on the idea of a descending classification (division of larger classes into subclasses, similar to the classification of inanimate objects), has evolved towards an ascending classification that consists of grouping related taxa into taxa of a higher order. The supraspecific categories (genus, family, order, division, class, branch) serve to take into account the degrees of relationship between taxa. The systems used presently are *phylogenetic hierarchy*, based on the evolutionary relationship of groups derived from common ancestors, and *phenetic hierarchy*, based on the similarity of forms or characters between species. Phenetic hierarchy, based on the similarity of appearance, does presuppose that there is a common ancestor and may give different results from phylogenetic hierarchy. An example illustrates this possible divergence: the lung-fish is phenetically closer to a salmon than to a cow, but it has a more recent common ancestor with the cow than with the salmon. The evolutionary line connecting lung-fish to cows was transformed so rapidly that cows do not resemble their aquatic ancestors at all (Ridley, 1989).

7.2.2. Biotic assemblages: from individual to biocoenosis

Within ecosystems, we can distinguish hierarchical sets and functional levels of increasing complexity in collections of living organisms.

a) The individual

The elementary unit of the living world is the individual, carrier of a genetic make-up. An individual is made up of a set of structural elements—molecules, cells, organs, etc.—that are themselves organized in a hierarchical manner. The set of genes constitutes the *genotype* of an individual that is, in a way, its initial capital that it will be able to exploit in different ways according to the constraints of its environment. Each individual belonging to a species is slightly different from others genetically. A bacterium contains around 1000 genes and some mushrooms contain around 10,000. There are some 30,000 genes for human species.

b) The population

A population consists of all the individuals of a single biological species inhabiting a particular environment. It is also the favoured topic of research on evolution of living organisms because it is at this level of organization that natural selection operates. A population is called *panmictic* when it is made up of a unique set in which all the individuals can freely exchange their genes during the course of reproduction. But most often it is subdivided into subsets whose geographic distribution is a heritage of the environmental history. For these fragmented populations, called metapopulations (see Chapter 9), their existence and dynamics are functions of possible exchanges between isolated subsets.

c) The community

Ecological research rarely concerns the entire set of species present in the ecosystem, but rather multispecies sets selected as a function of the issue, or, most often, as a function of the available taxonomic competence. These groups are commonly called assemblages or communities, which are usually defined as a set of species. When there are species belonging to a single systematic group (birds, fishes, termites, etc.), they are bird or fish communities, or *taxicoenoses*. The planktonic or benthic communities, on the other hand, includes several taxonomic groups. Southwood (1987), in an attempt at a synthesis, defined the ecological community as "a group of organisms, generally of wide taxonomic affinities, occurring together in a location; many of them will directly interact with each other within a framework of both horizontal and vertical linkages."

The central question of community ecology is the following: is the set of populations present in an environment the result of chance (i.e., a random collection of all the populations that have succeeded in reaching the site and staying there) or the result of selection on the basis of coevolution between the species themselves and between the species and their abiotic environment, in such a way that there exists a network of interdependence between these species? Many ecologists lean toward the second hypothesis and estimate that the community has properties (including a spatial and temporal organization) and a functional structure.

According to Wiens (1984), "the goals of community ecology are to identify patterns of ecological communities, to determine the causal processes that underlie these patterns, and to generalize these explanations as far as possible."

In reality, there is a question of scale: in ecosystems that have a long life span, there may be a coevolution of species, and thus a degree of interdependence that is greater between the species than in young ecosystems.

A quite popular level of organization among ecologists is the *guild* (from the mediaeval Latin *gilda*, corporation), a set of systematically similar species that exploit the same type of resource within a single ecosystem. This concept is defined in functional terms, unlike that of community. For example, Cummins (1973) subdivided the benthic communities of rivers into functional groups: browsers, scrapers, filter feeders, etc. Within each of these guilds there are species that actively compete for resources. On the other hand, between individuals belonging to guilds that have different functional roles, the interspecies competition is weaker.

d) The biocoenosis

The biocoenosis (from *bios*, life, and *koinoein*, to have something in common) is the set of populations of animal and plant species that live in a given environment. Often, the term *zoocoenosis* is used to designate the animal part of a biocoenosis, and the term *phytocoenosis* to designate the plant part. The term *biocoenosis* was introduced in 1877 by the German zoologist Mobius, who worked on banks of oysters. He applied the notion of biocoenosis to a community of plants and animals. According to Mobius, science did not yet offer a word to designate such a community of living things, in which the sum of species and individuals, being mutually limited and selected by external average conditions of life, reproduce and continue to occupy a given territory. He proposed the word *biocoenosis* for such a community. Each oyster bank was analysed as a community of living things (a biocoenosis) that found in this precise place all the conditions for their reproduction and survival: a suitable soil, sufficient food, and a suitable salinity, as well as temperatures that were favourable to their evolution. Mobius invented the concept but did not develop the theoretical bases of the discipline that would develop into biocoenotics (or synecology) in the 1920s (see Chapter 2).

e) Ecosystem-biosphere

The most inclusive levels of integration, *ecosystem* and *biosphere*, are the final terms of biological hierarchy. They are treated in detail in Chapters 2 and 14.

7.2.3. Trophic hierarchy

The food chain, from producers to consumers, including decomposers, is another hierarchical system. The concept of trophic level is useful in grouping roughly the populations of various species into large categories, thus allowing us to estimate the flows of matter and energy that circulate in the ecosystem. However, considering the difficulties often found in assigning a trophic level to a species, the trophodynamic approach could not be used as a general organizing principle for ecosystems (see Chapter 11).

7.3. EXAMPLES OF HIERARCHICAL STRUCTURES IN THE FUNCTIONING OF ECOSYSTEMS

The question of hierarchical structures does not concern only biological systems. There are hierarchical organizational levels in factors that control the functioning of ecosystems. They are illustrated by the two examples given below.

7.3.1. The rule of stream order

Water courses, fed by runoff or by springs, are organized into hierarchical hydrographic networks (Fig. 7.3). Hydrologists and geomorphologists have established a classification of water courses by using a simple rule known as the Strahler rule: the meeting of two water courses of the same category gives rise to a water course of higher rank. Each of the levels identified corresponds to a set of mesological and biotic characters that structure the river ecosystem.

This system of classification has been used mostly by North American biologists, in preference to other approaches such as the distance from the source. It has the merit of being quite objective and easily applicable, but like any simple method, it has its limits. In particular, it is not always easy to identify

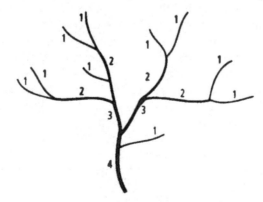

Fig. 7.3. Hierarchical classification of water courses (Strahler's classification). Conventionally, water courses of order 1 are defined as those without tributaries. The confluence of two streams of order 1 produces a stream of order 2, and the confluence of two streams of order 2 produces a stream of order 3. On the other hand, a stream of order 2 receiving a stream of order 1 remains a stream of order 2.

streams of order 1 when the flows are intermittent. This hierarchical structure of water courses corresponds to a biological structure that has been formalized by the concept of fluvial continuum (see Chapter 9).

7.3.2. Hierarchy of physicochemical variables controlling the metabolism of inland water systems

The dynamics of a lake ecosystem, whatever its geographic location or nature, is under the control of several abiotic parameters that depend on the climate and the geological nature of the substrate. A hierarchy of these factors can be proposed (Fig. 7.4) showing simultaneously their inclusion in a vertical structure and their interactions in a horizontal structure. The geographical situation directly controls the type of climate to which the system is subject as well as the form of the basin, and indirectly the physicochemical quality of the environment. The climatic regime determines parameters such as precipitation, temperature, and insolation. The shape of the basin determines a certain number of morphometric characteristics. These factors interact to control parameters such as the water balance, the vertical structuration of the water column, the turbidity, the inputs of nutrients, etc., which in turn limit the context in which various species can develop as a function of their ecological needs.

This type of hierarchical structure of abiotic factors resulting in the control of the functioning of the ecosystem is known as *bottom-up* control or control by resources. In fact, the presence and abundance of species is estimated to be

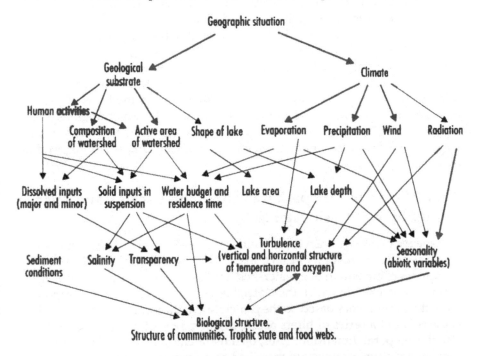

Fig. 7.4. Abiotic parameters controlling the functioning of lake systems (Lemoalle, 1999)

under the direct influence of abiotic factors, in opposition to *top-down* control, which puts forward the biotic interactions, notably predation and trophic cascades (see Chapter 11).

Asymmetry of control

The study of the dynamics of hierarchized systems shows that there is a strong relationship between the duration of processes, or their frequency, and the spatial extension of those processes. Generally, long-term processes act over vast areas and exert a determining control on processes that occur more rapidly over smaller areas. The latter actually have little influence on the former. The result of their relationship between different hierarchical levels is that processes that act at higher levels constitute constraints over lower levels, is called *asymmetry of control*.

Just as with lakes, the chain of factors that control the ecological functioning of a river can be summarized.

On the regional scale, the diversity and productivity of living communities depend on primary determinants such as the geology, the relief, and the climate. The soils and vegetation of the watershed also intervene in the water and sediment balance. On the scale of a stretch of the river, the spatiotemporal dynamics depends on the hydrological and sedimentological regime and the geomorphology of the valley that determines the form and stability of the bed, as well as the structure of the riparian vegetation and the connectivity of the water courses (interactions with the flood plain).

Finally, on the local scale, diversity and productivity depend on key factors such as the physical habitat (substrate, hydraulics), the hydroclimate (temperature, light, dissolved gases), and the food resources.

From this conceptual framework, a process could be developed leading to the regionalization of ecosystems of running water on the basis of primary determinants (geology, relief, climate). In this way, hydro-ecoregions can be delimited within which the water courses are supposed to present similar physical and biological characteristics, and a similar process of longitudinal evolution.

7.3.3. Hierarchy of spatial and temporal scales and species composition of freshwater fish communities

The species richness of fish communities in a watershed is the result of many historical and contemporary factors. To explain the existing relationship between the species composition of a watershed and the set of species observed in the region or continent in which the watershed is located, Tonn (1991) proposed a theoretical framework based on the principle that the local species composition is the result of a series of filters acting at different spatial scales, considering also the temporal dimension. Thus, on a hierarchical process of structuration, community structurations are inscribed as a function of the scales on which they are studied.

In fact, the species richness observed locally carries the imprint of historical processes that are manifested sometimes on a global scale. For the different continents, for example, the continental filter began with the separation of Gondwana, and the present fish fauna is the result of phenomena of speciation and extinction that occurred subsequently on each continent. These phenomena varied from one continent to another. It can be shown that the species richness of a river system is correlated to the continental richness (Obserdorff et al., 1995). For example, Europe and North America, which have directly and repeatedly suffered the impact of glaciations, have a poorer fish fauna than African and South America, in which the impact of glacial periods was less significant. As a result, the species richness in rivers of equivalent size is greater in African than in Europe (Lévêque and Paugy, 1999).

On the regional scale, the qualitative and quantitative composition of fish fauna result from historical events that enabled the dispersal of species and colonization of hydrographic basins or, on the contrary, led to the extinction of species present in these basins. The succession of many climatic events and the existence of refuge zones, in which some species could survive during unfavourable periods, played a major role at the regional level.

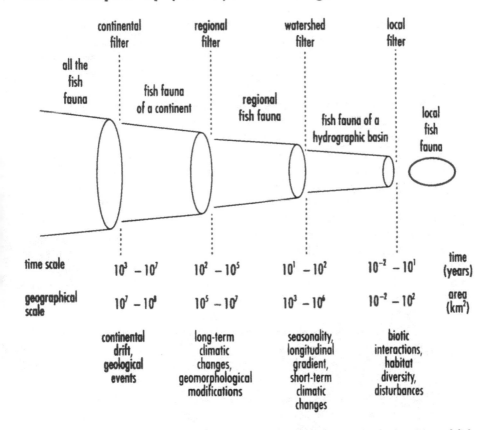

Fig. 7.5. Different spatial and temporal scales from which the species composition of fish populations in watersheds can be interpreted (modified from Tonn, 1991)

In each region, different watersheds can be identified that are separated from each other by barriers that fish cannot cross. These watersheds could have communicated in the past (and could perhaps do so in the future), because of rivers capture as well as orogenic events, or climatic changes that could cause exceptional floods, for example. The number of species in a basin is the result of an equilibrium between processes of colonization and extinction, on the one hand, that depend on the history of the watershed, and on processes of speciation that depend on the evolutionary potential of existing fish families and the duration of isolation.

Finally, at the local level, it is the contemporary factors, whether abiotic (physical, chemical, climatic) or biotic (competition, predation, diseases), that have a critical impact on the composition of bio-communities, each of them being a subset of the regional pool of species. The species richness is a function of the diversity of habitats, and the biotic interactions play a much more important role at this scale than at other scales (Fig. 7.5).

7.4. TYPOLOGY AND ECOLOGICAL CLASSIFICATION

Description, characterization, and differentiation are various elements that, once integrated, can be used to construct a typological system. The units identified can be grouped and classified into "types", which are then organized hierarchically to result in a classification (Godron, 1975; Nilsson and Grelsson, 1995).

There is no single and universal method by which to classify ecosystems; the success of an approach is measured by its capacity to meet the scientific or management objectives for which it is used. This calls for attention to the scales of the study, the aims of the research, the definition of selected criteria, the choices of principal characters, and their hierarchy (Barnaud, 1997). The improvement of typologies and classification of natural areas is in great demand and there is an abundant literature about it to respond to cognitive or targeted questions, in order to identify ecosystems considered representative in a context of protection, restoration, research, monitoring, and other objectives.

In a syncretic work, Klinj (1994) compared various theoretical points of view on the classification of ecosystems as a function of environmental management objectives. The inductive process uses quantitative methods that allow the emergence of similarities and differences (patterns) between the elementary units. These methods do not have an inherent behaviour. The deductive process supposes the construction of a theoretical framework from hypotheses on the manner in which nature functions, and a subsequent positioning of the entities analysed in that framework on the basis of several criteria.

In the inductive process that is often favoured, ecological classification is first of all a tool for organizing information. The possible groups are looked for on the basis of common attributes, from units of observation. These groups may involve biological elements (distribution of species, species richness, structure of communities, etc.), processes, or physical characteristics (types of soils, frequency of fires, precipitation). The final product is a set of groups or ecological units. Within each of these groups, the units of observation share more common attributes than with units belonging to other groups. In operational terms, we can expect, for example, that all the representatives of an ecological unit respond in a predictable manner to natural changes as well as to management practices.

Ecological classification is thus an iterative process that involves field observations, statistics, and modelling. On the small scale, ecological units are often identified in the field, by direct observation of the flora, fauna, soils, and geomorphological criteria. On larger scales, aerial photographs and satellite images, associated with field observations, are useful in identifying the ecological units. Once the data are collected, various methods can be used to identify patterns and to reduce the number of variables. Multivariate methods can be used to detect patterns in the canopy of forest systems or the distribution of plankton, for example. These patterns must, if possible, be confirmed by field observations or by comparison with other sets of data. Multivariate methods are highly useful for such ecological classification because of the large number of variables and observations that are usually analysed and the difficulty of identifying patterns because of the complexity of interactions at work. However, they are insufficient to develop a classification of systems.

Generally, the present trend is to look for closer integration of biological and physical information in ecological classifications to better respond, in particular, to the objectives of integrated and sustained management of environments. The theoretical bases of the classification must thus refer to the following:

— *The recognition of patterns.* It is necessary for spatial and temporal relationships to be well defined.

— *The dynamics of systems.* Many ecological classifications have static patterns or structures, but the possible trajectories of ecosystems must also be taken into account.

— *Hierarchical organization.* The hierarchy theory offers a conceptual framework to study multiscale systems.

— *The biotic component.* The patterns observed in time and space can be useful to define the biotic component of the ecosystem in two ways: according to the concept of continuum, the species assemblages are temporary and unstable; according to the concept of community, reproducible species assemblages are observed in limited habitats presenting characteristic properties.

In theory, ecological classification is independent of mapping. But mapping is an effective way of representing the spatial distribution of ecological units and is an essential tool for discussing management measures.

The CORINE programme

In order to apply the European Community Council directive on the conservation of wild birds (1979), it was necessary to make an inventory of wetlands and to determine which ones were the most important for the conservation of these species. This was the starting point for thematic data bases: CORINE-Biotopes and CORINE-Land Cover. CORINE-Biotopes is a standardized repertoire of important sites for nature conservation in Europe (Devillers et al., 1989). The level of resolution chosen corresponds to that of the needs of small vertebrates, large invertebrates, and vascular plants. The hierarchy was drawn up as a function of geomorphological characteristics, the composition of plant communities, and finally biogeographical or ecological factors controlling the composition of animal communities.

The European programme CORINE-Land Cover involves the mapping of ecozones or homogeneous units of landscape analysis based on characteristics of the territory studied from remote sensing images (Blandin, 1989). This is a classification centred on land cover, with a partition into significant major entities in terms of land management.

Chapter 8

Spatial and Temporal Scales and their Consequences

If roses, which last only one day, had histories and left memoirs for each other, they would draw a portrait of their gardener in a certain way. The roses that came later would leave that portrait unchanged for those that followed. Of their gardener they would say, we have always seen the same gardener; from the memory of rose we have seen only him; assuredly, he does not die like us; he alone is unchanging.

Fontenelle (1657-1757)

Entretiens sur la pluralité des mondes, "Cinquième soir"

"Four blind men are led into a courtyard to experience an elephant for the first time. The first grasps the trunk and declares that elephants are fire hoses. The second touches an ear and maintains that elephants are rugs. The third walks into its side and believes that elephants are a kind of wall. The fourth feels a leg and decides that elephants are pillars..... . Like elephants, ecosystems can be viewed as many perspectives. Our conclusions are biased by the way we observe ecosystems."

O'Neill et al. (1986)

In theory, the ecosystem is a functional unit of the biosphere that has abstract limits. However, the spatial dimension of ecosystems is the subject of recurring debate among ecologists. It was taken into account explicitly in the theory of island environments of MacArthur and Wilson (1967), according to which the nature and species richness of island communities depend on the size of the island (see Chapter 10). *The concept of ecosystem as an independent and discrete entity is now essential in the opinion of many scientists, and they immediately pose the problem of the scales at which ecosystems must be studied.* These scientists refer increasingly often to ideas of spatial and temporal scales to untangle the complexity of ecological systems that are often fragmented, heterogeneous, and changing.

On the one hand, we must understand how scales of space and time interfere with the perception of ecological processes. "The spatial and temporal scales of community investigations determine the range of patterns and processes that may be detected, and therefore, the level of understanding and explanation that can be achieved" (Giller and Gee, 1987). For example, some processes may lead locally to the definitive or temporary disappearance of a species, while the same species expands rapidly in other places.

On the other hand, the question of change in scale is one of the most significant obstacles to the development of general laws or global models. In fact, extrapolation of observations from one spatial or temporal scale to another remains a particularly delicate exercise.

One of the main advantages of concepts of scale is therefore that they oblige ecologists to question the adequation between the phenomena they study and the scale at which they observe them. This question concerns not only the treatment of observed data but also experimental approaches. The structures detected sometimes do not correspond to the processes that are being studied, or they accumulate the effects of different processes relating to different scales (Wiens, 1989). The prolonged neglect of the interdependence between scales and nature of phenomena has been the source of futile debates between scientists, since models constructed from data collected at different scales are not comparable. *There are no good or bad scales of observation. There are only scales that are adapted or not adapted to the questions being asked.*

There is often a confusion between the concepts of scale and organization level. Scale is linked to the measurement of space or time. To go from one scale to another is to change the unit of measurement and to use an instrument adapted to observation at that scale. The idea of organization level is not linked to the instrument but results from the fundamental properties of matter and the living world (Pave, 1994).

8.1. TIME

The measurement of time is based on cyclical phenomena such as the earth's rotation around the sun (the year), the earth's revolution on its own axis (the day), or the cycle of the moon (28 days). Time is a continuum, but it can be arbitrarily divided into scales of different durations depending on the nature of the phenomena we wish to study. For example, we can distinguish the time required for a biological response (a displacement, a growth) or the time corresponding an individual's life span, or the time needed for evolutionary phenomena.

8.1.1. Time scales

The intervals of time that serve as a reference for each of the scientific disciplines involved in the study of ecosystems may vary considerably. In addition, there is a relationship between time scales, physical events, and biological phenomena (Table 8.1).

a) Time scales of geologists and palaeontologists

For geologists, the elementary unit is a million years (Bernard et al., 1995). This time scale is actually that of the establishment of continents and their major orographic units. It is also the time scale of major erosive and sedimentary processes that are the origin of morphogenic systems. Biologically, this time scale is that of the establishment of flora and fauna on different continents. The origin of life dates from around 2000 million years ago, and the separation of

Table 8.1. Theoretical relationship between time scales, physical events and biological phenomena (modified from Magnuson, 1990)

Time (years)	Ecological equivalent	Physical events	Biological phenomena
100,000	100 millennia	major tectonic movements	species evolution phenomena of speciation
10,000	10 millennia	glacial cycles	extinction of species
1000	millennium	wet and dry periods communications between basins	exchanges of fauna and flora between continents
100	century	climatic changes	life span of vertebrates
10	decade	impacts of human activities El Niño	evolution of communities life span of invertebrates
1	year	hydrological cycle seasonal cycle	reproductive cycle
0.1	"month"	lunar cycle	planktonic succession
0.01	"day"	diurnal cycle tornadoes, tides	daily feeding cycle daily migrations
0.001	"hour"	changes in temperature	physiological processes

the present continents (the fragmentation of Pangaea) began around 150 million years ago. The disappearance of the dinosaurs, around 65 million years ago, was concomitant with the development of mammals. One characteristic of phenomena studied at this scale is their irreversibility: e.g., degradation of rocks and radioactive materials; reduction of the internal energy of the earth, or biological evolution.

b) Time scales of climatologists and palaeoenvironmentalists

Periodic climatic cycles, related to the orbital parameters of the earth, evidently have a periodicity of 20,000, 100,000, or 400,000 years. They are one of the causes of the major cycles of glaciation and deglaciation observed over around 3 million years. This time scale, in which the interval is 1000 or 10,000 years, is that at which the establishment of flora and fauna in regional palaeoenvironments is studied. At this scale, in fact, variations in the sea level, as well as climatic changes of large amplitude, have led to significant modifications of the physiognomy of ecosystems and the distribution of species (see Chapter 17).

Northern Europe, which lay under ice for 20,000 years, is for example poorer in species than the regions further south. The time scale of geomorphology ranges from the study of palaeoenvironments to that of present environments. It lies between the long time scale of geology and the shorter time scale of present phenomena.

c) The time scale of ecologists

The unit of time for an ecologist ranges from a day to a century. This is the scale of periodic cyclical processes over the short term (diurnal, seasonal, and other rhythms) and the medium term, and of non-periodic phenomena (disturbances). Certain species of fish, for example, are active during the day and inactive during the night, or the other way around, so that they occupy certain habitats only at particular times of the day. Similarly, many species base their biological behaviour (the reproductive cycle) on seasonal variations in temperature, rainfall, or hydrology. At the scale of 10-100 years, it is also possible to observe changes in the processes of runoff and river flow linked to climatic events (droughts) or human intervention.

d) The time scale of biologists and physiologists

Biologists and physiologists are concerned with the time scale of biochemical reactions and physiological processes such as nutrition, growth, and reproduction. This is also the time scale of interactions between individuals and of behavioural studies. We can include in this the studies of life cycles, keeping in mind that these could vary from some days to some years depending on the species. The scales range from a second to a year.

8.1.2. Interactions between time scales

Phenomena commonly interact at different time scales simply because reaction times ares not the same for different processes or for different species.

a) Synchronization between processes: match/mismatch theory

There are critical periods in the lives of organisms. One such period in the fish is that at which the larva, having reabsorbed its vesicle, must feed on prey that it catches from the natural environment. These prey must be present when the larva needs them. Thus, when fish lay eggs at the same time each year, and the cohort of young larvae appear at the time of peak phytoplankton production (match), a large quantity of food is available and the growth rate and survival rate are high. If, on the other hand, the eggs have been laid too early or too late, there is no coincidence between the phytoplankton production and the arrival of young fish in the system (mismatch). These fish may suffer famine and the recruitment is low. This match/mismatch theory, developed by Cushing (1966, 1996), emphasizes the requisite synchronization of two independent biological processes, the favourable result being the successful recruitment of juveniles, which determines the abundance of adults.

b) The "invisible" present

In hydrosystems, a decade corresponds to the average life of many fish species. Many ecological studies are of a shorter duration, while the ecological and socio-economic processes that determine the dynamics of these ecological systems are played out most often over much longer time scales. Over these short time scales, and in the absence of major disturbances, scientists have great difficulty in detecting very slow changes. They are thus forced to believe that the world is static, whereas it actually is changing when we analyse it at a larger time scale. This is the "invisible present" concept articulated by Magnuson et al. (1983), which serves as an argument in favour of long-term research on ecosystems.

c) Delayed effects

It is often very difficult to interpret observations because of the time gaps that may occur between a disturbance and the response of the ecosystem. Because of the existence of such *delayed effects*, it is more difficult to distinguish causes and effects, which complicates the interpretation of phenomena observed and the search for ecological laws (Magnuson, 1990). Effects are delayed simply because physical and biological processes take time and because organisms have different reaction times: the effect of a disturbance may manifest itself in a few days in planktonic communities or in a few years in fish communities, while the consequences for trees with a longer life span will be visible only after a few decades. During a research programme, we may end up observing the response of an ecological system to a disturbance that occurred before we began to make observations and therefore has unknown characteristics. Similarly, when we carry out experiments on an ecological system, it is possible, if the duration of observation is too short, that we have seen only the transitory period and cannot observe the new state towards which the manipulated system is slowly evolving (Magnuson, 1990).
Delayed effects explain the existence of relic communities that have successfully survived despite substantial changes in their environment.

d) The retrospective approach

Ecologists have gradually realized that ecosystem structure and functioning cannot be explained only on the basis of observations made in the present. For observations made on a site to be correctly interpreted, they must be placed within a wider spatial and temporal context, especially in terms of the dynamics of the principal factors of disturbance.
The local or regional ecological diversity thus carries the imprint of historical events of a climatic, geological, or biological nature, which are largely responsible for the present distribution of species and the structure of communities. In other words, *the retrospective approach, by which the history of ecosystems is reconstructed, is important in explaining the composition and dynamics of present ecosystems.*

8.1.3. Long-term ecological research

Long-term studies have a double advantage when we seek to resolve, at least partly, the problems of interactions between spatio-temporal scales:

— They are essential in studying slow phenomena or in attempting to discover the consequences of disturbances that may produce delayed effects. For changes occurring over a period of several decades, the major difficulty is in obtaining basic observations, which supposes the establishment of a long-term observation system.

— They are useful in recording less frequent, episodic, and/or unpredictable events. This is particularly true for natural events such as tornadoes or fires, or even for biological invasions or the occurrence of epidemics.

Long Term Ecological Research (LTER)

The LTER network was established in 1980 in the United States in order to lead ecological research to larger temporal and spatial scales than those that were commonly possible with the funds usually available for research. A set of 20 sites was selected in which the research was oriented around five principal problems:

— patterns and control of primary production;

— spatial and temporal distribution of populations selected as being representative of the trophic structure;

— patterns and control of organic matter accumulation in the surface layers and sediments;

— patterns of inorganic inputs and movement of nutrients through soils, ground water, and surface water; and

— pattern and frequency of disturbances.

An international network (ILTER) was established in 1993 with the participation of several countries. This network was particularly concerned with promoting comparative studies of sites as well as the standardization of methodologies and technology exchange for observation and experimentation in order to facilitate such studies. The objective was also to help develop bases for a scientific management of ecosystems and to participate in training and education in the domain of ecological research.

8.2. SPATIAL ECOLOGICAL UNITS: A QUESTION OF SCALES

Reflection on notions of spatial scales in ecology traces its roots to maps of vegetation around the 16th century. The first topographical map of France was drawn by Cassini at the end of the 18th century and provided a precise representation of the land cover. In the 19th century, an ecological approach to vegetation was developed by means of maps drawn at various scales. Mapping flourished during the 20th century, following efforts made by phytosociologists to classify plant associations. Maps concretely pose the question of choice of scales of observation and representation, indication of spatial agencies, and the consideration of hierarchies in the classification of vegetation (Decamps and Izard, 1992).

The question of spatial scales is also important with regard to flows in ecosystems. For a long time, ecologists did not quite realize that the ecosystems they studied were integrated within a larger whole. Some ecologists sometimes even forgot that ecosystems could not be reduced to their individual biotic components. Many studies were thus focused on sites or zones judged to be representative of a whole system without considering that biotic and/or abiotic interactions with neighbouring systems were possible. This is what was sometimes called the invisible place, by analogy with the invisible present (Swanson and Sparks, 1990).

It is true that the lack of means of large-scale investigation, such as aerial or satellite photography, which are now available, limited the ability of ecologists to spatialize their observations. Spatial analysis has given rise to many studies for redefining organization levels in space by varying the scale of analysis or defining the possible regularities in the spatial structure. Spatial analysis reveals recurrent structures and forms of spatial organization (patterns). It looks for and analyses the processes that are at the source of these structures by means of statistics, modelling, and various other methods.

In the history of ecology, there has actually been an epistemological process leading to the identification and definition of basic ecological units characterized by a power of integration that is always larger. To put it more simply, we have passed from the elementary plant association to the biotic community, then the ecosystem, the landscape, and finally the biosphere. In all such transitions, each time we identify a new level of integration, we observe after some time that it is insufficient and that we must refer to a higher level of integration (Bergandi, 1995).

8.2.1. Ecosystems

An ecosystem is fundamentally made up of a physicochemical environment called the *biotope* and a set of living species, animal and plant, called the *biocoenosis*. This is the *basic ecological unit, characterized by its structure and functions*. At first it was an abstraction rather than a concrete and autonomous entity. Ecologists often speak of the "object" ecosystem, a discrete and functional unit. This is the entity that allows us to integrate the relationships between the different biotic communities (plant and animal) and the different abiotic factors.

In theory, the biotope that occupies a certain volume in space is relatively homogeneous, as is the biocoenosis. However, the concept of ecosystem is not synonymous with a homogeneous environment (see Chapter 2). It is not a fixed whole but rather a dynamic structure in which there are constant exchanges and interactions between the biological components themselves, and between the latter and the physicochemical (or abiotic) components. According to the theory of systems, such wholes manifest properties that are different (or emergent properties) from simple addition or superimposition of the properties of constituent elements. The task of the ecologist is to look for the overall new properties that emerge from these organized wholes.

8.2.2. Macro-ecosystems: landscapes, geosystems, ecocomplexes, biomes

Over the last few decades, in response to environmental problems such as climatic changes, changes in land cover, or habitat fragmentation, many ecologists have been looking for an ecological unit larger than the ecosystem. The idea gradually took shape that a set of ecosystems interacting in space and in time (and not just mosaics of juxtaposed ecosystems) may itself form a functional ecological entity. This is a level of integration that is higher than that of ecosystems, a level at which new properties appear linked to the structure of mosaics, to the network of interfaces they form, and to the flows of matter, energy, and organisms the structure favours or hampers. It is in this context that concepts such as landscape or ecocomplex were proposed. Although their theoretical status may still be debatable, they have an obvious practical importance in responding to anxieties about the impact of human activities. Advances in technology such as satellite imagery or geographic information systems that allow us to tackle the dynamics of ecological systems at different scales has contributed to the operational development of these concepts.

In reality, ecological units intermediate between the biosphere and the ecosystem were identified largely under the influence of geographers, whose approach then attracted the interest of ecologists. The concepts of landscape, geosystem, and ecoregion are in fact products of physical geography. For some authors, such as Bailey (1996), it is only a question of scale: the site or ecosystem corresponds to the micro-scale, the landscape to the meso-scale, and the ecoregion to the macro-scale.

The geography of ecosystems

The geography of ecosystems is the study of the mode of distribution, the structure, and the processes of differentiation of ecosystems considered as spatial units in interaction, and at different scales. The geography of ecosystems is similar to landscape ecology, but it places much greater emphasis on mapping, as well as the regional and global dimensions. According to Bailey (1996), the climate modified by geomorphology forms the logical basis of delimitation of large and small ecosystems.

When we try to map ecosystems, we must pose two fundamental questions: what are the most important factors for delimiting the ecosystems, and how do we determine their limits? Geographers have developed different methods of classification for identifying these "frontiers", methods that take into account several abiotic and biotic factors (Vogt et al., 1997; Bailey, 1996). The ecological classification and mapping that result suppose that we identify limits between ecosystems, limits that correspond to significant changes in the nature of relationships between the constituent elements. One classification method consists of selecting a factor that is estimated to exert a critical impact on the ecological functioning of the environment studied, and using it to divide the landscape into ecological units at different spatial scales. Conventionally, vegetation has been used (and more rarely fauna) to identify the ecological units. Some authors have drawn the limits on the basis of the nature of soils or the existence of hydrographic basins or used multivariate analyses resulting in groups or hierarchical classifications. Methods based on the theory of fractals have also been applied to the identification of geographical and ecological units.

a) The landscape

In 1939, the German biogeographer Troll associated the term *ecology* with the term *landscape* to come up with *landscape ecology*. He defined the landscape as a spatial and visual entity proper to the human species that integrates the geosphere, the biosphere, and all human artefacts. This approach has developed greatly in Europe as well as in the United States, but it has also attracted controversies, given that the association of the two terms has been judged awkward. It has even been called "unnatural", in that the term *landscape* has a strong evocative power, appealing to the imagination of individuals who have aesthetic notions, while ecology pretends to be a rational science.

When we consider the landscape as a level of organization, we recognize a certain autonomy, perceiving it as a self-organized system. From the ecological point of view, the landscape can thus be defined as an organized mosaic of ecological units in interaction, whose spatial heterogeneity is, at least partly, the consequence of human activities. Still, many researchers who use the term *landscape* recognize that it is ambiguous because it has a wide variety of uses. With the emergence of concern about the environment, the landscape has become the object of public policy, while remaining a classic term of geographic research and gradually penetrating the fields of agronomy, ecology, history, and law.

For some scientists, landscape ecology has contributed to the reunification of natural and social sciences in considering that human beings are an integral part of ecosystems (Burel and Baudry, 1999). This new "invasive species" can considerably modify ecosystems, which are thus the reflection of human-nature interactions. Animal and plant populations must adapt to this landscape dynamics to survive. A strong trend at present is the fragmentation of habitats and populations, which leads to a functioning of the patch dynamics type.

To each "his own" landscape?

For a geographer such as Bertrand (1978), the landscape is a social interpretation from an economic and cultural system of production, an object that exists outside the observer. Cloarec (1984) considers the landscape the cultural expression of the relationships of the subject (individual or social group) to the space that surrounds it. For an agronomist (Lizet and Ravignan, 1987), the landscape constitutes the mirror of ancient or present relationships of human beings with the nature that surrounds them. According to Barrue-Pastor et al. (1992), the landscape is an indispensable notion in which the ecological and social components of humans' influence on their environments and resources have been integrated. Phytosociologists have also attempted to develop the scientific analysis of the landscape on the basis of the vegetation, by studying the "associations of associations" within the spatial units identified. According to Beguin et al. (1979), the plant species and, more so, the plant associations are considered the best integrators of all the ecological factors responsible for the distribution of the vegetation and of landscapes.

It appears that, depending on the practices of his discipline, each author defines his "own" landscape. In these conditions, no science can pretend to explain landscape on its own, since it is a product of multiple interactions of an ethical, philosophical, geographic, sociological, and ecological nature.

b) The geosystem

The concept of *geosystem*, resulting from Soviet geography, includes not only the ecosystem, but more widely the interrelations between the biophysical environment and the activities of human societies (Barrue-Pastor and Muxart, 1992). A geosystem is characterized by:

— a *morphology*, i.e., spatial, vertical, and horizontal structures;

— a *functioning* that comprises all the transformations linked to the energy cycle, the cycle of water and minerals, and the processes of geomorphogenesis; and

— a *specific behaviour*, for example by changes of state that intervene in the geosystem for a given time sequence.

In a way, the geosystem begins from the same integrative principles as the ecosystem, but it explicitly includes anthropic activities. Besides, it is a territorial concept that envisages a high degree of integration, so that the ecosystem thus becomes a part of the natural geographic system. According to Rougerie (1988), the geosystem generally has a larger territorial extent than the ecosystem and its limits are much more clearly defined. It is often made up of the joining of several ecosystems under the influence of a dominant factor, the network of a watershed, for example, or a mode of land exploitation.

The concept of geosystem has attracted the interest of geographers and has probably advanced thinking with respect to the integration of biophysical compartments and human activities, which some also call the ecosystem approach. But it remains a naturalistic concept (Barrue-Pastor and Muxart, 1992). The notion of landscape, also adopted by ecologists, seems to be more open and flexible.

c) The ecocomplex

The term *ecocomplex* was proposed by Blandin and Lamotte for the use of decision-makers (Lamotte, 1985; Blandin and Lamotte, 1988). Here, a level of integration higher than that of ecosystem was emphasized. According to these authors, what is important is to consider spaces individualized by an original set of interactions between ecosystems. Whatever the reasons to delimit them, it was convenient to designate these systems of ecosystems by a single term, and they proposed the term *ecocomplex*. It evokes the ecological nature, i.e., the spatial, temporal, and relational nature, of these assemblages of ecosystems, which represent, on the territorial scale, a higher level of integration. A hydrosystem with a river, its upstream and downstream dependences, its flood plain, and its watershed is a good example of an ecocomplex.

Structurally, the ecocomplex concept emphasizes phenomena of interfaces, the relative arrangement of various ecosystems, and phenomena of connectivity. Functionally, it draws attention to the flows between ecosystems and allows us to analyse coherently the species that depend on the simultaneous use of different ecosystems. According to its advocates, the word *ecosystem* avoids the ambiguity associated with the word *landscape*, which corresponds more to spaces that can be perceived by an observer with a particular sensitivity.

Moreover, the ecocomplex concept refers less directly to the terrestrial environment and can thus be applied to other types of environments such as the oceanic environment.

d) The ecoregion

Intuitively, we recognize the existence of large, relatively homogeneous ecological regions on the basis of the combination of geological, climatic, pedological, phytosociological, and other characteristics. Off-hand, we can cite the Pyrenean Massif, the plateaux of the Massif Central, or the flat countryside of Flanders. The term ecozone is used in Canada. This idea can be used to identify regional systems on the basis of their geography. It is very useful in daily life, but it has been relatively rarely used operationally by ecologists.

e) The watershed

Schematically, a *watershed* is a territory that receives the rainfall that feeds a water course (Burel and Baudry, 1999). The fluvial system, as conceived by geomorphologists, comprises the entire hydrographic network and its watershed (Petts and Bravard, 1993). These are complex geomorphological systems made up of subsets, each with its own history in the case of major rivers.

Limited by a dividing line of water, the watershed is a naturally spatialized ecosystem. With respect to matter and energy, it is an open system with entries, transfers, and exits. The network of channels of the hydrographic basin concentrates the water flows and dissolved or suspended matter from the slopes along the water course that receive the rain to the principal course and the estuary. The geographic unit *watershed* is particularly easy to understand, for scientists as well as administrators, and it is therefore popular.

f) The biome

The *biome*, a term invented by Clements, corresponds to vast sets presenting some homogeneity as to plant and animal associations, such as the tundra, the steppes, the boreal forests, the Mediterranean forests, and deserts. The distribution of these biomes is controlled strictly by the macroclimate (temperature, humidity), in such a way that they are found in bands more or less parallel to the equator (see Chapter 6).

8.2.3. The biosphere

The biosphere strictly speaking comprises all the living organisms, animals and plants, found on the earth's surface. Nevertheless, it is most often defined as the superficial layer of the planet that encompasses living organisms, in which life is always possible. This space comprises the lithosphere (the earth's crust), the hydrosphere (all the oceans and continental waters), and the atmosphere (gaseous envelope of the earth) (see Chapter 14).

8.3. INTEGRATION OF SPATIAL AND TEMPORAL SCALES

Ecological systems are dynamic in time and in space. The integration of spatial and temporal dynamics is a virtual challenge for ecologists. What are the consequences of spatio-temporal dynamics on communities and ecological processes? How does land use evolve over time? Such questions have led to a reconsideration of a long static approach to ecosystems.

We now understand that the dynamics of communities depends essentially on the dynamics of habitats, which control the distribution of organisms as well as the distribution and accessibility of resources. In fact, we can distinguish among ecosystems a series of entities and processes enclosed within one another, like Russian dolls, forming a continuum on spatial and temporal scales. According to the hierarchy theory, in which such structures are made up of successive levels of organization, the higher levels "constrain" the behaviour of lower levels. For example, in the ecology of a river, the geomorphology of the watershed and the hydrological regime will largely determine the ecological conditions that might be found in a stretch, while the inverse is not true.

8.3.1. The erosion-transport-sedimentation cycle

The evolution of the morphology of river channels and their valley over time is the result of processes of erosion, transport, and deposit. On a section of a river, if erosion is greater than sedimentation, the bed is dug out, and this is called an *incision*. Inversely, if the sediment deposited is greater than the erosion, the height of the river bed increases and this process is called *silting*. The erosive capacity of water courses is considerable, and rivers with a rapid course have a significant impact on the evolution of a landscape. They may hollow out deep valleys, sometimes virtual canyons edged with steep cliffs, as in the gorges of Verdon, where the cliffs are 700 m high. Hydrological episodes play a critical role in the evolution of erosion processes. In a rainy period, the runoff waters, which contribute to the erosion of the watershed by their mechanical action, bring suspended matter to the river.

Along the course of a river, there is a marked evolution of geomorphological characteristics (Fig. 8.1). In the zone furthest upstream, water courses are narrow and often have a steep slope. Despite the low flow, the processes of erosion are intensive because of a strong current, and the beds are stony. This part, the upper course, is sometimes called the "production zone" or the "erosion zone" because a number of these small streams supply the river system with water and sediment. Further downstream, the slope and the speed diminish, while the flow and the width increase. The beds are made up of pebbles and gravel. As the slope diminishes, the processes of sedimentation predominate over those of erosion. This part of the river system, the middle course, is also called the "transfer zone" of water and matter towards the alluvial plain. Finally, in the plains, where the slope is low, the current slows considerably and a large part of the transported matter is deposited to form wide alluvial plains, or deltas. This is the "storage zone" or lower course in which there is an accumulation of alluvia resulting from the demolition and mechanical and chemical transformation of rocks on the surface.

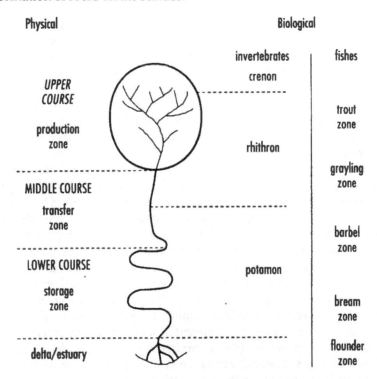

Fig. 8.1. Major geomorphological entities and longitudinal zonation of invertebrate and fish communities

From upstream to downstream, we can also distinguish changes in the morphology of river valleys. In the upper course, where water breaks down matter, we observe V-shaped valleys, with a straight course. When the slope diminishes in the transfer zone, the course with a single channel gives place to a course with multiple channels, flowing around stony islands in which high mobility is associated with an active transport of matter.

These are *tress* water courses. Further downstream, in the alluvial plain, the rivers with a slower course again form a single channel but a more curved one, forming more or less pronounced loops. These are meanders in which islands and abandoned channels are frequently seen.

Variations in the base level may lead to modifications of the river bed. If the base level drops, there is an increase in the fall and therefore of the current, which leads to a resumption of erosion that works back from downstream to upstream. Inversely, when the sea level rises, the river loses its speed and sediments. This regressive erosion became very active in water courses during the last glaciations, about 15,000 to 20,000 years ago, when the sea level was about 120 m below the present level. This explains the existence of rias in the Brittany rivers; rias are estuaries carved by erosion that were invaded when the sea level rose again.

8.3.2. Fish habitat

During its life cycle, the size and weight of a fish change considerably. Ontogeny is the process by which development stages are differentiated in the life cycle of the organism. These development phases are present simultaneously and correspond to different ecological, physiological, and biological needs, which implies in many cases the use of different biotopes. In fact, the biological cycle can be completed only if the individual finds the conditions needed for its development at each of the stages of ontogenesis. The ontogenic niche thus comprises all the habitats and resources needed for the proper progress of the life cycle.

In a highly variable environment such as that of a river system, individual fishes will constantly look for a compromise so as to optimize three principal functions: to feed themselves, reproduce, and protect themselves from predators. Their habitat (by definition the geographic environment suitable to the life of an animal or plant species) will result from this search for compromise between the variability of the environment and the optimization of their vital needs (Lévêque, 1995). In the course of its life, and as a function of its daily or seasonal activity, a single species can thus successively occupy several types of habitats.

Habitat is an essentially dynamic concept. It is a spatial and temporal reference: the position occupied by an individual at a certain stage of development seeking to optimize the necessary compromise between different biological and ecological constraints (to feed, reproduce, and protect itself) in an environment that is itself highly variable. For a species to be able to live and reproduce in an ecosystem, it must have access to different habitats needed for the completion of its life cycle. This requires an excellent synchronization between the ontogenesis and the changes involved in the physicochemical environment. One of the practical conclusions for the management of aquatic species and spaces is that we must consider all the environments a species needs during its development. It is not enough to preserve the biotopes required by adults to ensure the survival of the species, we must also ensure that it can find conditions favourable for reproduction and the growth of larvae and juveniles.

For practical and operational reasons, the need to recognize a typology of habitats has brought certain authors to propose a hierarchical approach to habitats for the study of fish communities at different spatial and temporal scales.

Bayley and Li (1992) initiated this approach by distinguishing four major types of spatio-temporal organization. According to these authors, the microhabitat corresponds to the zone of daily activity: feeding, selection of the best abiotic conditions, and social behaviour (gregariousness, territoriality). On the scale of the month, the domain of activity extends to the river (home range), while on the seasonal scale it may involve the hydrographic basin if the species makes large-scale migrations. Finally, the regional scale is that of evolution (speciation) and the establishment of fauna under the influence of climatic and geological events (extinction, colonization).

Taking a cue from this approach, Lévêque (1995) also recognized four major units:

— The *resting zone* is the smallest scale at which a fish responds to a more or less complex set of biotic and abiotic stimuli. First of all, it must find a shelter from unfavourable environmental conditions and/or predators. By frequenting this resting zone, the fish minimizes its energy expenditure.

— The activity zone corresponds to the *home range* within which the biological and behavioural rhythms are determined by diurnal or lunar cycles. The territory, for territorial fish, may be the spatial scale of reference for this category. For the others, there is a set of refuge or resting zones as well as zones in which the fish will feed, which supposes small-scale migrations.

— The scale of *ontogenic niche* corresponds to the set of environments a species needs to complete its life cycle. The spatial limits are the geographic limits of various habitats occupied by the development stages, including egg-laying zones towards which the species makes sometimes significant migrations during reproduction. While the preceding scales essentially concern the individual, the ontogenic niche concerns the population as a whole.

— The regional scale, that of the *metapopulation*, corresponds to different hydrographic basins in which the species is present. These basins are geographically isolated, except during exceptional periods on the geological scale.

The four scales identified above correspond to an increasing complexity at the level of land use for biological functions (Fig. 8.2).

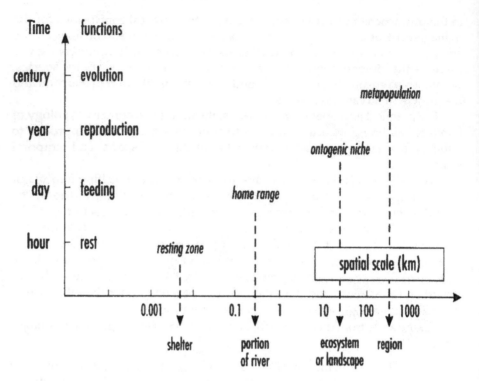

Fig. 8.2. The four major typological sets of habitat based on fish activity

Fishes of the genus *Diplodus* occupy various habitats during their life cycle (Harmelin-Viven et al., 1995). The juveniles of *Diplodus sargus, D. puntazzo,* and *D. vulgaris* occupy the slightly inclined sandy bottoms of less than 2 m depth. During their growth, the young Sparidae widen their home range moving deeper, with a preference for rocky zones. The habitats occupied by the adults at a depth of a few tens of metres are more varied, with soft or rocky beds.

The adults migrate towards a depth of 40-50 m to reproduce in groups and release pelagic eggs. The phase of pelagic development lasts around a month and a half and the transition from the pelagic to the benthic phase occurs when the juvenile is 8-9 mm long.

For migratory animal species, the notion of habitat may cover various ecosystems, sometimes very far apart. The life cycle of the European sturgeon (*Acipenser sturio*), which is no longer found in the Garonne basin, while it was present at the beginning of the 20th century in several rivers of western Europe, occurs between fresh water and sea water. The sturgeons reproduce in fresh water in the lower part of the Garonne or the Dordogne. The juveniles pass most of their lives in the sea but make frequent incursions into the brackish waters of the Gironde estuary, where they feed. The area of distribution of Gironde sturgeons in the sea extends from the Gulf of Gascogne to Scandinavia. One cannot help making a comparison with migratory birds that winter in the tropics and return north during the summer to reproduce.

8.3.3. Spatial and temporal dynamics of river systems

The dynamics of inundation and exundation of the major bed of a river during a high water period or a hydrological cycle is a good illustration of the spatial and temporal dynamics of an ecosystem. The use of space-time along a river system can be illustrated by some fundamental processes of their functioning.

a) High water regimes and ecological functioning of the alluvial plain

In the major river systems having flood plains, the variations of flow in the river regulate the intensity and duration of exchanges between the different geographic elements of the river-alluvial plain system. At the beginning of the high water period, the river overflows its low flow channel and invades the flood channel and flood plains, where it creates new aquatic habitats that can be colonized by living organisms (Fig. 8.3). Depending on the duration and amplitude of the high water period, the characteristics of these habitats (e.g., depth, speed of current) are modified with time, as well as the biocoenosis they shelter. Then, with the lowering of the water level, the habitats gradually dry out before the river recedes to its low flow channel. Apart from this, there is some variability from one year to another. No one year is exactly like another

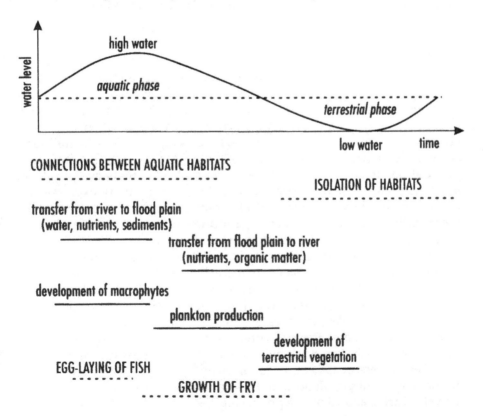

Fig. 8.3. Relationships between high waters and some biological and ecological processes

with respect to the hydrology, and exceptional events may also affect the water course in an episodic manner.

The system functions in pulses (flood pulse) (Junk et al., 1989), the rhythm of which is regulated by oscillations of the hydrological regime and the impact of which depends on the amplitude, duration, frequency, and/or regularity of the high waters. In particular, the time at which connections are established between the river and its annexes has important consequences for the functioning of the system and the biology of the species. These systems are mostly represented in the tropics, because generally the European rivers, mostly modified by man, no longer have flood plains.

In relatively large rivers, a predictable and long-term high water period favours the development of organisms that are opportunistic in their adaptations and strategies, effectively exploiting the transition zone between the aquatic and terrestrial environments, rather than depending only on the resources of the permanently aquatic environments. Some organisms, such as fish, leave the principal course during high water to use the resources available on the flood plain and return to the principal course during low water. If the high water period is exceptionally short, some organisms cannot complete their life cycle.

Many studies have shown the existence of a correlation between the duration of the high water period and the size of fish catches. In the Danube, for example, the annual catch rises to 500 t when the floods last 20 days and may reach 1500 t after 200 days of high water.

b) Connectivity

Connectivity is a parameter of landscape functioning. *It consists of all the processes that ensure that the subpopulations of a subdivided population can be interconnected to constitute a single demographic whole called the metapopulation* (Merriam, 1984). A high connectivity is manifested by a lesser isolation of populations and thus a greater demographic stability. This concept of connectivity, taken in a functional sense, is different from that of structural connectedness, which refers to the physical possibility of exchanges between ecological units by episodic or permanent connections (Amoros et al., 1993). Connectivity expresses the processes by which the different elements of landscape are integrated functionally. It is a parameter of the organization of complex systems. The river hydrosystem offers a good illustration of this concept.

According to the local geomorphological characteristics, a cross-section of a water course shows the existence of many habitats in the flood channel (Fig. 8.4): the river bed, the riparian zone adjacent to the channel with forests of hardwood and riparian trees, the flood plain with large or small, fairly permanent oxbow lakes, marshes, abandoned channels, secondary channels, and grassy formations. These environments, which occupy an intermediate position between the terrestrial environments and the deep-water environments, are called *ecotones*, or sometimes *fluvial annexes*. Each of these zones harbours animal and plant populations and communities that interact passively or actively. During annual or exceptional high water periods, the waters spread

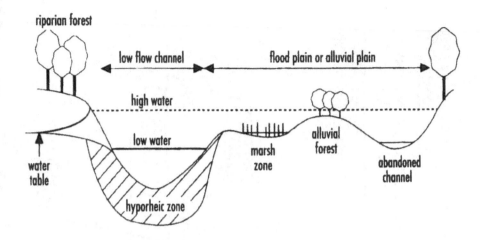

Fig. 8.4. Cross-section of an alluvial plain

across the flood plain, temporarily flooding some environments or linking the principal course of the river with abandoned channels or wetlands that survive on the flood channel, and sustaining exchanges of nutrients and organic matter.

Each piece of the mosaic plays a role in the equilibrium of the whole. The marshes and wetlands, for example, generally facilitate the percolation of surface waters and the refilling of groundwater. The alluvial forests produce organic matter that will be decomposed in the river and serve as a natural filter with respect to the surface and ground water, which moves toward the major bed. The abandoned channels, secondary channels, and oxbow lakes of the flood plain are virtual biological reservoirs and spawning zones for some fish species, which find here conditions more favourable for the development of the fry (slower current, richer and diverse food).

c) Spiral flow of nutrients

In a river, nutrients such as carbon, nitrogen, and phosphorus are not transported uniformly and continuously but are stored in the biomass, cycled, and distributed from upstream to downstream following a helix-shaped route, passing alternatively into organic and inorganic forms. This *spiral flow* involves relationships with the riparian storage zones of organic matter (abandoned channels, pools), which increase the accumulation of nutrients in the flood plain and tend to increase the efficiency of the river system in producing living matter (Fig. 8.5).

d) Drift

Drift is the downstream movement of organisms of a water course that benefit from the force of the current. It follows a distinct daily cycle and reaches a

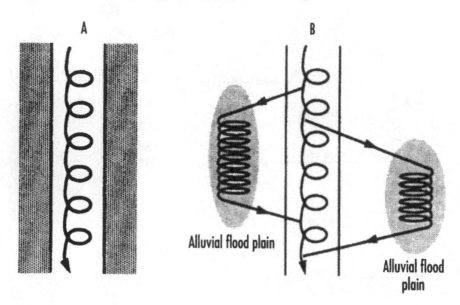

A   B

Alluvial flood plain

Alluvial flood
plain

Fig. 8.5. Spiral flow of nutrients in a water course deprived of a flood plain (a) and in a water course in which floods maintain regular exchanges between the river bed and the flood plain ecosystems (b) (Amoros and Petts, 1993)

maximum during the night, which corresponds to the maximal activity of many benthic invertebrates.

It can be explained partly by the behaviour of organisms. Some benthic organisms in water courses live in the shelter of stones during the day but come to feed during the night on the periphyton and organic matter on the upper surface of the stones. This nocturnal behaviour is a biological strategy to escape visual predators. The periodicity of the drift can thus be explained partly by the biological rhythm of the search for food, so that the invertebrates are at high risk of being led accidentally by the current (*passive drift*) when they are active during the night (Allan, 1995). But it also seems that certain aquatic organisms allow themselves voluntarily to be led by the current (*active drift*). In this context, drift seems to be a way for aquatic organisms to quit highly populated upstream zones or zones that have become less favourable to (re)populate downstream zones. Active drift is thus a significant means of colonization of new habitats created by the rise of the water level.

8.3.4. The ergoclines theory

The paradox of vertically stratified aquatic environments (such as oceans or deep lakes) with respect to primary production is the following: light reaches only the upper layers of the water, in which plants and chlorophyll are found, while mineral salts are concentrated in the deep water, where photosynthesis is not possible. In a distinctly stratified environment, planktonic algae multiply in the well-lit surface layer but do not sink to the bottom. The result is an exhaustion of the available nutrients, which limits production. If the water layer is vertically turned over, the reserve of nutrients is constantly renewed, but part of the algae are pulled down into less favourable light conditions. The optimal situation is an alternation between periods of stratification and periods of homogenization. The periods of homogenization may be caused by wind, for example. If windy and calm periods alternate with a suitable rhythm, plant cells will have at their disposal a sufficient quantity of nutrients and light energy, favourable conditions for significant growth. If the alternation occurs at intervals that are too long, production is limited by the deficiency of nutrients.

In reality, the turnover of water masses that supplies the upper layers with nutrients is ensured by an input of mechanical energy, which is of solar origin but takes indirect routes: differential heating of water masses or action of the wind, which itself results from differential heating of soils and waters. Some authors, such as Margalef, have used the term *covariance* to describe this auxiliary energy that allows contact between elements that must "vary together" in order to interact (Frontier and Pichod-Viale, 1998). For photosynthesis to take place, there must simultaneously be light energy, carbon dioxide, and all the essential nutrients (phosphorus, nitrates, silicates, and trace elements). It is the movement of fluid masses that allows these elements to be present or renewed once they are exhausted. This auxiliary energy is manifested particularly in the movements of water masses in an aquatic environment, as well as in movements of air and water in terrestrial systems (Frontier et al., 1992). This auxiliary energy has been calculated to be 10 to 25 times as high as photosynthetic energy. But in reality it is not the quantity of auxiliary energy that ensures the precision with which the biological rhythms and rise of nutrients remain in phase. The auxiliary energy has a qualitative rather than quantitative importance (Frontier and Pichod-Viale, 1998).

This dependence of the biology of water masses on the injection of auxiliary energy, which highlights the fundamental role of spatial scales and the rhythms of hydrodynamic phenomena, has been generalized in the form of an *ergoclines theory* (Legendre and Demers, 1985). The essence of the theory is that primary production occurs basically at the level of spatio-temporal transitions between conditions of stability and conditions of instability: e.g., hydrological fronts, abrupt gradients of nutrient salt levels, the water-sediment interface, and the temporal transition between periods of vertical stability and instability in the water layer.

8.3.5. The refuge zones theory

The variability on the annual scale, which can be observed in the river basin, with its alternation of high and low water and the dynamics of the mosaics that constitute the river system, is observed in scale over much larger scales of time and space. The alternation of glacial and interglacial periods that characterize the Quaternary era was the cause of significant spatial variations of many terrestrial ecosystems (see Chapter 17).

The refuge zones theory (Haffer, 1982; Prance, 1982) suggests that the dense rainforest has undergone periods of regression and extension, depending on the wetness or dryness of the climate. This is why, at the period of its greatest extension, the dense rainforest must have stretched without major discontinuity all along the Atlantic coast of Africa. In the dry period, on the other hand, the dense forest was located at certain zones in which the edaphic conditions still allowed it to survive. These zones, called *refuge zones*, served as a refuge not only for plants but also for animals that depended on the forest. During the periods of regression, many species ought to have become extinct. However, the isolated refuge zones may have served ultimately as matrices of allopatric speciation. When the forest again extended to its maximum, these surviving zones became the present zones of great biodiversity and high endemism. In theory, the species richness and rates of endemicity decrease as we move away from the refuge zones towards recently colonized zones. But the repetition of periods of alternation, during the Quaternary, helped complicate the rather simplistic process described above. Still, this theory may be relevant on the scale of entire continents and a time interval that is close to a hundred thousand years. In a way, the refuge zones constitute a phenomenon similar to that of patch dynamics (see Chapter 9), although the time scales are very different (Fig. 8.6).

The refuge zones theory is by nature impossible to demonstrate experimentally. It is really only a hypothesis based on an interpretation of observations. Many studies have nevertheless relied on the arguments of this hypothesis with respect to plants, vertebrates, or invertebrates, in Africa (Maley, 1996; Hamilton, 1988), South America (Prance, 1982; Meggers, 1994), and other regions of the world.

A similar situation exists with aquatic systems in the northern hemisphere because of the advance and retreat of the ice cap over the course of glaciation-

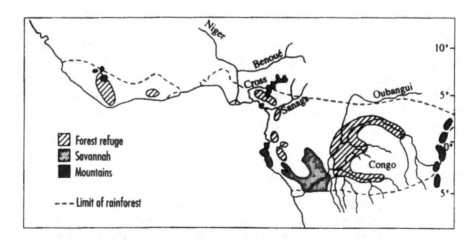

Fig. 8.6. Distribution of major forest refuge zones in equatorial Africa during the last arid phase (18,000 years before present) (Maley, 1989; Leveque and Paugy, 1999)

deglaciation cycles. In Europe, for example, the Danube served as a refuge zones for aquatic fauna during the ice ages. It is from this refuge zone that the aquatic fauna was able to recolonize the waters when the ice melted. These also are pulsations of great amplitude, in comparison to seasonal cycles.

8.4. SCALES OF OBSERVATION

The existence and interaction of multiple scales are no longer contested in ecosystem ecology. However, to progress from the descriptive and conceptual stage to an operational stage is another story. *The transfer of scales—effect of a phenomenon on a small spatio-temporal scale on a large-scale phenomenon, and vice versa—is today one of the most delicate questions in ecology and a major factor in the understanding of functional processes.*

For technical reasons, models cannot integrate several spatio-temporal scales. It would be more appropriate to invent new formalities that can explicitly include these scales and transfers of scale. In a seminar entitled "Predicting across Scales", the organizers concluded: "Plenary speakers identified the challenges of extrapolating between scales, but did not propose procedures or principles" (Dale et al., 1989).

Although some approaches seem promising, the question is far from being resolved. In some cases, the concept of transfer of scale may create the impression that the knowledge acquired on the small scale can be "transferred" to larger scales by simple extrapolation. This is a trap. In hydrology, for example, the experimental results obtained in northeastern Brazil on micro-plots of 1 m^2, parcels of 100 m^2, and microbasins of 1 ha have shown that in average conditions of soil humidity, and for the same level of rainfall, the runoff diminishes by around 5% each time the area doubles. This effect of the scale of the area on the runoff is known to hydrologists and makes it difficult to extrapolate the results obtained from small areas to large areas (Molinier et al., 1991). A possible explanation for this phenomenon is the following: on a micro-plot the runoff inherently depends on the mode of recording it, while on larger parcels a proportion of the water that runs off infiltrates the soil before reaching the outlet, and that proportion has a probability of being more significant as the area covered becomes larger. However, the heterogeneity of soils may also be cited as a reason.

In reality, there are many situations in which the experiments and models developed on the small scale have a limited value on larger scales. We may even say that extrapolation of physical laws that are valid at one scale to a higher level has consistently been a source of disappointment and paradox (Lebel, 1991). It is the question posed that determines the scale of the study, and it is a matter of identifying the variable or variables that best characterize a hydrological system at a given scale.

The complexity of ecosystems, especially the overlapping of scales of functioning, lead logically to the consideration of several scales of observation to analyse the structures of a single system. This approach allows us to break down the overall structure into structures appropriate to each scale and, thus, to more easily generate hypotheses as to the relationships between the structures observed and the processes that cause them.

The choice of a scale of observation is a strategic process in ecology because it will influence the structures detected and the conclusions we can draw from a research programme.

The scale of observation needs to be defined in two dimensions: the resolution, i.e., the level of detail or *grain* of the information, and the extent of the observation period (Wiens, 1989).

8.4.1. Macroecology

Ecologists have long been aware of the advantage of taking into account phenomena acting on a large scale. But they have usually avoided this issue, which considerably complicates their efforts to study the relationships between organisms and their local environment (Brown, 1995). In particular, constraints linked to the experimental approach have pushed ecologists to work on reduced scales of time and space, which has considerably limited our understanding of ecological phenomena.

Brown (1995) recognizes also that the results of microscopic studies cannot generally be extrapolated to a large scale to explain macroscopic patterns and phenomena that are observed. In reality, we learn more about the microscopic organization of complex ecological systems, and we need more macroscopic studies to place this knowledge in a larger context.

Macroecology is thus an ecological discipline that is certainly new but based largely on earlier traditions, particularly biogeography. *The principle is to seek the statistical properties emerging from large series of ecological data in order to identify patterns that can be generalized to scales at which the hypothetical-deductive process cannot be applied.*

Theoretical explanations or hypotheses proposed to explain the patterns observed must be confirmed by complementary observations that allow these hypotheses to be tested. In fact, this process emphasizes statistical analysis and not the experimental approach characteristic of other approaches in ecology. It is more a question of operationality than of philosophy, according to its practitioners. It is often difficult, if not impossible, to conduct controlled experiments and to duplicate them on large spatio-temporal scales. Thus, alternatives must be found to make deductions about the functioning of nature.

The macroecological approach thus advocates the need to study phenomena at different spatio-temporal scales, especially the macroscopic scale. It is a matter of seeking out how local ecological phenomena interact with the processes operating at larger spatio-temporal scales. This process, as it is now presented, attempts to take into account the results obtained in other disciplines, such as biogeography, evolution, palaeobiology, and earth sciences in general, which must be studied on the large scale. Macroecology confirms the fact that natural ecosystems, open to exchanges of energy and matter, are subject to climatic and geological factors acting on the large scale that are impossible to manipulate experimentally. Such a multi-scale approach is essential if we wish to comprehend what are the phenomena at the origin of existence and maintenance of biodiversity, at all the spatial scales (from the local to the global) and temporal scales (from the annual cycle to the scale of the history of life on earth).

8.4.2. The biosphere's primary production

Issues of change in scale, or the spatialization of local results, are at the heart of the problem of major international programmes such as the Global Change Programme of the PIGB: How can the functioning of the carbon and water cycle be predicted on the biosphere level (highest hierarchical level) from the knowledge acquired at the lower levels of ecosystems and organisms? How can studies conducted on parcels or small regions of a few square kilometres be generalized to the biosphere?

Operationally, we have often treated the issue of change of scale by simple transposition, on the scale considered, of models of functioning indicated at the local level, or by reiteration of the local model in every respect. More elaborate processes, taking into account problems of scale, have been used with respect to terrestrial primary production, for example (Saugier, 1992):

— A first stage involves the transition from the chloroplast to the leaf. Photosynthesis is quite a rare example in plant physiology in which a very complex set of reactions can be summarized in a simple model. Now we have satisfactory leaf models for plants growing in normal conditions.

— A second stage involves the transition from the leaf to the cover. It is easier to define primary production by unit of area of soil than by unit of leaf area, and the same applies to the needs of plants (e.g., light, water, nutrients). We customarily define a cover by its *leaf area index*, which is the ratio of the area of all the leaves (counting one side) to the soil area. Different models have been proposed, associating light penetration and leaf photosynthesis to evaluate the photosynthesis of the plant cover.

— A third stage involves the transition from the plant cover to the biosphere. From the information obtained by satellite about the distribution of the plant cover (vegetation index), the measurements taken on some sites can be extrapolated spatially. We can thus calculate the production of each biome identified, and then the overall production of the biosphere.

The problem of changes in scale in the case of terrestrial primary production is thus relatively "easy" to resolve, in particular by means of instruments that can be used to measure photosynthesis in natural conditions (enclosures, for example), satellites for observation of the earth, and increasingly efficient models that take environmental conditions into account. Still, we must not assume that this situation is the rule.

8.4.3. Spatial scales and soil functioning

The soil plays a key role in ecosystems. It is the medium of exchanges and transformations, a filter for the circulation of water and dissolved elements. This function of the soil takes specific forms depending on the spatial and temporal levels considered, which poses various types of questions:

— What is the representativity of a measurement taken on a small area in evaluating the average value on a large area?

— What is the possibility of generalizing laws characterizing processes studied over small areas to large areas?

— Are there similarities between structures represented at different scales?

To study the role of soil in the functioning of ecological systems, it is necessary to make use of a tool of spatial analysis. Fournier and Cheverry (1992) propose five scales.

- The *point* scale at which elementary physical or biological mechanisms occur, for example, that of the rhizosphere, in which phenomena of the soil-plant interface and root adsorption occur.

- The *station* scale, which is that of the soil profile considered in its three dimensions, about a square metre in area, about a cubic metre in volume. This is the scale at which most *in situ* measurements of functioning are made: e.g., hydric functioning by the use of neutron probes, evaluation of the solute composition by means of porous candles. This scale is most suitable in helping us understand the adaptation of the entire plant to the soil. It also allows a functional approach in controlled conditions of biological activities such as that of earthworms and their role on the structuration of soils.

- The *slope* is the scale at which the topography, pedogenic characters, and state of the soil surface combine to influence the functioning of the soil-water-plant system. This slope is in fact frequently composed of segments, each of which functions according to a logic appropriate to the combination of slope, soil, and surface condition that characterize it. The overall functioning of the slope will thus be the result of the combined effect of the functioning of each of the segments. The spatial scale is variable, but the segment of the slope is frequently based on the dimension of the cultivated field or landscape altered by man. The bocage landscape with the establishment of a hedgerow network is an illustration of this scale.

- The *watershed* is the spatial unit at which the measurements taken at different scales can be synchronized. It allows the transition from flows estimated from spatio-temporal means (local measurements) to average flows at the meso-scale (flows at the outlet). Fairly small watersheds are the elementary units in which input-output balances can be drawn. When we go to large watersheds by successive framing of elementary watersheds, the weight of local factors (nature of soil, type of plant cover) diminishes and that of more general factors such as the climate increases.

- The *biogeographic zone* is the last spatial scale at which the role of the soil can be considered. This is the scale of specific landscapes resulting from a dynamic equilibrium of soil, vegetation, and climate. It is at this level that functions of the soil develop that may be significant on the global level.

Depending on the questions posed, we decide on the initial scale for the spatial analysis and the study of inter-scale relationships. For example, the starting point of the study of hydric erosion is the watershed, whose functioning is explained by the processes involved at the scales of point and station. Inversely, it is from the watershed that we explain what happens at regional and biogeographic scales. If we are concerned with nitric pollution, the starting point is the scale of the station.

8.5. THE DYNAMICS OF SPATIAL STRUCTURES: FRACTALS

Euclidean geometry describes the world in terms of lines, areas, and volumes. Living organisms, on the other hand, often have a more complicated geometry. According to the mathematician Benoit Mandelbrot, classic (Euclidean) geometry only caricatures nature. He therefore elaborated in the 1960s a non-Euclidean geometry, the theory of fractals (fractured as opposed to the classical "smooth" objects), which could take into account the irregular forms of nature: for example, clouds are not spheres, mountains are not cones. Fractals can be defined as objects or motifs that have non-whole dimensions. In Euclidean geometry, a point has zero dimension ($D = 0$), a line has one dimension ($D = 1$), a plane has two dimensions ($D = 2$), and a volume has three dimensions ($D = 3$). But natural objects, such as landscapes or trees, present non-whole dimensions, between 2 and 3, which characterize their complexity. These complex spatial motifs have two major characteristics:

- A scale dependence in the measurement of length: Mandelbrot (1967) showed that the length of a rocky coastline measured on a map depends on the scale of that map and that, from one scale to the other, the lengths may vary by several orders of magnitude. For example, a coastline measured at a scale of 1/500,000 will be shorter than a coastline measured at a scale of 1/20,000, because the latter will show much more detail and irregularity in the coastline, which significantly increases its length. Similarly, the number of islands surveyed in an archipelago increases when we study smaller sizes. In other words, the scale of observation has a fundamental importance in evaluating lengths, objects, or areas.

- An internal similarity: Nature offers many examples of self-similar forms. A head of cauliflower represents a motif that is repeated thousands of times, in smaller scale, within each of its branches and sub-branches. Looking

172 **Ecology:** From Ecosystem to Biosphere

again at the example of coastlines, the enlargement of a portion of the coastline makes new circumlocutions apparent. This is a phenomenon of self-similarity. The explanatory hypothesis is that a single process or a single set of processes organizes the form at all the scales. For example, the same physical forces carve out the coastal landscape into bays and peninsulas and shape the rocks. We can hypothesize that different processes will be the source of different fractal dimensions.

Without entering into the details of the theory of fractals, which are quite clearly explained by Frontier and Pichod-Viale (1998) we will give here some examples of its application in the context of ecological systems.

a) To increase the areas of contact

The theory of fractals has specific applications in biology. Organisms are crossed by flows of matter and energy that penetrate or come out of these organisms through areas of contact with the external environment. For organisms of equivalent form, the area of contact is proportionally less when the volume increases. The reduction of the ratio of external surface to the volume with increasing size thus reduces the significance of the flows, so that in theory, from a certain size onwards, an organism no longer receives enough matter and energy to maintain its metabolism and continue to grow. It must thus use various means to increase the areas of contact. One such means is the acquisition of a puckered or branched geometry. A human lung that has only a few dm^3 in volume has a contact area with the atmosphere of about 400 m^2. Similarly, in a tree or a plant, the branches, stems, and roots constitute complex arborescences according to a geometry called "inextricable" (Frontier and Pichod-Viale, 1998), which ensures especially a better contact with the atmosphere or soil. For example, as biomass grows it becomes organized and ramified, acquiring a fractal geometry whose properties cannot be really expressed in Euclidean geometry. However, the stages of fractionation are necessarily limited. From the trunk of a tree to the last twig, the number of ramifications is less than 10. Beyond this appears a plan of organization that is different: the leaf. Thus, there is a finite geometry whose transitions are abrupt and involve changes of nature.

The major advantage of interfaces with a fractal geometry is thus that they can resolve one of the essential problems of the biology of a living organism: to increase the areas of contact of an organ or organism (and thus increase the exchange capacity) without substantially increasing the initial volumes and metabolic costs inherent to their maintenance (Lode, 1994).

b) Hierarchical structures

A direct application of the fractal concept is the characterization of the arborescence of a hydrographic network (Taborton et al., 1988). To measure the fractal dimension of a water course we use the curve of cumulative frequencies of the length of the branches of the arborescence. Fractal analysis aims to define the properties of scale of water courses; it can be used to differentiate them according to the factors that control the irregularity of their outline (Beauvais et al., 1994). The morphology of the network depends thus on the influence of geomorphological factors such as the nature of the bed rock and the soils, the tectonics, the hydrological factors, and other aspects. An application to the water courses of the Central African Republic showed that their fractal dimensions are of the same order of magnitude as those of the oceanic coasts. Besides, in the framework of this study, the fractal analysis has indicated two types of sinuosity: one controlled by the lithological and pedoclimatic factors that interact on the local scale on decennial periods of time, and the other controlled more by the tectonics evolving over much greater time periods. The principal criticism levelled at fractal modelling in hydrology is that its utility goes little beyond the descriptive stage (Lebel, 1991).

c) Fractal representation of the soil

Knowledge of the dynamics of water in soils is fundamental for agronomists and pedologists. However, to understand the path of water in a heterogeneous soil, we must know the number and morphology of the particles, aggregates, and spaces that make up the soil. Generally speaking, a porous medium with a complex structure such as soil actually carries very robust elements of organization that are expressed in a fractal structure (Rieu and Perrier, 1994). The size class distribution of the number of aggregates of the soil or the number of pores is often described in terms of exponential laws, the exponent of which may be explained by a fractal dimension of soil. For example, we observe that the cumulative number of solid elements related to successive size classes varies according to an exponential law of the form $(Nr > L) \sim L^{-D}$, where $(Nr > L)$ represents the number of elements of linear size greater than L and D is the fractal dimension of all the elements. From an analysis of these laws of scale, theoretical models of fractal structures can be elaborated that represent, in a way, an ideal and simplified abstraction of the soil as a whole.

d) Fractals and landscape ecology

Landscape ecology is an especially suitable field for the application of fractal geometry. Baudry (1993) measured the fractal dimension of some elements of an agricultural landscape in Normandy, France, comprising grasslands, ploughed fields, and bramble patches resulting from the decline in grazing pressure. While the distribution of the bramble patches in the landscapes may appear random, it is actually remarkably self-similar for grains of analysis ranging from 0.25 to 16 ha. However, the use of fractal geometry remains limited because there are many methodological problems and the interpretation of results is delicate, especially with respect to the functional significance of descriptions (Leduc et al., 1994; Milne, 1997).

Chapter 9

Spatial Heterogeneity and Temporal Variability

"The real lake is not a basin with two vertical sides as in the textbook. One that is like that, Loch Ness, is so out of line that it harbours monsters."

Hynes (1975)

"An attractive, promising, and frustrating feature of ecology is its complexity, both conceptual and observational. Increasing acknowledgment of the importance of scale testifies to the shifting focus in large areas of ecology. In the rush to explore problems of scale, another general aspect of ecological systems has been given less attention. This aspect, equally important, is heterogeneity. Its importance lies in the ubiquity of heterogeneity as a feature of ecological systems and in the number of questions it raises— questions to which answers are not readily available."

Hamilton and Milbrook (1991)

The organization and structuration of ecosystems in space and time are fundamental questions. Starting from the essentially static notion of homogeneous environment, ecologists have gradually integrated notions of spatial and temporal heterogeneity to describe the dynamics of ecosystems and landscapes. Now we find a consensus that nature is heterogeneous. This evolution of ideas, marked by an increasing consideration of complexity, is partly the result of advances in means of observation and treatment of information.

Conventionally, we speak of spatial heterogeneity and temporal heterogeneity. *Spatial heterogeneity* can be perceived from a static or dynamic point of view. An environment is heterogeneous if a qualitative or quantitative variable, such as plant cover or air temperature, has different values in different places. However, in functional terms, heterogeneity is also apparent when there is a change in the intensity of functional processes in response to variations in the structure of the environment. Despite the efforts of ecologists, the quantification of spatial heterogeneity still remains problematic because of the difficulty of the task and because the components of heterogeneity are often poorly defined (Kolasa and Rollo, 1991). *Temporal heterogeneity*, more often called *temporal variability*, can be defined simply by the different values taken by a variable in a single point of space as a function of time. In reality there is a close relationship between these two dimensions, spatial and temporal, so that *structural heterogeneity can be defined as the complexity and variability of a property of an ecosystem in time and in space.* This property can be a measurable entity such as plant biomass, annual precipitation, or the concentration of nutrients in the soil.

9.1. FROM THE PARADIGM OF HOMOGENEOUS SYSTEMS TO THE RECOGNITION OF HETEROGENEITY

From the origins of ecology, the question of spatial and temporal homogeneity and/or heterogeneity was at the core of the debate (McIntosh, 1991). At the end of the 19th century, scientists were still driven by the traditional conception, arising from natural history and Christian belief, that there was an intrinsic order in nature and that nature was in equilibrium (see Chapter 10). When we have an image of the world based on notions of equilibrium and homogeneity, heterogeneity may seem an obstacle to scientific progress. It was thus not surprising that the words *homogeneity* and *similarity, constancy*, and *stability* were frequently found in the ecological literature from the early 20th century. However, they were used without an explicit definition while quantitative methods were still in their infancy. The term *homogeneous* was often used (and perhaps still is) in the sense of uniform, which may be expressed as regularly spaced, or by the absence of change, or even by similar composition when samples are taken. The ambiguity of the word *uniform* is obvious in the definition of plant association proposed by the Third International Conference of Botany in 1905: an association is a community of defined composition, having a uniform physiognomy, and growing in uniform conditions.

The idea of homogeneous environment marked the science of ecology until recently. With the development of theoretical ecology in the 1960s, some ecologists even lost sight of the fact that spatial homogeneity and temporal equilibrium are only simplifications of reality (Wiens, 1999). During the 1960s, the ecosystem could still be described as a homogeneous biocoenosis developing in a homogeneous environment (Duvigneaud, 1980). It was on this basis that Bourliere and Lamotte (1978) referred to the notion of homogeneity. The ecosystem, according to these authors, could extend over vast areas, such as a forest or savannah, and may therefore be difficult to delimit, but it presents a structure that is repetitive in space and may thus, at a certain scale, be considered homogeneous. The plant fraction of a community, the phytocoenosis, gives a preliminary idea of its homogeneity. Botanists were thus the first to establish rules that delimit and define homogeneous groups that are plant associations (Bourliere and Lamotte, 1978).

In other words, plant associations, which may appear virtual integrators of climatic and edaphic conditions, were used by terrestrial ecologists to visualize the space occupied by ecosystems and serve as a basis for an ecological partitioning. As early as the beginning of the 20th century, the principal activity of the botanist F. Clements was to identify units of vegetation corresponding to natural areas having identical or similar climatic characteristics, including a "plant formation" with dominant characteristic species. It was the same spirit that, in the first decades of the 20th century, presided over the development of various schools of phytosociology that exerted a virtual hegemony in ecology for some time, animal communities being defined simply by reference to "homogeneous" spaces identified by botanists (Lefeuvre and Barnaud, 1988).

For all that, the notion of the homogeneity of ecosystems has rapidly been demolished by profound field observations. Even at the end of the 19th century, researchers had raised the question of representativity of the sampling. In the 1880s, the oceanographer V. Hensen calculated plankton abundance and biomass in a large volume of sea water by extrapolating the results obtained from a sample at some points, stating the hypothesis that the plankton was distributed uniformly over space. He was vigorously attacked by Haeckel (the father of the term *ecology*), who showed that the plankton was actually distributed irregularly, often forming swarms, and concluded that Hensen's extrapolation was an error. These debates had the merit of drawing attention to the spatial distribution of organisms. The importance of problems relating to the measurement of spatial heterogeneity became increasingly obvious during the 1940s and 1950s. Many studies were carried out to explore the statistical properties of samples and their relationships with the modes of distribution of organisms. The concept of scale was looked at to tackle the question of spatial heterogeneity (Greigh-Smith, 1964) and it soon became apparent that sample size had an influence on the conclusions that could be drawn from the analysis of samples (Frontier, 1983). In reality, ecologists, who were imbued with the notion of homogeneity, but conscious of the heterogeneity of ecosystems, have sought to determine, through a statistical approach, the extent to which the area studied is far from a homogeneous situation. Jaccard, a plant ecologist, introduced in 1902 a numerical coefficient called "similarity" to compare samples and to determine to what extent they share the same species or other attributes. This search for similarities using numerical methods, a sort of homogeneity test, would develop further during the 1930s and 1940 and reach a peak during the 1950s and 1960s with the establishment of more elaborate mathematical methods. Manual data processing and limitations in the means of calculation partly explain why coefficients of similarity, relatively easy to calculate, were popular during the 1960s.

More generally, recognition of heterogeneity raises the question of sampling strategies. For example, research on the Cape Frehel moor dominated by a few species of gorse and heather have shown that this moor, which is apparently homogeneous and at first sight easily distinguished from the neighbouring formations, is actually composed of 16 units of vegetation (Lefeuvre and Barnaud, 1988). Subsequent research indicated a virtual organization of these units as a function of topographic profiles and of the soil thickness and moisture content. Such an observation certainly challenged the notion of random sampling practised until then on a supposedly homogeneous environment.

As Barbault (1992) has emphasized, heterogeneity is so obvious a characteristic of ecological systems that it is now difficult to imagine that ecological theories have been able to underestimate its importance for so long. Because of this, ecologists have long felt that biological reality violates the laws of ecology established on the basis of homogeneous environments, but they did not have the conceptual and methodological tools, till recently, to integrate heterogeneity in an operational manner in their research. In the domain of inland waters, the study of rivers has long been more rare than that of lakes, perhaps because running waters are among the most heterogeneous natural systems and are therefore difficult to study. Many ecologists also consider that heterogeneity was an obstacle in the search for general laws of functioning of ecosystems, so the concept of homogeneity (and the associated notions such as equilibrium, stability, uniformity) has long been favoured by field ecologists, who attempt to overcome the additional difficulties posed by heterogeneity by working on uniform spaces. It was only during the last few decades, partly because of advances in tools of data management and treatment, that the consideration of spatial heterogeneity and temporal variability developed afresh.

Distribution of plankton into aggregates

An important source of heterogeneity in the pelagic environment is the spatio-temporal distribution of zooplankton (Angeli et al., 1995). It has been observed that zooplankton communities are not distributed randomly but have a strong tendency to constitute aggregates in the pelagic environment, at different spatial scales (Pinel-Alloul, 1995). Consequently, a good knowledge of the spatio-temporal distribution of zooplankton is necessary to understand its interactions with the other biological compartments of pelagic systems, as well as its role in the overall functioning in terms of productivity and biogeochemical flows.

For a long time, it was thought that the aggregates or swarms were the result of large-scale physicochemical processes (Pinel-Alloul, 1995). Various studies have shown a strong relationship between the spatial distribution of plankton and the advective processes of the water mass, such as the Langmuir circulation or upwellings. Besides, the physicochemical gradients generated by the vertical stratification of water masses may limit the distribution of planktonic species. We now know, thanks to the development of investigative techniques (acoustic systems, videos, etc.), that biological processes also contribute, in a highly significant manner, to the formation of swarms (Folt and Burns, 1999). The heterogeneity of zooplankton distribution was shown to result from four major biological mechanisms: daily vertical migrations, avoidance of predators, the search for food, and the search for sexual partners.

Thus, the distribution of zooplankton into aggregates, in a marine or continental environment, is the result of the interaction of several physicochemical and biological processes. For example, the effect of light on the formation and maintenance of swarms is best understood when we simultaneously consider the predators, endogenous rhythms, and the tides. These various processes are manifested more or less according to the scale considered: the physical processes are more important for the large spatial scales, while the biotic processes become predominant on smaller scales (Fig. 9.1).

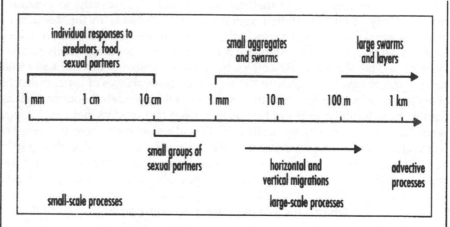

Fig. 9.1. Various aggregative behaviours operate at different spatial scales. The figure gives qualitative indications of the biological processes at work at different scales: from 1 mm to 10 cm individual behaviours are predominant; from 10 cm to 1 m relations with sexual partners are predominant; from 1 m to 10 m small aggregates are observed; from 10 m to 100 m vertical and horizontal migrations structure the processes; beyond 100 m, migrations combined with advective processes are at the origin of major aggregates and formations of layers (Folt and Burns, 1999).

9.2. THE DISCOVERY OF FRONTIERS: ECOTONES

To recognize that heterogeneity is a component of ecosystems does not resolve the problems that arise. In practice, whether for ecological research or for administrative purposes, it is often essential to identify the limits of ecosystems or any other ecological entity that is the subject of attention. And the means of doing so are not unanimously agreed upon because it easily identified limits between ecological entities are rarely found. In some cases they may be physical. A lake is often intuitively considered an ecosystem with contours fairly well identified by a shore that constitutes the demarcation from the surrounding terrestrial environment. However, in reality the water level is always rising up against the banks, and in shallow lakes the shore is a transition that is sometimes geographically extensive and/or poorly defined spatially between the aquatic and terrestrial environment.

A pragmatic approach is to recognize that there are zones of discontinuity, or *ecotones*, between different and neighbouring systems, and those ecotones have specific ecological characteristics. The ecotones are often perceived by the observer as a transition zone between systems that have markedly different physiognomies. Examples are the forest-savannah contact or the river-ocean contact. In reality, the concept of ecotone covers a set of terms such as interface, border, transition zones, frontier zone, or contact zone.

In the logic of an approach that has long been based on the study of homogeneous environments, the "discovery" of ecotones has been one way for ecologists to take heterogeneity into account: homogeneous entities continue to be recognized, but the interfaces, which pose a problem to the extent that they cannot be attached to either of the major entities, can now be ranged in a clearly identified category for which an explanatory theory can be attempted.

The notion of transition zones between ecological units is not recent. In 1897, Clements defined the ecotone as a frontier between two associations or two plant communities in which the processes of exchange or of competition between neighbouring formations can be observed. Subsequently, Odum (1971) suggested that the community of the ecotone contains not only organisms of adjacent communities, but also organisms that are characteristic of, or even exclusive to, the ecotone. In any case, the concept of ecotone is now based on three principal ideas:

- First of all, an ecotone is not a static zone of contact between two communities, but a dynamic zone that changes with time and has particular properties. The spatial limits of an ecotone vary with time.

- Second, ecotones are not isolated entities but are an integral part of the functioning of a landscape (Risser, 1990). According to Holland (1988), they are transition zones between adjacent ecological systems, having a set of characteristics defined uniquely by temporal and spatial scales and by the force of interactions between the adjacent ecological systems.

- Finally, ecotones are compared to a permeable membrane such as that found in physical and biological systems. They can modulate the flows of energy, matter, or organisms between the adjacent ecological units. This role of chemical and biological filter has particularly been studied for riparian forests.

Ecotones in a river environment obviously involve the horizontal interfaces between the components of the landscape. Riparian tree formations occupy the interface between the aquatic environment and the terrestrial environment. Depending on the hydrological and morphological processes, they have their own dynamics. Many studies have indicated their role in the fixation and stabilization of banks, as well as their role as filter of the nutrients and pollutants that come from the terrestrial environment (Naiman and Decamps, 1990).

But there are also vertical ecotones at the level of exchanges between the ground water and the river or between the ground water and soils. For example, research conducted over the past two decades has shown the existence of an underground compartment of the river that may reach thicknesses of a few decimetres to about a hundred metres. This hyporheic zone, made up of a combination of gravels, pebbles, sands, and clays, occupies the entire fluvial plain. The underground layer flows between the interstices, following the direction of the river, but much more slowly. In the sector of the Rhone close to Miribel, the river flows at an average speed of 1 m/s while the layer in the hyporheic zone flows at a speed of a few mm/s. This underground environment harbours in its interstices an entire living world from bacteria to invertebrates that has been unknown for a long time (Gibert et al., 1996). These hypogeal organisms (as opposed to the epigeal organisms that live on the surface) never come to the surface and belong to various zoological groups: ciliate protozoa, Oligochaetes, Annelids, Crustaceae Syncarids and Harpacticoides, and so on. They have a thin, long body adapted to movement between grains of sand. These are depigmented and blind species. Besides, they have a slow metabolism in comparison to their epigeal homologues and they can tolerate a long period without food. The amphipod *Niphargus* may fast for four months and resist lack of oxygen by using metabolic pathways that allow ATP synthesis in the absence of oxygen. Many epigeal species (e.g., *Baetis* sp., *Caenis* sp., *Gammarus*, Leptoceridae) are able to migrate into the underground waters and remain there temporarily when the conditions at the surface become unfavourable.

Estuaries and the muddy stopper

Functionally, an estuary is an interface between the land and the sea, between continental waters and sea water, a kind of ecotone. This place in which sea water and fresh water mix is also a zone of transit and accumulation because the joint action of the tide and gradients of salinity create a trap for clay particles carried by the rivers. This accumulation of suspended matter constitutes the *muddy stopper* that represents around 5 million t in the estuary of the Gironde. This muddy stopper is generally located in the central part of the estuary, from upstream of the saline intrusion, and moves with the flow of the river and the tide (Paskoff, 1985). The ecology of estuaries is closely linked to the existence and the dynamics of the muddy stopper. During tides of dead water (low coefficients of tide), the currents are too slow to keep the sediments in suspension. The particles of the muddy stopper sediment toward the bottom, contributing to the rapid filling up of the bottoms and channels as well as the muddying of the banks. On the contrary, during high tide coefficients, the sediments are suspended again. Ecologically, the suspended matter limits the penetration of light and thus the primary production. Besides, a large part of the organic matter of elements in suspension is mineralized in the estuary, which sustains a consumption of dissolved oxygen that may cause anoxia. Finally, the muddy stopper is a place at which pollutants from upstream concentrate as they are adsorbed on the fine particles. These pollutants are transformed in the estuary, sometimes releasing toxic substances.

9.3. GRADIENTS

The concept of community that has gradually become modified as a function of changes in the environment is opposed in theory to that of discontinuity between well-contrasting entities. In simple terms, a gradient is a continuous variation of an ecological factor in space and in time. This factor, biotic or abiotic, must be sufficiently structuring for us to observe the modifications of the composition and the organization of communities along the gradient. It is thus a form of organization of nature that allows us to study more easily the empirical relationships between the environmental conditions, the distribution, and the abundance of species, and the biological traits of these species. These are favourable places to articulate and quickly test hypothesis or to conduct experiments in order to improve our ability to predict the responses of ecosystems to environmental changes.

Some spatially continuous gradients present a relatively clear pattern of zonation of organisms or communities. Mountain slopes, the banks of lakes, and coastal environments are examples of gradients in which an environmental factor (here the altitude or the depth) plays a major role in the organization of communities. An example of gradient at the scale of an ecosystem is that proposed by the concept of fluvial continuum (Vannote et al., 1980).

Elaborated originally for aquatic invertebrates in temperate rivers, the fluvial continuum concept attempts to establish the relationships between the continuous longitudinal gradient of physical factors, on the one hand (morphology, hydrology), and the biological strategies (trophic guilds) of benthic invertebrates and the dynamics of organic matter, on the other hand. It is an attempt to take into account the functioning of the river from the inputs and flows of matter and energy along the longitudinal gradient.

The energy required locally for biological production has three possible origins: inputs from outside the system from the vegetation on the banks, the autochthonous primary production in the river, and the transport of organic matter from upstream. The importance of these three sources of energy varies, however, along the water course. In the upper course, the riparian vegetation is important and the allochthonous plant debris (leaves, twigs, branches) is used by the invertebrates of the "shredders" guild, which break them down into fine particles used by invertebrates of the "collector" guild. Besides, the riparian vegetation intercepts light, strongly limiting the aquatic primary production during the summer. This heterotrophic system functions essentially with external inputs.

Downstream, the water course becomes wider, so there is better penetration of solar energy and development of the algae or plants fixed on the higher plants or rocks. The autochthonous primary production takes on an increasing importance while the relative importance of allochthonous inputs diminishes. The system thus becomes autotrophic, and the autochthonous primary production is exploited by the "grazers". In a relatively slow, large river, the turbidity of waters may limit the development of rooted plants or algae that are fixed on supports. The significant inputs of fine particulate organic matter from upstream thus creates new conditions of heterotrophy, comparable to those of the streams upstream. This theoretical process is to be modulated as a function of the geographic and climatic context (Statzner and Highler, 1985). In particular, the existence of tributaries and human interventions may modify this continuum (Fig. 9.2).

Along the longitudinal gradient, we can also observe, in theory, an increase in species richness with the size of the water course, which reaches a maximum in the intermediate zones, to decrease again downstream. Among the factors

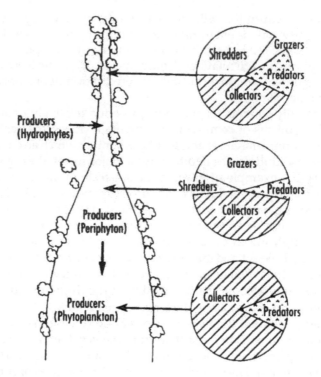

Fig. 9.2. Representation of the fluvial continuum for water courses in the temperate zone. The upper courses are heterotrophic, the middle courses autotrophic. Coarse particles of organic matter predominate upstream, while fine particles dominate downstream. The macroinvertebrate communities succeed one another along the water course as a function of the nature of available resources.

that may explain this pattern are hydrological conditions, temperature, the composition and diversity of substrates, and the abundance and type of food (Minshall et al., 1985).

9.4. TEMPORAL VARIABILITY: A FUNDAMENTAL CHARACTERISTIC OF ECOSYSTEMS

Ecologists are primarily concerned with the role of variability of the environment in the dynamics of ecosystems. Is such variability an obstacle to any prediction about the characteristics of communities, or is it possible to develop hypotheses that take this reality into account? How can this apparent disorder be expressed in the form of a coherent discourse, because the variability is expressed at various scales of space and time, from the diurnal cycle to climatic variations of great amplitude (Barbault, 1992)?

Like many other terms, *variability* can have different meanings. It involves genetic variability as well as changes in time and space of abiotic characteristics of the environment, or fluctuations of the abundance and composition of natural communities. The variability of abiotic factors of the environment and their consequences on the dynamics of communities are the focus of this discussion. Two kinds of variability must be distinguished: what is more or less regular and thus predictable, such as the diurnal and seasonal rhythms, and what is irregular and thus unpredictable, such as inter-annual climatic variations or random events. The vulnerability and adaptive responses of organisms to these different types of variability are not identical: communities living in ecosystems with a range of limited and predictable fluctuations are more sensitive to disturbances of anthropic origin than communities from highly variable environments that are, in a way, pre-adapted to changes (Begon et al., 1990).

9.4.1. Diurnal cycles

The diurnal rhythm is expressed in an alternation of day and night that has consequences for the other parameters of the environment, such as temperature or oxygen production. In aquatic ecosystems, for example, plants produce oxygen during photosynthesis only in a layer of water that receives enough light to enable photosynthesis to take place, i.e., the *euphotic* layer. We can easily understand that oxygen production linked to light energy is not continuous and follows the rhythm of sunlight, which itself varies according to the time of year. The oxygen level in the euphotic layer thus depends on the equilibrium between oxygen reduction by photosynthesis (P), essentially diurnal, and oxygen consumption (R) by the respiration of living organisms, including those that mineralize decomposing organic matter, which is a constantly recurring process. All these processes are dependent on temperature. At any moment and at any point in the ecosystem, the oxygen level will thus be the result of the balance of biological activities and atmospheric exchanges, keeping in mind that the solubility of oxygen is greater in cold waters than in warm waters. Another consequence of the day/night alternation involves the vertical migrations observed in many zooplanktonic populations in a marine or freshwater environment, which concentrate on the surface during the night and at dawn sink to the bottom, where they remain during the day. The role of light (phototaxy) as a signal triggering and controlling the speed and amplitude of these migrations has been experimentally demonstrated, even though the phenomenon is apparently more complex (Angeli et al., 1995).

9.4.2. Annual variability

On the annual scale, variability is largely the result of seasonal variations of climatic factors. Even if the level of predictability remains high, there is a margin of error to be considered that is not negligible, because climatic events are not similar from one year to another: the quantity and temporal distribution of rain may favour or limit the development of certain species and communities in the terrestrial environment.

At the individual level, the first reaction is behavioural: the individual modifies its level of activity or migrates. For example, after rain in the tropical savannah, various amphibians or insects start multiplying, radically changing the structure of communities and food webs of terrestrial communities. Simultaneously, the vegetation begins to grow, also completely modifying the food webs and the structure and habitat of the dry season. On the population level, it is trivial to speak of fluctuations of abundance in the seasonal scale, and a physiological cycle of reproduction and growth is followed.

On the level of ecosystems, the seasonal variability is part of the normal functioning of these environments. Various studies have indicated that changes in the flow of rivers, which are the consequences of rainfall, are simultaneously a particularly important physical disturbance and an essential ecological factor with respect to the biological cycle of fishes. High waters that last a short time and are often difficult to predict in upper stretches of the water course are rather comparable to disturbances. However, in large rivers, with flood plains, high water is not a catastrophic event; it is a recurring event that is part of the annual ecological cycle. The regular long-term flooding creates temporary lentic habitats that are essential to the completion of the life cycle of river species. The organisms that inhabit these environments are adapted to (or have been selected for) these spatio-temporal fluctuations.

In the concept of *flood pulse*, Welcomme (1979) and Junk et al. (1989) recognized that high water is the principal factor controlling the productivity

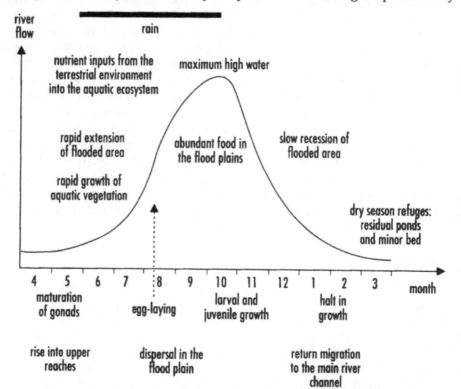

Fig. 9.3. The cycle of seasonal events in a river with a flood plain in relation with the biology and ecology of fishes (modified from McConnell, 1985)

of rivers. The system actually responds to the amplitude, duration, frequency, and regularity of pulsations. The seasonal cycle of events in a tropical flood plain has been summarized by Low-McConnell (1985) (Fig. 9.3). When a savannah is flooded at the time of high water, the water is enriched by nutrients resulting from the degradation of terrestrial organic matter or the excrement of large wild and domestic animals. The result is a rapid development of bacteria, algae, and zooplankton, and a rich fauna of invertebrates with a short cycle, while the vegetation grows quickly. There is eventually a sudden increase in the biomass and production of various organisms that may serve as food for fish. Among these organisms, many lay eggs during the high water flow at the limit of the flood plain or penetrate this zone (Welcomme, 1979). Some species even migrate long distances to find favourable conditions for reproduction. Fish larvae or juveniles find in the flood plain abundant food in the form of small invertebrates and in the vegetation they find refuge against predators. When the water recedes, the fish go back to the low water bed of the river or are trapped in the residual ponds and lakes, many of which will later dry up. The stretches of water that last throughout the year serve as a refuge zone for aquatic organisms that recolonize the system during the next high water period.

Variability of the oxygen level in aquatic ecosystems

Oxygen is soluble in water but its solubility depends greatly on the temperature. It is 14.5 mg/l at 0°C, 10 mg/l at 15°C, and only 7.5 mg/l at 30°C. This difference in solubility as a function of temperature partly explains the existence of seasonal variations of the oxygen content of waters, and the fact that the warmer tropical waters are overall less oxygenated than temperate waters.

In aquatic environments, oxygen has two origins: the production of oxygen by aquatic plants and dissolution of atmospheric oxygen. The gaseous exchanges with the atmosphere depend on the turbulence of waters and the degree of saturation. In stagnant waters, these exchanges are limited to the surface layers, while in turbulent waters the turnover ensures good exchange between the atmosphere and all the water masses.

Unlike in terrestrial ecosystems, the distribution of oxygen is not uniform in aquatic systems such as the oceans and deep lakes. The oxygen level in waters can even vary considerably in time and in space, with important consequences for the distribution of organisms. In lake systems we observe generally a marked stratification. In the well-lit upper layer, oxygen production is high and there may be temporary supersaturation especially during the day, in the absence of turnover of the water by the wind. In deeper zones, on the other hand, the oxygen level is lower because of respiratory processes bringing about photosynthesis and we very often observe even anaerobic conditions (total absence of dissolved oxygen). The impact of stratification of oxygen on the aquatic fauna is important: species with aerobic respiration can only survive in the well-oxygenated surface waters, while the deeper waters are inhabited only by anaerobic organisms.

9.4.3. Inter-annual variability

Inter-annual variability is highly uncertain because it is difficult, if not impossible, to predict the evolution of the climate in the years to come. It is thus at this temporal scale that we observe the greatest fluctuations in the dynamics of ecosystems.

While the marine environment may seem at first sight to be more stable than terrestrial systems, it has been demonstrated that the stocks of pelagic coastal fishes fluctuate greatly. The decline of many fisheries has presented a profound challenge to the way in which scientists study the dynamics of living resources in relation with the environment. In particular, the instability of certain stocks of pelagic coastal fishes that are developed in highly productive upwelling zones has been indicated (Cury and Roy, 1991). In the history of fishery, the anchovy in Peru or the sardine in Japan are well-known examples of stocks that have fluctuated greatly, with dramatic impact on the regional economy. It has been shown that in the upwellings generated by wind, there is a dome-shaped ratio between the success of recruitment and the intensity of the upwelling. This is the theory of the optimal environmental window of Cury and Roy (1989) (see Chapter 6). Environmental fluctuations thus seem to be the essential factor in understanding the variability observed in the dynamics of pelagic fish populations in upwelling zones (Cury and Roy, 1991).

Fishery studies have long used models to describe and predict changes in catches of an exploited stock: global or synthetic models and structural or analytical models. The former, the archetype of which is the Schaefer model, takes only the fishing effort into account as an explicative variable. In various, more or less improved forms, it has resulted in a set of models in which the variability not linked to fishing is considered as a background noise that is supposed to be random. In other words, these models are based on the principle of equilibrium: each year, the quantity of fish caught is compensated for by the recruitment of juveniles. If the catch becomes too high, the stock diminishes and, at worst, disappears. There is an ideal situation, the "maximum equilibrium yield", at which the largest number of fish can be caught while the stock is preserved.

The example of pelagic coastal fishes illustrates a situation in which the need to exploit these unstable resources leads to a concern with variability, instability, change, and management methods. Fishery managers must learn to live with that variability, instability, and change to act, decide, and react in a situation of uncertainty. We are far from a conventional dogma of stock management that is based on the concept of equilibrium and the postulate of predictability.

In reality, it is instability that has now become the central idea, a virtual object of research. It is a difficult field in which scientists progress slowly. It is also an approach that conforms poorly with economic planning. To be objective, we are right to question the credibility of a discipline that for a long time has promised and used models that are poorly adapted to real situations, with the ambition to play a critical role in decision-making on fishery issues.

9.5. LANDSCAPE ECOLOGY: HOW TO INTEGRATE SPATIAL AND TEMPORAL HETEROGENEITY

The term *landscape ecology* was introduced in 1939, three years after the concept of ecosystem was introduced, by the German biogeographer C. Troll, who wished to combine geography and ecology (Troll, 1939). Landscape ecology is another way of tackling issues of spatial heterogeneity and of discovering how modes of spatial organization (patterns) control ecological processes. A landscape is presented as a mosaic of interactive elements whose arrangement gives it a spatial structure. It is thus a heterogeneous but nevertheless organized, spatially structured system in which we can observe flows of species and energy. In this framework, landscape ecology is seen as the spatial expression of the ecosystem (Richard, 1975). In fact, the two raisons d'etre of landscape ecology are that nature is heterogeneous, and that we must be attentive to questions of scale (Wiens, 1999).

The term *landscape* has been recomposed and reappropriated by ecologists who shifted their object from ecology in the landscape to landscape ecology (Lefeuvre and Barnaud, 1988). There are few landscapes, at least in the temperate regions, that have not been influenced by human beings, whether recently or historically. Landscape ecology has thus incited ecologists to integrate human activities in their ecological models, and to approach other disciplines that arise from the social sciences, such as geography.

Within the field of landscape ecology are grouped various orientations. One of the traditional currents of thought is linked to analysis of the land cover and its evolution. This is a sector with important applications and impacts, some relating to the phytosociological approaches. Another more recent current of thought, influenced by the ecological approach, involves the structure, functions, and genesis of the study object "landscape": a landscape, according to Forman and Godron (1986), is a portion of heterogeneous territory made up of sets of interacting ecosystems that are repeated in a similar fashion in space. According to geographers such as Bailey (1996), a landscape is made up of geographic units with identifiable limits: ecosystems. This point of view has the advantage of considering the landscape as a level of integration higher than the ecosystem. This is also the point of view of Burel and Baudry (1999), who add that the landscape is characterized essentially by its heterogeneity and its dynamics, governed in part by human activities. It exists independently of our perception of it. These authors restrict landscape ecology to a spatial scale corresponding to human activities (from a few hectares to a few hundreds of square kilometres).

9.5.1. Principles of landscape ecology

Landscape ecology has given rise to a specific terminology. *All landscapes, even those that are different, are made up of a similar structure, an "ecological mosaic" made up of various types of elements that are more or less fragmented and connected with one another by flows of minerals, species, and energy.* This landscape mosaic constitutes a spatially heterogeneous whole. Without going into detail, specialists in landscape ecology distinguish three major types of elements: the *matrix*, the *patches*, and the *corridors* (Forman and Godron, 1986).

The *matrix* constitutes in a way the canvas, the structuring element that gives it its general physiognomy. It could, for example, be a prairie, a forest cover, or an extent of water.

In this matrix there are occasional elements called *patches* or parcels: e.g., groves, ponds. These patches are comparable to islands that are more or less isolated from one another. They are the result of a spontaneous development of the vegetation influenced by natural or anthropic disturbances. Their type, number, form, and dimensions are variable but determine the structure and species richness. All these patches together constitute a *mosaic*. Each landscape element may have relationships with all the others. The flows of matter and energy are determined by vectors such as wind, water, animals, and human beings.

Disturbances play a fundamental role in the origin of patches and their dynamics. They may consist of fragmentation of the habitat due to human activities (deforestation, cultivation), natural fragmentation of a type of plant cover resulting from soil heterogeneity, or heterogeneity resulting from regeneration after a severe disturbance (fire or drought). Many agents that end in the formation of patches are episodic, so that the mosaic is modified permanently. The process of creation and evolution of patches, as well as the changes that occur in the patches, themselves constitute what is called patch dynamics.

Patches may be linked to one another by linear elements, *corridors*, which ensure the circulation of species and energy between the patches. All these corridors together constitute a *network*. The corridors may differ in nature and function. The linear corridors (hedgerows, roads) are narrow and represent the favoured habitat of ecotone species of the patches or the matrix. The corridors bordering water courses, formed of a band of vegetation, help regulate the movements of minerals and species between the patches, the matrices, and the water course itself.

To the diversity of the constituents of the landscape is added the complexity of their spatial relationships. The concept of *connectivity* expresses the intensity and regularity of processes by which the different landscape elements are related and integrated functionally with the whole. Connectivity is higher when the landscape elements are better connected to one another. The modalities of exchange and interaction fluctuate in time and in space. *Connectedness* refers to the fact that two patches of the same type are connected in space, and *connectivity* refers to the processes by which isolated sub-populations, each having its own characteristics (density, mortality, recruitment), may be interconnected to form a genetically viable demographic whole called a *metapopulation* (see Chapter 10). The perenniality of these populations depends on the possibility of connection of subsets to enable genetic exchange.

Geographic Information Systems (GIS)

The term *Geographic Information Systems* means several distinct things. Simply put, a GIS is a data base coupled with a graphic software.

Geographic information is the representation of a portion of a territory by a drawing (geometric figure) or by a realistic copy in the form of an image (photograph). This representation is georeferenced, i.e., the data are located in a single system of coordinates. The GIS is thus a statistical tool that allows geographically referenced data to be stored, managed, manipulated, structured and synthesized, and finally visualized. Each theme of information is represented by a layer, a set of elementary objects of the same kind. A layer relates the cartographic representation of spatial objects with the table of statistical information that is associated with it. These layers of information may be of different kinds (e.g., satellite images, maps by the national geographic institutes, and ecological data).

The GIS allows us to combine information contained in the data base in different ways. For example, we can superimpose different layers of information in order to relate the information derived from layers that are inherently incompatible (remote sensing, field surveys, etc.) and of different nature. We can also, by intersection and inclusion, try to delimit spatial units that respond to certain criteria.

9.5.2. Landscape structure and functioning

The inventory of constituents of a geographic entity is an anatomical approach (to identify the structure), which does not necessarily allow us to understand how the elements organize themselves or function. In reality, any ecosystem presents a *vertical structure* that corresponds to the integration of different constituents over a defined area.

For example, a forest ecosystem comprises not only a community dominated by trees but also a soil, surface waters, a water table, a rocky substrate, etc. However, an ecosystem is never isolated and its limits are not impermeable but open to transfers of energy and matter. Each ecosystem within a landscape thus interacts with the neighbouring ecosystems over its entire vertical dimension. The nature and intensity of these exchanges determines the manner in which the landscape "system" will function. The *horizontal structure of a landscape* is thus the spatial association of ecosystems that are structured vertically and in interaction.

If we hypothesize that the structure regulates and controls many ecological functions, landscape ecology provides a formal framework to study nature, the causes and effects of spatial heterogeneity and temporal variability. But appropriate scales must be selected for different ecological phenomena for us to understand the processes that structure the landscape. A lake, for example, can be seen on a macroscopic scale as a relatively homogeneous unit, while on a smaller scale it may appear to be a heterogeneous but horizontally structured environment (pelagic-littoral) or vertically structured environment (epilimnion-hypolimnion). Simultaneously, the recognition of these appropriate scales for ecological processes leads us to identify discontinuities in scale, discontinuities beyond which the scale is no longer pertinent for the process studied.

In fact, landscape ecology has become the preferred domain to study the crucial problem of scales in ecology, in showing how processes corresponding to different spatial and temporal scales can interact. In a way, it has allowed us to recognize that heterogeneity is a fundamental element of the functioning of systems, rather than an embarrassing source of errors (Pickett and Cadenasso, 1995).

In sum, ecological systems are dynamic and the heterogeneity perceived at a given moment, in a given place, is the result of various ecological processes and disturbances of anthropic and natural origin (Fig. 9.4). Over the spatial heterogeneity is superimposed a temporal variability that in turn helps create and maintain the spatial heterogeneity. This temporal variability is the result of ecological processes that are more or less cyclical (e.g., seasons, high water periods) or natural and/or anthropic disturbances that may be physical in origin (e.g., storms, floods, fires), chemical (pollution), or biological (epidemics).

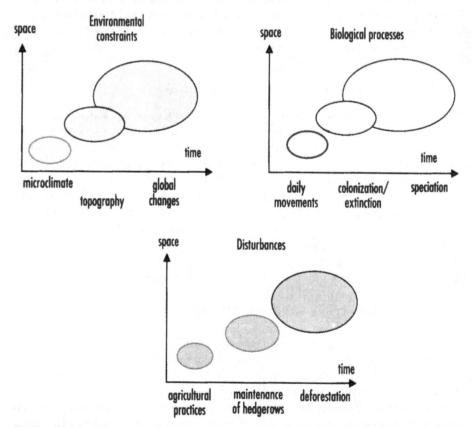

Fig. 9.4. The landscape mosaic is the result of environmental constraints, disturbances, and biological processes, each operating at their own spatial and temporal scales (Urban et al., 1987; Burel and Baudry, 1999).

In reality, the time factor is a major element of landscape dynamics and we must know the history of environments in order to understand their present structure and attempt to predict their future evolution. In particular, the distribution of species may be the heritage of historical events (climatic changes, fires, droughts) that have profoundly marked the landscape and the effects of which continue to be felt with a delay that varies according to the species studied.

9.5.3. The hydrosystem: a heterogeneous landscape

The conventional notion of a spatially well-defined ecosystem is difficult to apply to fluvial systems, which are stretched out over space and for which interactions with the surrounding terrestrial environment have long been observed. Studies of fluvial systems during the 1980s (Amoros and Bravard, 1985) showed that spatial heterogeneity and temporal variability are essential components of the functioning of large rivers. We thus progress from the concept of the ecosystem to that of a fluvial system or hydrosystem, which finally brings us back to the idea of an aquatic landscape. The term *fluvial hydrosystem* (Amoros et al., 1987) applied to large water courses designates a complex ecological system. It is made up of an overlapping mosaic of running-water ecosystems (principal channel), stagnant-water ecosystems (backwaters, oxbow lakes), semi-aquatic and terrestrial ecosystems (marshes, alluvial forests), systems that are superficial as well as subterranean, that are established in the alluvial plain. Their functioning depends directly or indirectly on the active course of the river. A good number of these ecosystems in fact result from one another by

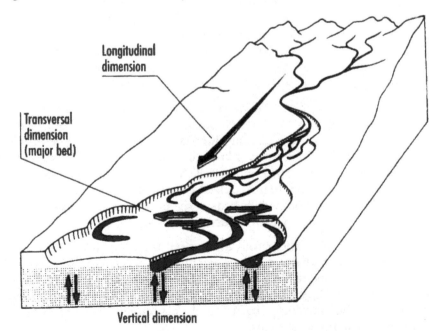

Fig. 9.5. Flows of hydrosystems according to longitudinal, transversal, and vertical dimensions (Amoros and Petts, 1993)

transformation of biotopes along the river dynamics and as a function of ecological successions. This interactive set of natural or artificial environments, subjected to the same regime of disturbance, is characterized by multidirectional exchanges. It has become common to speak of the four dimensions of the hydrosystem (Fig. 9.5).

- The *longitudinal dimension*. Rivers are systems that are stretched out spatially but continuous in their longitudinal dimension, to the extent that the dynamics of systems located downstream depends on physical, chemical, and biological processes that occur upstream. This longitudinal axis expresses the evolution of principal hydraulic, geomorphological, and ecological characteristics of water courses following a more or less continuous gradient from the source to the mouth.

- The *vertical dimension*. There are close relationships between the surface waters and ground waters that are manifested particularly by exchanges of water and dissolved elements such as mineral salts, nitrates, and pollutants. The water table ensures the maintenance of flow in a low water period, while in the high water period the water table is recharged in the flooded areas. The water table serves temporarily as a refuge zone for many benthic invertebrates when conditions at the surface become unfavourable.

- The *transversal dimension*. The river functions together with its alluvial plain, which is made up of a mosaic of ecosystems. The role of flood plains and fluvial annexes such as backwaters has been indicated in the functioning of running waters, as well as the role of riparian forests on the cycling, retention, and distribution of nutrients. The riparian zones provide the running waters with most of their supply of organic matter in the form of detritus. Over the entire watershed, the land use (woods, cultivation, organization) influences the runoff and hydrological functioning as well as the inputs of particulate and dissolved substances.

- The *temporal dimension*. The present aquatic systems are a heritage in geomorphological as well as ecological terms of climatic and tectonic changes that have taken place in the past. To understand their structure and functioning, it is necessary to know this history as well as the conditions in which the flora and fauna were established. In a more recent past, various types of developments (dams, dykes) or activities (removal of sand and gravel) have often modified the beds of rivers. At each time scale, there are phenomena that have had an influence on the present dynamics.

In reality, we must add a fifth dimension to hydrosystems, that of the exchanges of the river and its watershed with the atmosphere: e.g., evapotranspiration and rainfall, inputs of dissolved substances by rain, gaseous exchanges (N_2, CO_2, O_2), the fallout of atmospheric dust. These exchanges control the elements of the hydrological balance and, consequently, the driving force of the functioning of fluvial systems.

9.5.4. Lakes as part of a landscape: regional ecology

For a long time, lakes were considered the archetype of the ecosystem, with spatial limits that were relatively clearly identified by the shore line. Subsequently, it was perceived that these ecosystems were not actually closed but kept up exchanges of matter and energy with neighbouring aquatic or terrestrial environments, as well as with the atmosphere. The lakes are thus part of a larger landscape comprising the lake itself and its watershed, characterized by the local geology, the land cover, the topography, and other factors. In other words, the spatio-temporal dynamics of lake environments is partly influenced by the external environment.

Some authors have proposed the concept of *regional ecology*, a more general term than landscape ecology, to designate mixed terrestrial and aquatic systems. In the context of regional ecology, what occurs outside the systems studied is focused on in order to understand the respective role of endogenous and exogenous factors in the functioning of lake systems. According to the terminology of landscape ecology, the lakes as in a way considered as patches and their dynamics is studied in a network of interactions at the regional level (Magnuson and Kratz, 2000). This process refers to some extent to that of island biogeography, which integrates the regional and local determinants of the biological structure of lakes (Tonn et al., 1990).

a) Temporal coherence

An important aspect of regional ecology is *temporal coherence*, i.e., the more or less synchronous way in which the different patches will react in time. To test this coherence, we can use the comparative method and study, for example, the behaviour of a series of adjacent lakes in a relatively homogeneous region. Two lakes may fluctuate in a similar and synchronous way if the external factors, such as climatic factors, exert a determinant effect on their dynamics.

On the other hand, if it is the internal factors such as the availability of food or interactions between species that are predominant, the lakes react independently or asynchronously (Kratz et al., 1997) *Thus, temporal coherence will be the degree to which different systems behave, in a similar or dissimilar manner, as a function of time*. Temporal coherence has been measured on a series of Canadian lakes (Magnuson and Kratz, 2000). On the one hand, coherence is greater between lakes located in comparable climatic conditions. On the other hand, there are differences according to the type of variable considered. The physical variables are well correlated, but with the chemical variables the coherence is stronger for those that are controlled by atmospheric inputs than for nutrients such as silica or phosphorus. The biological variables are the least coherent.

b) Position in the landscape

Comparisons between lakes have also shown that neighbouring lakes of the same climate, geology, and pool of species have different dynamics controlled essentially by factors such as hydrology and geomorphology. The concept of "position in the landscape" has been used to attempt to explain the differences observed. There is the position in the hydrological system as well as the nature and importance of the connectivity between the lakes and the hydrological systems. This position in the landscape, in the case of Canadian lakes, is a geomorphological heritage of glaciations.

To test this concept, a study was conducted in North America on nine chains of lakes connected by ground waters and/or by a surface network. With respect to the *spatial dimension*, the study indicated the existence of patterns in the variables structuring the lake systems, as a function of their position in the landscape (Sorano et al., 1999). Along a chain of lakes, for example, the ratio of the area of the watershed to that of the lake increased, while the time of residence of the water decreased. Physicochemical variables such as alkalinity, conductivity, and calcium level generally increased along a chain of lakes. The position in the network of flows was thus an important factor of the organization of lake systems.

With respect to the *temporal dimension*, the criterion of reference used was synchrony. In the present case, synchrony at the annual scale was high but was not linked to the position in the landscape. However, the greatest synchrony was observed among lakes in which the waters stay for a short time, while patterns were discerned in sets of lakes in which the waters remained for a year or longer. Some of the results needed to be validated, but one thing was certain: lakes, which were long considered independent units, are actually connected and organized within a landscape. The concept of "position in a landscape" may be compared to that of *river continuum*: hierarchically organized systems of rivers or lakes are the result of processes that act on large spatial scales.

9.6. FRAGMENTED COMMUNITIES

In an environment characterized by spatial heterogeneity, and/or by the fragmentation of ecosystems, populations of a single species are generally fragmented and more or less isolated from one another. Following in this respect the island biogeography theory (see Chapter 10), the first studies on fragmentation of forest habitats showed that the size of wooded islands was a good indicator of the species richness of birds (Forman et al., 1976). However, the island theory gave way from the 1980s onwards to the theory of metapopulations that had been stated by Levins in 1970, which served as the basis for much theoretical and empirical research on the effects of habitat fragmentation on populations (Hanski and Gilpin, 1997). Presently there is an interest in these question of fragmentation of habitat, metapopulations, and metacommunities, largely because land development tends to transform landscapes into virtual mosaics.

9.6.1. Metapopulation concept

A *metapopulation* is a set of sub-populations that are geographically more or less isolated but interconnected by exchanges of individuals that help maintain a gene flow between different sub-populations. A relatively simple case is that of populations occupying islands of the ocean or continental type, between which there may be constant or occasional exchanges. The exchanges depend obviously on the aptitudes of species to disperse: insects and birds have a better ability to move from one island to another than earthworms or tortoises. These exchanges can be regular or occasional. Freshwater fish populations, presently isolated in different watersheds, have been in contact in the past (and will possibly be in contact in the future) during exceptional events. Exchanges can thus involve different spatio-temporal scales, depending on the species and the environment.

In a metapopulation, sub-populations in which natality predominates over mortality serve as "sources" from which individuals disperse to other places. Inversely, sub-populations living in constrained environments in which the mortality exceeds the natality constitute "sinks". A metapopulation is thus a dynamic system characterized by migratory flows and processes of extinction in subsets of a fragmented habitat. Conventionally, a metapopulation survives owing to an equilibrium between local extinction and the establishment of new populations in unoccupied sites (Hanski, 1998). The concept of metapopulation can be extended to multispecies communities and a metacommunity may be defined as a set of ecological units having the same biotic components and between which exchanges are possible.

9.6.2. Patch dynamics

The emergence of landscape ecology in the 1970s offered theoretical ecologists a simplified view of heterogeneity by defining space as a mosaic of patches arranged in an ecologically neutral matrix. This spatial model gradually became more complex with the recognition that patches can be different, and then with the taking into account of the role of corridors as well as the connectivity or permeability of landscapes.

In landscape ecology, a patch is a spatially delimited structure, at a given time, within a matrix. The concept of patch dynamics makes the link between the distribution of communities in a mosaic and the spatio-temporal dynamics of patches: they can in fact disappear or extend themselves over time as a function of fluctuations of environmental factors. Besides, and even though located in a single matrix, each patch and the communities it harbours may have a totally different dynamics. In this matrix there may also be patches that tend to disappear or, on the contrary, to spread out, and new patches may be created. Depending on the trend, the associated communities may thus be senescent or pioneering, or they may correspond to a stage in a succession.

To understand the interactions between patches, we must observe changes at different scales of time and space. The seasonal inundation and exundation of the major bed of a river constitute a good illustration of this patch dynamics to the extent that the variations at the level of the water create and/or modify the heterogeneity (Denslow, 1985). If the species are capable of exploiting this heterogeneity, the disturbances can favour the coexistence of several species having varying ecological needs in patches that are at different degrees of evolution. We can add a little more complexity to the model by considering species that use different landscape elements to complete their life cycle. In these conditions, the survival of populations depends on the spatial distribution and the composition of the landscape mosaic, and its dynamics over time. It is necessary, particularly, that the individuals find the habitats they need for each stage of their life cycle at the appropriate time.

According to Townsend (1989), the concept of patch dynamics provides a unifying principle in stream ecology, where we observe a high spatial heterogeneity of ecological characteristics such as the speed of the current, the substrate, and the availability of food resources. Patch dynamics, in a holistic sense, implies the following:

— Natural or anthropic disturbances act on the ecosystem to modify the distribution of habitats in time and in space. A simple example is a high water period, which creates new habitats in a river while modifying the environmental parameters in the flooded habitats: e.g., speed of current, depth. The inverse is true during the receding of waters.

— In this heterogeneous system, the pioneer communities are established as soon as the habitats are available and normally evolve towards a more mature stage. Consequently, we can find at a given time patches at different stages of the ecological succession, as a function of the chronology of floods.

— Such a dynamic system in time and in space allows a greater biological diversity than systems evolving monotonously towards a climax.

In this context, heterogeneity creates diversity. *Spatial heterogeneity and temporal variability are thus key elements of the functioning of ecological systems and not a simple "background noise" that may disturb the dynamics of communities, as has long been believed.*

An example of patch dynamics: the tiger bush

The expression "tiger bush" refers to the peculiar appearance of some plant formations in a semi-arid zone characterized by an alternation of parallel bands of vegetation with bare zones having very sparse plant cover, or even crusted bare ground. The bands of dense vegetation are arranged perpendicular to the line of the greatest slope. They represent 20 to 30% of the total area and their dimension varies from 5 to 50 m width for 20 to 400 m length. It is called "tiger bush" because the whole landscape looks like a tiger skin from aerial photographs. These formations are found in all the arid zones of the globe, including western and eastern Africa, Australia, the Middle East, and Mexico. There is a variant, the "patch bush", which is commonly found on a crystalline substrate (Casenave and Valentin, 1989).

The mosaic structure of the plant cover as well as its dynamics are the result of various factors:

• A relatively flat relief in which the hydrographic network is not hierarchical (Leprun, 1992).

• Surface reorganization of soils due to hydric and wind erosion. The mechanical action of rain on bare soil and the humidification-desiccation alternation create a crusty organization of the surface that reduces permeability.

• Runoff, given that the basic mechanism is the redistribution of water and nutrients that create and maintain the zones of vegetation. The bare crusted zones correspond to dry soil on which the water runs off, while the wooded zones, located downhill, receive and capture the runoff, which infiltrates easily because of the high porosity of the surface soils.

• Biological processes such as seed dispersal and competition between species.

Tiger bushes are not static ecosystems (Aguiar and Sala, 1999). The uphill part of the patch tends to spread because of the input of sand and runoff, while the downhill part wastes away and the soil erodes (Fig. 9.6). Thus, dynamically, there is a continuous transformation of soils and vegetation under the combined effects of hydric and wind erosion and the system progresses upwards at a rate that varies from 20 to 70 cm a year in the Sahelo-Sudan region (Leprun, 1992). This example of patch dynamics is interesting because it is not the result of large-scale disturbances, as is the case with the effect of fire: here it is the endogenous factors that are predominant.

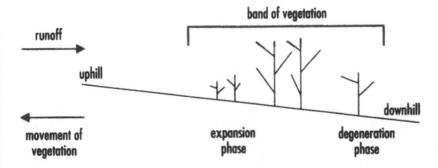

Fig. 9.6. Patch dynamics of vegetation in a tiger bush (inspired from Aguiar and Sala, 1999).

Part III
Functioning of Ecosystems

Dynamics of Communities and Ecosystems: from the Balance of Nature to Self-regulated Systems

Let us remember, my children, that nothing is constant but change.

The Buddha

The paradigm of the equilibrium of communities and ecosystems, derived from the myth of the balance of nature, is one of the most ancient in ecology. Equilibrium, stability, and resilience were common terms in the ecological vocabulary that implied a certain constancy, even perenniality, in time. Still, we must agree on what we mean by equilibrium. Is it an intrinsic property of the ecosystem or a property that is applied more particularly to some of its components? Ecologists most often apply notions of equilibrium and stability to biotic components of ecosystems (Golley, 1993). It is thus a way of investigating the nature of processes that regulate the organization and dynamics of populations and communities subject to abiotic and biotic constraints (competition, mutualism, parasitism, etc.). But what meaning is implied in this for ecosystems? What meaning should be attributed to notions of equilibrium and stability for entities that are dynamic, heterogeneous, and variable over time?

10.1. BRIEF HISTORY OF THE IDEA OF EQUILIBRIUM

10.1.1. The dogma of equilibrium

The idea that ecological systems tend to perpetuate themselves in a self-same manner marked the beginnings of ecology (Egerton, 1973). The paradigm of equilibrium was one of the founding principles of scientific ecology with, as a corollary, the statement that systems are not constituted by chance, and that there is an organization that maintains a cohesion. In fact, if we recognize that there is an order in nature it becomes possible in theory to identify general laws of structuration and functioning of ecosystems and to predict their future trajectories. If on the contrary ecological communities are random collections of species, the search for laws becomes useless and ecosystem ecology loses its rationale. This subjective, somewhat deterministic conception of nature is expressed in the famous concept of "balance of nature", which is often the basis of theories and models that address the regulation of populations and communities and the stability of ecosystems.

The notion of equilibrium comes from mechanics and was developed through classic thermodynamics "at equilibrium" throughout the 19th century and the first half of the 20th century (see Chapter 5). It was invented to describe and characterize systems in which the status variables do not change, or practically do not change (Pave, 1994). The desire to identify a state of equilibrium in ecological systems has more of aesthetic or metaphysical motivation than reasons of convenience of analysis (Drouin, 1993). According to Prigogine and Stengers (1986), from the end of the 19th century, the study of equilibrium states, mechanical and then thermodynamic, found fruitful resonances in biology and the social sciences. Here also, it is pointless to emphasize that the intellectual and affective load of the concept of equilibrium came not from mathematics but from ideas of order and harmony from entirely different fields of study. And these were the same ideas that would lend their weight to physical and mathematical methods when such methods finally entered the fields of biology, economics, and sociology.

We should probably add that the notion of equilibrium also made it easier for ecologists to study life, at a time when technologies for information processing were still limited or non-existent. The quest for general ecological laws led ecologists to work by preference on environments that offered the most stable conditions, and the most homogeneous possible, so that they could treat data with the statistical and mathematical tools available at the time.

10.1.2. Controversy: the paradigm of non-equilibrium

Two major trends brought ecologists to refute the notion of equilibrium. The first was the accumulation of field observations showing that ecosystems are not isolated but open systems, crossed by flows of matter and energy, and that there are many exchanges between systems (theory of dissipative systems, see Chapter 5). Besides, it is evident that the notions of equilibrium and stability cannot be applied strictly to ecosystems when the abiotic and biotic components are known to vary on annual or multi-annual scales. It is thus preferable to admit that the dynamics of the ecosystem comprises an important part of variability and that the constitutive parameters oscillate more or less regularly around average values with greater or lesser amplitudes. A simple example is that of a river with an annual high water period: we may statistically know the variability of the high water level from one year to the next (every 10 years, every 100 years) and propose limits of variation within which the system is estimated to be "stable". The second trend was that ecologists acknowledged the need to tackle ecosystem dynamics at different spatial and temporal scales (see Chapter 8).

The result is that ecology has developed a totally different paradigm—that of non-equilibrium; it is not that equilibrium does not exist in nature, but that it does not necessarily appear at all scales or for all phenomena. The paradigm of non-equilibrium recognizes that ecosystems are open thermodynamic systems, regulated by non-deterministic internal and external factors without an equilibrium point. The disturbances are a fundamental element of their dynamics. This paradigm, as opposed to the preceding one, is based on the idea that population density is not constant in time in various parts of the space considered.

The theory of patch dynamics (see Chapter 9) clearly illustrates this idea that spatial heterogeneity and temporal variability are key elements of the functioning of ecological systems.

10.1.3. The mutation is not complete

We can also consider the evolution of paradigms involving the notion of equilibrium as part of the vast debate that animates ecologists concerning the respective roles of biotic or abiotic controls in the dynamics of ecosystems. Some authors, such as Lack (1954), in the sphere of density-dependent regulation, have favoured biotic relationships such as competition, mutualism, and predation. Other theoreticians have taken the other position and affirmed the primacy of physical factors, especially climatic factors, in fluctuations of abundance in populations. The studies of Andrewartha and Birch (1954) were accepted as authoritative on the subject and sustained, with advocates of density-dependent regulation, a sterile debate, because abiotic factors obviously also play a role in population dynamics, especially in extreme situations: catastrophic events, such as severe winters, or prolonged droughts, may lead to considerable mortalities.

It is now known that according to the situations, the species concerned, and the questions posed, the relative importance of these different factors in the control of ecological systems may evolve over time or space. It is also known that several factors most often act simultaneously. Still, we must experimentally test the hypotheses proposed and thus improve the predictions about these multi-factor controls. Most often we cannot practically do this.

The evolution of ideas summarized above has taken a great deal of time in actuality, and the mutation is probably not complete. As early as 1930, Elton said that "the balance of nature does not exist and perhaps never has existed. The numbers of wild animals are constantly varying to a greater or lesser extent, and the variations are usually irregular in period and always irregular in amplitude. Each variation in the numbers of one species causes direct and indirect repercussions on the numbers of others, and since many of the latter are themselves independently varying in numbers the resultant confusion is very remarkable."

However, Elton received hardly any attention in his time. Ecologists later developed other concepts that could serve explanatory functions. Thus, in borrowing certain concepts from physics, let us talk of self-regulated or auto-organized systems. The idea of equilibrium still persists. The Gaia hypothesis, based on the principle of an self-regulation of the planet earth, is implicitly based on the ideas of an order and regulation of nature (Fig. 10.1).

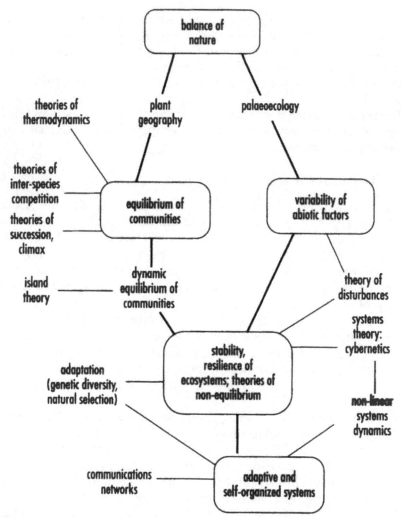

Fig. 10.1. Diagram of the evolution of ideas about equilibrium and regulation of ecosystems

10.2. THE BALANCE OF NATURE

The notion of the equilibrium of ecosystems probably originated from the notion of balance of nature, which was the basic concept in natural history from ancient times, where it emerged implicitly in the discourses of the Greeks to describe the functioning of the world (Egerton, 1973). These ideas are partly the reflection of a good popular sense since at the scale of human life, at least, we have the impression that nature perpetuates itself, renewing itself in more or less the same manner each year, according to the seasons. This leads us to consider, perhaps mistakenly, that certain natural systems that we do not see the movement of are at equilibrium. This is true, for example, for forest or marine systems.

According to Popper (1962), almost all scientific theories originate from myths and myths are, in a way, anticipations of scientific theory, whereas, in one form or another, the concept of the balance of nature is part of many cosmologies. This equilibrium comes from the will of the gods, in the face of which humans intercede by prayers, sacrifices, and rituals so that they will have a good harvest, a good hunt, a good catch. Till the Renaissance, the western world was in the grip of the Christian religion, which considered that God managed the world. In the 16th and early 17th century, scientists still had an Aristotelian view of natural beings. They believed that all living things formed an uninterrupted and hierarchically ordered chain, from raw material (minerals) to the angels and encompassing plants, animals, and humans. This antique concept of the chain of beings acquired a theological significance in the Middle Ages. An all-powerful God could not tolerate vacuums in creation. This is why, at the end of the 17th century, Leibniz could say that all living things form a single chain, in which the different classes, in the form of rings, are so closely linked to one another that it is impossible for the sense and imagination to fix precisely the point at which any one begins or ends (Duris, 1994).

10.2.1. The economy of nature according to Linnaeus

The Swedish naturalist Carl von Linnaeus (1707-1778) is primarily known for the introduction of binomial nomenclature in the classification of living things. In the mid-18th century, it was still thought that God had created all the species of plants and animals, which were of a fixed number. The task of scientists was to inventory them. Two and a half centuries later, this inventory is still in progress and no one has attempted to fix a deadline for it.

But Linnaeus was also the author of a set of works that developed general principles of an "economy of nature" (Linnaeus, 1972). By economy of nature, he meant the very wise disposition of natural organisms instituted by the Sovereign Creator according to which these organisms had common ends and reciprocal functions. This natural order could be characterized according to four phenomena: propagation of the species by mechanisms of generation; its geographic distribution; the destruction that is the certain end of each natural body; and finally the propensity for conservation inscribed in the structure and mode of life of each. The economy of nature (otherwise called Divine Economy or Divine Wisdom) was thus essentially a concept according to which the interactions between natural bodies resulted in an intangible equilibrium that maintained itself throughout the ages.

In this fixed conception of the world, Linnaeus considered that minerals, plants, animals, and human beings lived in a harmony willed by the Creator. The environmental conditions and climate explained the geographic distribution of species. The author of nature, according to Linnaeus, procured for each animal a vestment perfectly appropriate to the region it inhabited and the body structure of the animal is artfully adapted to its mode of life and to the type of soil in which it lives. Lions, elephants, and rhinoceroses feed on plants that grow throughout the year only in warm regions and this is why they are assigned these places. Each species plays a role and fulfils a function in this balanced world and if any one element is lacking, he said, we can expect disastrous consequences for the universe. It is important to emphasize that the world of Linnaeus is immutable. It reproduces itself in an identical way from the beginning, in the same proportion of species, according to the will of God. The cultural environment imbued with Christian theology in which Linnaeus lived did not allow him to question the cause and effect relationships between environmental factors and the distribution or characteristics of species, since God had already decided it thus.

In the philosophical context of the 18th century, the work of Linnaeus helped rehabilitate the idea of a harmonious nature, work of God, in which his glory was manifest. This process was part of a current of thought initiated in great part by English scholars such as William Derham, whose writings at the beginning of the 18th century demonstrated how creation indicated the divine power. In their wake, other naturalists such as Reaumur saw in the anatomy and behaviour of the living things they studied the indisputable mark of the wisdom of their author (Duris, 1994). This conception of nature is highly anthropocentric: it is for the use of human beings that the nature we inhabit was created, and all the things created by God are useful to us, directly or indirectly (Limoges, 1972). Thus, what seems a nuisance to us, according to Limoges (1972), is most useful to us: aphids are eaten by larger insects that themselves feed sparrows, which are a pleasure to us and charm us with their song. In this world, human beings have a role to play: we appear, for example, to be a means of maintaining the proportion of the large predators.

10.2.2. Towards a more dynamic conception of nature: time as a dimension

At the end of the 18th century, the influence of the Catholic religion, still active in European science, would progressively be wiped out after the French Revolution, to the extent that a fixed idea of nature would give way to a transformist idea of nature. Simultaneously, the field of physics became prominent.

The French naturalist Buffon (1707-1788), who opposed the ideas of Linnaeus, was one of the first to consider the forest as not a collection of trees, but an entity in itself, a whole in which the individuals maintained particular relationships and reacted to one another, something that definitively announced what we today call an ecosystem. He pointed out even the role of birds in the dissemination of seeds, and that of field mice that store food for the winter (Deleage, 1992). Buffon glorified a nature cultivated by man, a civilized nature. One of his important contributions was that he considered that the order of nature as we observe it is a process subject to the grand worker of nature: time. The history of living nature is related to that of the earth. The mammoths, larger than the elephant, disappeared without descendents, he observed. There must be parallels for other species. This was a rupture with the immutable world of Linnaeus.

In his work *Principles of Geology*, published in 1832, C. Lyell, who is known to have influenced Darwin, proposed a global conception of the equilibrium of species that is a secularized version of the Linnaean notion of the economy of nature. Natural equilibrium is the result of antagonist causes that do not come from divine providence. This author gave a greater importance to competition among the causes responsible for extinction, thus opening up the route to ideas of the "struggle for survival". In proving the existence of an evolution of the world, Lyell moreover opened up the route to Darwinism.

Limoges (1972) underlined the importance of a historical and epistemological transition from the classic and theological paradigm of the economy of nature to the Darwinian revolution, which marked the emergence of a dynamic and properly naturalist conception. According to Limoges, ecology was made possible by the Darwin/Wallace theory, which in proposing the concept of a newly produced adaptation replaced the ancient and static conception of natural economy and introduced the division of labour in a new conception of the economy of nature. Because of this new theory, Haeckel in 1866 was able to announce the formation of a new discipline, ecology, defined by the body of concepts required to examine the complex interrelations that were the real conditions of the existence of the struggle for survival (Limoges, 1972).

Paradoxically, even though many naturalists in the 19th century began interpreting nature in a dynamic manner, few of them felt the need to challenge fundamentally the concept of the balance of nature. Science in the 19th century is always marked by the idea that there is an underlying stability in the universe. In biology, Claude Bernard based physiology on the theory of equilibrium, i.e., on the idea that the internal environment of the body was constant. The ecologists Clements and Forbes also applied these ideas to living communities. The notion of state of equilibrium also played an important role in the development of the ecosystem concept, given that Tansley made it a central element of his studies.

10.3. EQUILIBRIUM THEORIES BASED ON INTRA- AND INTERSPECIFIC RELATIONSHIPS

To observe that the density of many natural populations fluctuates around an average value is to recognize that these populations cannot have an unlimited growth and, consequently, that there is a certain stability in time. The identification of factors implicated in the regulation of population density has mobilized the efforts of ecologists. It involves discovering the exogenous factors limiting population growth, as well as the role of interactions between species in the maintenance of a certain equilibrium within communities. For reasons that are partly cultural, western ecology since the 1950s has emphasized competition between species. In this, it follows Darwin's ideas, which, probably imbued with the social and economic context of Victorian England, stressed competition and the *struggle for survival*.

10.3.1. The logistic equation and density-dependent regulation

In 1798, Thomas R. Malthus, one of the pioneers of demography, developed in his essay on the principle of population the idea that the growth of the human population is limited by the lack of food, which is the consequence of differing rates of increase: the geometric rate of increase of the population and the arithmetic rate of increase of production. For more than a century, this Malthusian paradigm would colour all the studies of animal demography. Darwin found it a source of inspiration.

For theoreticians imbued with the idea of the balance of nature, the density-dependent factors played an important role in the regulation of populations. The most simple hypothesis for the adjustment of numbers of a population to the capacities of the environment is based on the cybernetic concept of density-dependence introduced by Howard and Fiske (1911), who wrote with respect to insect populations that a natural equilibrium can be maintained only by the working of facultative agents that destroy individuals in proportion as they grow in number. Barbault (1981) remarked that it is more correct to speak of factors with density-dependent effects, if we are really talking of a factor (competition, predation, parasitism), or density-dependent phenomena, if we are talking of effects (mortality, natality, etc.).

The notion of density-dependence developed primarily around questions of mortality linked to density by the bias of intra-species competition, i.e., competition between individuals of a single species for resources such as food and space (Lack, 1954). When the density increases, the intensity of the competition increases and there is a reduction in the available resources per individual, which is expressed in a reduction of individual growth, fertility, and survival.

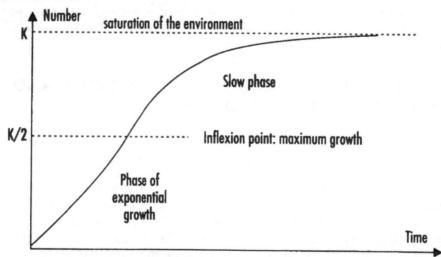

Fig. 10.2. Logistic growth of a population. This model expresses the fact that a population, after a phase of exponential growth, is limited by the availability of food resources and tends asymptotically towards a limit value K, which is the number of individuals that the environment can support as a function of the available resources.

The Belgian mathematician Pierre-Francois Verhulst stated the logistical law of development of populations (Fig. 10.2). The growth of a population, i.e., the increase in numbers, is the result of several antagonistic factors such as the rate of natality (b), the rate of mortality (z), and the rates of immigration and emigration. Generally, the intrinsic biological factors tend to favour population growth, while extrinsic factors proper to the environment exert negative or positive effects on the populations concerned. These antagonistic effects tend to buffer the variations in one direction or another. In an environment that is theoretically unlimited in terms of space and resources, the instantaneous growth rate of the population (r) can be estimated in a preliminary approximation as equal to b −z, and the size N of the population evolves as a function of time (t) according to the following formula:

$$dN/dt = rN$$

In reality, in natural conditions, the available resources (space, food) are limited. The growth rate of populations, which generally begins with an exponential phase, diminishes after a certain threshold, then becomes nil because of the resistance of the environment, which can only support a maximum number of individuals (K). If the resistance of the environment is represented by the term $(K-N)/K$, the growth of the populations is thus $dN/dt = rN(K-N)/K$. When $N = K$, the growth of the population becomes nil.

In practice this density-dependent logistic curve has been verified in a certain number of experimental populations of simple organisms, which has helped to maintain the idea of the balance of nature. But it presents limitations for general use. In fact, population growth is not a continuous phenomenon. The age structure may be variable, the biotic capacity of environments often fluctuates, and there may be delayed effects. The result is that population growth in a natural environment often presents fluctuations that appear random.

10.3.2. Equilibrium theories in relation to interspecific competition

The term *interspecific competition* refers to competition between individuals belonging to different species for the use of a limited quantity of resources, whether food resources or territory (Dajoz, 1996). The result may be, as for intraspecific competition, a reduction in growth, survival, and fertility for one or the other species present.

The principle of *competitive exclusion*, which originated in Darwin's *Origin of Species*, is one of the primary axioms of modern ecology. It seems to reflect the intuitive belief that two species that are very similar as to ecological characteristics cannot coexist over long periods of time (Chesson, 1991). The popular feeling that there is no room for two summarizes this idea that in a stable environment, when two or more species compete for the same limited quantity of resources, they cannot coexist and the most able will eliminate the others. This principle is based on the Lotka-Volterra equations (Lotka, 1925; Volterra, 1926), it being understood that the study of phenomena of interspecific competition proceed from mathematical analysis, from a logistical model, in studying for example the effects of the growth of a population A on the growth of a population B, and vice versa. The model recognizes that each of the two species has a logistic growth and that the maximal load of the environment is K_1 for the first species and K_2 for the second. If the coefficients of competition α and β represent respectively the inhibitor effect of species 1 on species 2, and that of species 2 on species 1, the following equation can be written:

$$dN_1/dt = r_1 N_1(1 - N_1/K_1 - \alpha N_2/K_1) \text{ and } dN_2/dt = r_2 N_2(1 - N_2/K_2 - \beta N_1/K_2)$$

It can be shown that when $K_2 < K_1/\alpha$ and $K_1 < K_2/\beta$, there is a stable equilibrium between the two species.

The model of competitive exclusion has been the object of a great many experimental demonstrations in the laboratory but it is fairly impractical in the field. The exclusion-coexistence alternative merits some closer study rather in terms of degree of coexistence or compromise. It is probable that the dominated species is rarely totally eliminated and many results seem to show that elimination of a species is perhaps a very long process that extends over many generations (Huston, 1994).

Some ecologists nevertheless have attempted to state that these models, which mimic the behaviour of real populations, should be realistic representations of population dynamics. The interspecific equilibrium theory elaborated on these bases thus postulates that different populations in a community will evolve in response to competition pressures. Even though it has been highly successful in the view of some ecologists, this theory manifestly corresponds to a reductive vision of reality because abiotic factors are practically ignored. The theory supposes the following:

— The characteristics of the life cycle of species can be correctly summarized by growth rates *per capita* of the population.

— Deterministic equations are applicable to model population growth.

— The environment is homogeneous in spatial terms and its temporal fluctuations are not taken into account.

— Competition is the only biological interaction that is really important (Chesson and Case, 1986).

Another approach consists of interpreting situations that, in nature, take up species that are supposed to be concurrent for certain common resources. This process, which often starts from a theoretical point of view (theory of ecological niche, hypothesis of competitive exclusion) may be complemented by field experiments. Hutchinson (1965), in particular, largely defended the idea that competition is the principal limiting factor of species diversity and that it leads to the emergence of "patterns" in the structure of communities. He also popularized the notion of "ecological niche" (see Chapter 12) by defining this niche as a hypervolume of n dimensions, each one corresponding to biological or ecological needs of the species considered.

There are some situations in which the existence of mechanisms of competitive interaction between two or more species has been shown (Barbault, 1997). However, given the methodological difficulties, it is not known what role these phenomena of interspecific competition actually play in the general dynamics of ecosystems. It is probable that they are not very important in relation to other factors, especially in ecosystems where disturbances create a strong spatio-temporal dynamics. In fact, over long periods, the environment is rarely constant, and changes in ecological conditions modify the competitive relations between species.

10.3.3. The role of predation

One way to limit the growth of a population is through predation. Theories about the role of predation in the regulation of communities are, in a way, a particular aspect of the general theory of competition. The dynamics of predator-prey systems is however complex and depends on local conditions, biotic as well as abiotic.

A famous model of the predator-prey interaction is the Lotka-Volterra model. It simulates the dynamics of the abundance of a prey and a predator in theoretical conditions in which the predator has only one prey and the prey population is not limited by the availability of food. Applied to the lynx and the Arctic hare, according to the number of furs recorded in the registers of the Hudson Bay Company over almost a century, this model takes into account prominent cyclical fluctuations observed in nature. One rarely finds such a simple situation. In particular, a predator may consume many varieties of fish depending on their relative availability, which does not necessarily imply that it exerts a limiting effect on each of those prey populations.

One hypothesis is that predators take the quantity of prey they need without touching their "capital". They thus have a constant supply of food. This is what Slobodkin (1986) called the *optimal strategy of predation*. But such a situation can no longer be generalized because there are many situations in which predators effectively limit the number of prey. Another theory suggests that in limiting the abundance of prey, the predators thus allow the coexistence of a larger number of species. In a famous experiment, Paine (1966) showed on the Pacific coast of the United States that the experimental suppression of a super-predator, the sea star *Pisaster ochhraceus*, led to a simplification of the biological system by allowing the expansion of a mussel (*Mytilus californianus*), whose population was normally controlled by the sea star, on the rocky shores. The biological community composed of 15 species at the outset was reduced to 8 species at the end of the experiment because the mussel eliminated several species by competition for space.

10.3.4. The MacArthur and Wilson model and the theory of dynamic equilibrium

The *theory of dynamic equilibrium* (also called the island biogeography theory) was proposed by Preston (1962) and then developed by MacArthur and Wilson (1963, 1967). It was applied at first to island environments. The basic hypothesis was simple: the species richness of island biological communities depends on the equilibrium between rates of immigration and extinction, these rates being expressed in number of species per unit of time. The rate of immigration of new species decreases as the number of species established on the island increases. Moreover, some species have greater capacity for colonization than others. The competitive interactions on the island tend to accelerate extinctions. From the growth of these two dynamic processes, we can explain the present richness of the community. We speak of dynamic equilibrium because there may be a replacement of species in this context. The community is not fixed.

The MacArthur and Wilson model predicts the following:

— that the islands close to a source-continent have more species than the distant islands;

— that the larger islands have more species than small islands for situations similar with respect to the source of invasion.

The relation that links the area (A) to the number of species (N) is expressed as follows:

$$N = kS^z \quad \text{or} \quad \log N = \log k + z \log S$$

This theory of island environments has many weak points. It ignores the fact that all the species do not have the same biology and thus the same opportunities for colonization, at least over fairly short periods of time. The theory can logically be applied only to species that can move easily or actively (birds, insects) or that can be transported by the wind or by other species. We must add that it does not allow for speciation *in situ*, that it ignores the relationships between species and environmental characteristics, that it does not take into account the size of the populations present. Despite all these limitations, this simplistic theory has a great heuristic value and a strong impact on research about the organization and dynamics of populations. It has been applied to continents as well: a mountain peak, an isolated forest, a river or lake basin can be considered "continental islands".

We can cite two examples that may appear to corroborate, at least partly, the island biogeography theory.

• The life-scale experiment conducted by Wilson and Simberlof (1969) in order to test the island biogeography model became a classic in ecology treatises. After the most exhaustive inventory possible of the arthropods inhabiting small islands of mangroves along the coasts of Florida, the fauna of these islands was exterminated. The recolonization was rapid, the number of

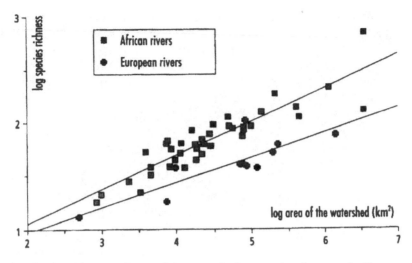

Fig. 10.3. Species richness and area of the watershed: comparison between the European and African rivers.

species stabilizing at the end of 200 days at values similar to those observed before the experiment. However, the taxonomic composition of the new communities was different from that of the original ones, which seems to indicate that the colonization was partly random, the first arrivals being favoured. Despite these taxonomic changes, the trophic structure of new settlements was similar to that of initial settlements, with the same proportions of different trophic groups (herbivores, decomposers, predators), which works in favour of the hypothesis that there are redundant species serving similar functions in the community.

• Relationships between the area of an island system and the species richness of a group were indicated several times. In the example presented in Fig. 10.3, which involves African and European rivers, a close relationship is indicated between the area of the watershed or the flow of rivers and the species richness of fish in these watersheds (Hugueny, 1989; Oberdorff et al., 1995).

The indication of a relationship between area and number of species is a result of an analysis that does not, however, explain its causes. Some authors believe that the small number of species in small islands corresponds to a higher rate of extinction because of the reduced number of populations. But a more convincing explanation is that habitat diversity is greater in a larger watershed, and that the settlements are richer when the habitats are more diversified (see Chapter 9).

10.3.5. Cooperation between species: the Japanese view

Ecology was largely dominated from the 1950s to the 1970s by the idea that competition was the principal factor of interaction between species. Since then, many observations have come to demonstrate that mutualist interactions were much more numerous than was earlier thought. The idea of mutualism was advanced by a Russian, P. Kropotkine, as early as 1906 in a work called *l'Entraide*. The author regretted that Darwin excessively emphasized the "struggle for survival" under the influence of the ideas of Malthus, without seeing that living things cooperate to better ensure their collective survival. A Japanese scientist, Kinji Imanishi, also proposed a theory of evolution that opposed the ideas of Darwin (Thuillier, 1988). He considered that it is the group and not the individual that constitutes the primary reality and that harmony is more frequent in the plant and animal world than competition and the struggle for survival (Ito, 1991). In nature, living things cooperate and collectively take over territories by practising mutualism rather than competition (Thuillier, 1988). Without going into great detail, we can cite the following examples (Faurie et al., 1998):

— Associations between plants and microorganisms, such as mycorrhizae on roots, which facilitate the transfer of minerals into higher plants (see Chapter 12).

— The role of insects in pollination and seed dispersal.

— The role of mammals and birds in the dissemination of fruits and seeds in tropical forests.

— Lichens formed by the association of an alga and a fungus.

— Protozoa capable of hydrolysing cellulose living in symbiosis with colonizing insects or in the stomachs of ruminants.

— The association between corals and unicellular algae of the zooxanthella group.

These mutualist relationships are as important as relations of competition in explaining the structure and the functioning of ecosystems. The disappearance of pollinators may lead to the disappearance of plants, just as the absence of certain vertebrates in the rainforests would make it impossible for seeds of tree species to spread and germinate. Plants that lack mycorrhizae grow more slowly than those that have mycorrhizae. We may thus consider that symbiotic relationships, products of coevolution, result in the following:

— They reinforce the cohesion of biological systems and improve resource use for the benefit of partners.

— They reinforce the relationships between species, thus helping to increase the resilience of ecosystems in the face of disturbances.

10.3.6. Ecological saturation and biotic interactions

If interspecific competition and other types of interaction actually act as structuring forces within biological communities, it must be possible to find situations in which these interactions limit the number of species that may coexist locally in a stable community. For example, in biogeographic zones rich in species, we may expect that the species richness of communities locally reaches a limit value, or a level of saturation, no matter how many species are present in the region (Cornell and Lawton, 1992). Theoretically, two types of results are actually expected.

- There is a linear relationship between the local richness and the richness observed at the regional level. In other words, the local species richness is independent of biotic interactions intervening at the local habitat level and proportionately increases with the regional species richness. There are non-saturated communities in which interactions between species are not sufficient to limit the local species richness.

- There is a limit value for the local species richness. This value may at first increase with the increase in regional species richness, and then stabilize and become independent of the regional species richness. In this case, the curve is asymptotic and we have a community that is considered to be saturated in which it becomes difficult to add species without eliminating other species (Fig. 10.4).

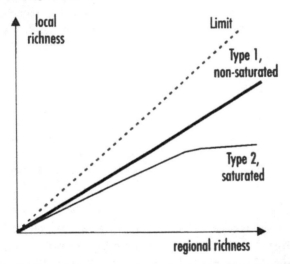

Fig. 10.4. Theoretical relationships between the local and regional species richness in saturated communities (asymptote) and non-saturated communities (linear relationship)

These hypotheses are nevertheless difficult to test directly in natural conditions and ecologists must use indirect methods (Oberdorff et al., 1998). A relatively simple test to verify the saturation hypothesis consists of comparing the local species richness to the regional species richness in similar habitats belonging to different geographic zones having a different pool of species. For example, the non-saturation hypothesis has been tested on fish settlements of the West African rivers belonging to the same biogeographic zone (Hugueny and Paugy, 1995) (Fig. 10.5). There is a close relationship between the area of the watershed and the species richness, which varies in this case between 18 and 95 species. The analysis of the local species richness observed in various

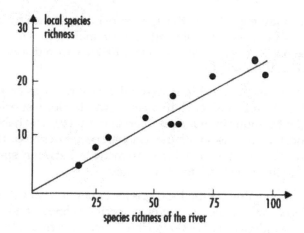

Fig. 10.5. Relationship between the fish species richness in the tributaries and the number of fish species present in the watershed of West African rivers (Hugueny and Paugy, 1995)

tributaries of these rivers indicates a strong relationship between the local species richness and that of the river, demonstrating that the communities are not saturated. In other words, the factors that control species richness on the scale of the river are the same as those that control the local structure of communities.

10.4. SUCCESSION THEORIES

An ecosystem is not a fixed structure determined once for all. Like a living organism (but there ends the comparison), it is a structure that is born, develops, acquires complexity and properties, and then eventually dies and disappears. Unlike living things, however, an ecosystem may rejuvenate itself or change in nature.

10.4.1. Successions

The emergence of the concept of succession in the early 20th century introduced a "time" dimension to the rather static perception of plant geography that had prevailed till then, thus opening the way to research on the temporal dynamics of communities. The notion of succession was used for the first time by the French author Dureau de la Malle (1825), but it was the American ecologist F. Clements (1916) who proposed a global system of description and interpretation of the dynamics of vegetation. The term *succession* referred to the process of colonization of a biotope by living things and changes over time in the floristic or faunistic composition of a station after a disturbance had partly or totally destroyed the pre-existing ecosystem (Lepart and Escarre, 1983).

A succession may, theoretically and simply stated, comprise the following stages:

— In a newly created virgin environment (*juvenile ecosystem*) or one that has just undergone a disturbance that eliminated a large proportion of the species in it, it is the pioneer or opportunist species that establish themselves. They are characterized by a rapid rate of multiplication (demographic strategies of the r type) and are poorly specialized. The food chains are simple.

— The biocoenosis diversifies with the appearance of species with a slower growth (demographic strategies of the *K* type), and the food chains become more complex.

— In the *mature stage*, there is a great species richness, and many of the species have slow growth and a long life span. The networks of interactions and the food webs are complex. Productivity is low and matter is for the most part cycled in place.

— In some cases, there is an *ageing* of ecosystems, notably when a small number of species predominate and eliminate the others. A particular form of this ageing is the disappearance of a lake in favour of a plant formation, when the lake gradually recedes and disappears.

"*r*" and "*K*" demographic strategies

Natural populations have developed many morphological and physiological adaptations to the environments in which they live. These *adaptive strategies* are associated with demographic characteristics: e.g., fertility rates, mortality rates, age at first reproduction, or life expectancy at birth. The objective is simply the survival of the species. Two extreme strategies have been identified, between which most species fall: the "*r*" strategies and the "*K*" strategies. These terms refer to two parameters of the logistic curve: coefficient of exponential increase "*r*" characteristic of the beginning of growth and maximal value of the biomass "*K*" admitting that this tends towards an asymptotic value.

Schematically, we can characterize these strategies as follows:

• "*r*" strategy: Small species with a short life cycle, having a high rate of multiplication and thus a rapid rate of renewal. These are expanding populations with a precocious reproduction that generally rapidly exploit a resource (food, space, etc.) and are limited only by the availability of the resource. They are in strong competition for resources with other species. These mobile and vagabond species colonize variable and unpredictable environments, i.e., "young" ecosystems.

• "*K*" strategy: Species with a long life cycle and a low rate of renewal. These are stationary populations with late reproduction that occupy specialized niches. The species present do not use the same resources and are not in close competition. The life expectancy is high and the juvenile mortality is low, often with protection of young ones. These species, which are economical in their use of energy, establish themselves after the pioneer and species and characterize the mature stages of the succession.

To put it simply, the "*r*" strategists are considered opportunistic, colonizing unstable environments, while the "*K*" strategists are climax species.

In functional terms, we can observe the following during a succession (Frontier, 1977):

— An increase of the total biomass up to the limit of the climax.

— A reduction in the ratio of production over biomass, i.e., the turnover rate of the biomass.

— A reduction in the energy flows per unit of biomass: less and less energy is required to maintain a similar quantity of living matter.

— An increasingly effective cycling of minerals: a young ecosystem cannot live without external inputs and constantly evacuates part of its matter. When the ecosystem becomes mature, the biomass increasingly tends to maintain itself by means of an endogenous cycling of essential chemical elements.

A fundamental characteristic of succession is its irreversibility. A disturbance may in fact lead to the disappearance of a part or all of the species present. If the disturbance is great and intervenes in a mature stage, the cycle of succession is initiated again. There is thus a "rejuvenation" of the ecosystem. *This succession can also occur repeatedly.* In fact, disturbances may take place in a cyclical or random way, with varying intensity. In other words, an ecosystem may undergo a form of rejuvenation several times, and each time the biocoenosis is brought to a stage that is more or less close to the pioneer stage. If the frequency of disturbances is very high, the ecosystem may even remain permanently close to the juvenile stage. This dynamics of succession is a function of time separating the disturbances in relation to the duration of the life cycle of the species that constitute the biocoenosis. Species with a short cycle ("*r*" strategy) rapidly recolonize the environment, while those with a long cycle ("*K*" strategy) have no chance of colonizing it if the interval between two disturbances is shorter than their life cycle.

a) Processes of colonization of a new environment: the case of gravel pits

Stone quarries create new ecosystems in the years following the end of the extraction process when the animal and plant settlements evolve rapidly. They are thus excellent models for the study of processes of colonization and succession.

The 20-year monitoring of bird communities in submerged gravel pits in Bourgogne (France) is an example of ecological succession (Frochot and Godreau, 1995). After the gravel pit was abandoned, birds quickly adopted the new water bodies according to a process that followed the general theories of ecological succession, with pioneer species that reached maximum density in the very first years (the lapwing *Vanellus vanellus* and the little ringed plover *Charadrius dubius*) and others linked to the mature stages (the great crested grebe *Podiceps cristatus*, the Rallidae, the warblers). In the case of nesting species, the total abundance of the population increased rapidly at the beginning of the succession and culminated after 7 to 8 years to decrease and subsequently stabilize at a lower level. The evolution of the species richness followed a similar path: growth until 7-9 years (up to 18 species) and then decrease to end at 8 species after 20 years. The maximum richness observed corresponds to a period of coexistence of pioneer species and species of the mature stages.

The conclusions of many studies on the recolonization of submerged gravel pits by plants and animals can be summarized as follows:

- Spontaneous colonization by flora and fauna was very rapid from pioneer species from nearby aquatic environments.

- Some species that move easily are very good colonizers (birds, insects, plants with wind-borne seeds, such as *Typha*). Other less mobile species such as gastropods or the large freshwater mussels (*Unio, Anodonta*) are also good colonizers. It is likely that birds, the wind, and anglers are responsible for the transport of eggs, larvae, or adults of some species.

- The successions observed in different plant and animal groups follow a graph similar to that of bird populations: a maximum at the end of a few years and then a decrease and stabilization.

Analysis of the plant succession shows a rapid development of pioneer plants, then belts of halophytes that culminate at around 5-10 years to stabilize and even regress, because of the competition exerted by the willow plantation.

b) Stages of succession of riparian woods of the Garonne valley

In the Garonne valley (France), the riparian woods, which succeed from the flood bed of the alluvial plain to the terraces that overhang the river, illustrate a distribution of the vegetation according to an environmental gradient and the phenomenon of succession as a function of time.

- In the flood plain, the processes of erosion and sedimentation interact with varying intensity and frequency, keeping the plant community at the pioneer stages. On the heights of the minor bed, the banks of alluvia are colonized by *Phalaris* on coarse sand and by trifid bur-marigold, and by knotgrass on the fine sand. Immediately above, and up to 3 or 4 m above the low water line, there are willow stands, especially with *Salix alba*, and alder stands with *Alnus glutinosa*. The area flooded during exceptionally high water is dominated by poplar, oak, and ash, including *Populus nigra* and *P. alba*, *Fraxinus excelsior*, and *Quercus robur*. In the flood plain, the plant successions evolve from a pioneer forest of softwood trees towards a hardwood forest in response to changes in the flood regime that modify the processes of erosion and sedimentation. These successions appear reversible, influenced by the high waters and subjected to dominant external processes. The time scale of the successions is about one century.

- The riparian woods of alluvial terraces begin from the low plain (5 to 20 m above the low water line) with mixed woods of deciduous trees, including *Fraxinus angustifolia*. In the higher terraces, the plant successions may

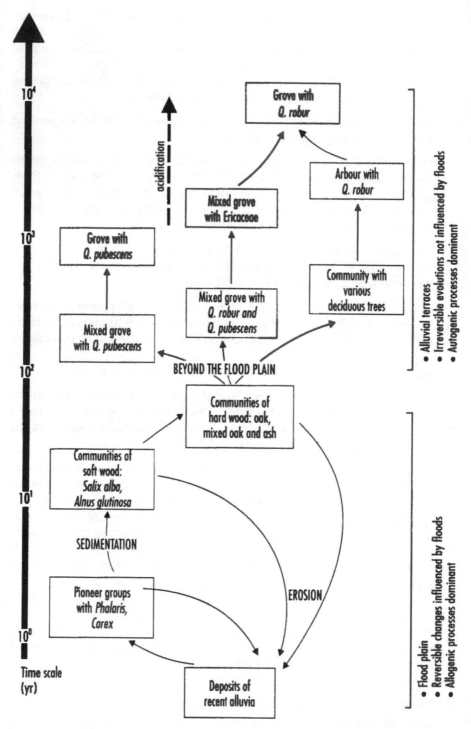

Fig. 10.6. Evolution of forest cover as a function of time in riparian woods of the Garonne valley (modified from Decamps et al., 1988)

proceed in different ways as a function of edaphic conditions (Fig. 10.6) to result in hornbeam and oak stands (*Carpinus betulus* and *Quercus robur*), mixed groves with *Quercus robur* and *Q. pubescens*, or groves with *Q. robur*. These successions are irreversible, not influenced by high water, and subject to predominantly internal processes. They follow a time scale of about a millennium (Decamps et al., 1988).

10.4.2. The concept of climax

In its original sense, *climax is the ultimate stage in the evolution of the vegetation in an ecosystem* (the mature stage) after a succession of intermediate stages, in the absence of natural or anthropic disturbance. It is a stage that is considered an equilibrium between the plant community and its environment, made up of the soil and the climate, toward which any terrestrial ecosystem tends. In the early 20th century, the American ecologist Clements made climax the central concept of ecological theory. The idea was not to classify plant groups according to their phytosociology, but to understand the dynamics that lead the plant formation towards a stage of equilibrium determined by the regional climate. This climatic climax in the temperate zones of western Europe corresponds to different types of middle-European and Atlantic oak stands, mountain beeches, and coniferous forests (spruce, larch, mountain pine) of the subalpine stage.

In the concept of climax applied to ecosystems, we find again the quest for equilibrium that inspired studies on the logistic curve in population ecology. Drouin (1991) notes that the term *climax* signifies "scale" or "staircase" in Greek, and it therefore connotes a gradation, a movement towards an objective. There is in fact an ambiguity because this trajectory exists in the notion of climax, but the term has often been used to denote the end point alone.

10.4.3. Holling's model

Since the 1980s, ecologists have challenged the concept of a well-ordered and unidirectional sequence of communities of species evolving towards a climax, the characteristics of which are determined by climatic and edaphic conditions. The studies conducted in various types of ecosystems make apparent the following:

— There is a non-negligible element of chance with respect to the species that colonize the ecosystem after a disturbance, or during the course of a succession.

— The pioneer species and those that are characteristic of the mature stage may be present constantly throughout the succession.

— Disturbances such as fire, wind, and herbivores are an essential element of the internal dynamics of systems and, in many cases, are at the origin of cycles of succession.

— Some disturbances may lead ecosystems towards different domains of stability, so that there is not just a single possible climax stage.

In sum, the notion of climax is useful but gives only a very static and incomplete perception of the phenomenon. Holling (1986) thus proposed a model that emphasizes the four primary stages in the cycle of a terrestrial ecosystem (Fig. 10.7). The traditional perception of the succession is under the control of two functions: an *exploitation* phase that corresponds to the rapid colonization of recently disturbed zones, and a *conservation* phase that corresponds to the slow accumulation and storage of energy and matter. There are two additional functions: a *distribution* phase during which the biomass and nutrients accumulated become increasingly fragile until they are abruptly released by factors such as forest fires, epidemics, or periods of intensive grazing. In the last phase, called *reorganization*, the processes of mobilization and immobilization minimize nutrient losses in the soils and make them available for a new phase of exploitation.

The progression in the cycle of the ecosystem proposed by Holling starts from the exploitation phase and evolves slowly towards the conservation phase, then very rapidly towards distribution and reorganization. During the transition from exploitation to conservation, the stability of the system increases and a "capital" of biomass and nutrients slowly accumulates. It is abruptly destroyed and released, to be ultimately reused.

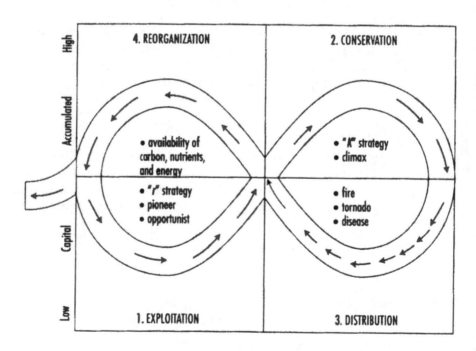

Fig. 10.7. The four functions of an ecosystem and the relationships between them. The closest arrows indicate that the system evolves rapidly. Those further apart indicate that it evolves slowly (Holling, 1986).

10.5. ECOSYSTEM STABILITY AND RESILIENCE

Changes in the state of ecosystems interest ecologists particularly since they are concerned with the consequences of human activities on natural environments. The respective roles of natural factors must be understood in relation to the anthropic factors in the changes observed, and we must explore the capacities of ecosystems to "resist" impacts, i.e., to absorb them and maintain them in a more or less identical (or stable) state over the long term.

In reality, many studies on the impacts of human activities are developed from simple ideas.

- The existence of a reference situation, an ecosystem that is presumed to be perennial, in relation to which the nature and significance of these impacts can be estimated. This simplistic vision, which has the advantage of being easy for non-specialists to understand, is evidently based on the old principle of the balance of nature. It is widely challenged today.

- The possibility of following in a simple manner the changes in the state of an ecosystem, measuring a certain number of variables that are supposed to be relevant, to characterize the state of the system and its trajectory over the long term. However, this idea, which is also a simple one, is not really operational. It is difficult to indicate the precocious changes and small-scale phenomena, so that when large-scale changes become visible the ecosystem is already engaged in major transformations.

In any case, ecologists have used a wide variety of terms to characterize the state of a system as a function of time, without always defining them precisely (Fig. 10.8). Even though we now know that variability and long-term changes of communities and ecosystems are the rule, the discourses of ecologists still remain mostly centred on the notion of stability. In reality, the myth of the reference system, based on arguable premises, tends to persist.

10.5.1. Ambiguous terms

It is hardly surprising that the term *stability* has never been satisfactorily defined with respect to ecosystems because it covers different notions. For many ecologists, the notions of stability and resilience essentially concern biological diversity. They thus speak of the stability of communities without explicit reference to abiotic factors. Others are concerned with biogeochemical flows and biological productivity. Still others are interested in the ecosystem as a whole through the bias of abiotic factors.

Another ambiguity is that stability is not necessarily a static state but can be defined in dynamic terms. It results in a way from a dynamic equilibrium of populations, communities, and abiotic factors (Peters, 1991). In reality, we often consider that a community or ecosystem is stable when we cannot detect significant changes in the species composition and relative sizes of the populations it consists of, at least during the observation period. This

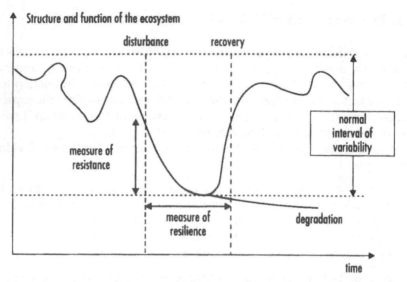

Fig. 10.8. Theoretical representation of responses (resilience and resistance) of an ecosystem to a disturbance (modified from Vogt et al., 1997)

immediately poses the question of the temporal scale at which observations are made, in relation to the duration of biological cycles of the constituent species and/or the intensity of global changes. The notion of stability in ecosystem ecology is necessarily relative to a time scale.

Stability

Attempts to define the term *stability* have concluded that it is too complex to be defined precisely and must therefore be broken down into smaller concepts. Thus, we distinguish the following:

• *Resilience* or *homeostasis*: A term borrowed from mechanics that expresses the capacity of an ecosystem to regain its primitive structure after having been affected by a disturbance. This resilience is a function of the intensity and frequency of disturbances, the more or less great isolation of the system, the presence of toxic substances, and so on. If a system has been greatly affected by a disturbance, it possibly will never return to a state similar to its initial state. Sometimes the term *elasticity* is used to describe the speed with which a system returns to its original state after a disturbance, and *amplitude* to describe the maximum scope of the disturbance that still allows a return to the original state. The implicit admission is thus that, beyond that scope, the system topples into another state.

• *Resistance, persistence*, or *remanence*: The capacity of an ecosystem to remain constant or to manifest a limited response to environmental variations. In the same order of idea, the term *inertia* is used to describe resistance to disturbances.

• *Reversibility/irreversibility:* When an ecosystem cannot return by natural processes to its earlier state, we say that it has crossed a "threshold of reversibility". This concept is complementary to that of resilience, because it defines the limits of a system's reaction capacity.

The concepts of resilience and resistance are thus a modern version of the notion of equilibrium. They attempt to describe how a system can respond to a disturbance and return to its initial state. Unfortunately, none of these concepts has been precisely defined so as to be really operational.

At present, the term *stability*, like *niche* or *environment*, is one of the elements in a network of more or less interlinked and sometimes redundant concepts, in which the real meaning often is based on the intuition and subjectivity of the user. In the absence of a precise definition, the meanings evolve in the field in conceptual and terminological terms. The words that express these concepts and belong to a common language may be interpreted differently by scientists and the users of ecology, which inevitably leads to incomprehension. Moreover, the concepts have now become commonplace to the point that they are accepted as inherent properties of nature (Peters, 1991). This example illustrates the difficulty that ecology has in following a path similar to that of the physical sciences (Fig. 10.9).

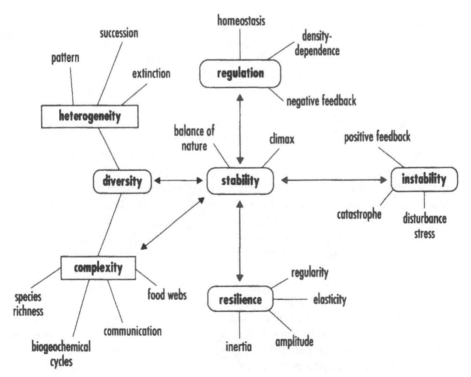

Fig. 10.9. Subjective and non-exhaustive representation of relationships between different concepts associated with the notion of stability (inspired by Peters, 1991)

In critically examining several long-term studies from the ecological literature, Connell and Sousa (1983) concluded that there are hardly convincing data to support the hypothesis of stable communities that can result in states of equilibrium. They also proposed to replace the concept of stability by the concept of *persistence*, which is more appropriate to ecological systems, on the condition that it is based within the limits of variation established in a stochastic manner. It is a notion very similar to that of the *permanence* of a system, which is applied to systems that perpetuate themselves in time following more or less complex dynamics, without any of the components disappearing (Pave, 1994).

10.5.2. Complexity and stability

A subject of animated debate among ecologists is the postulate that ecosystems are "stable" to the extent that they are diversified. The more species there are, the more interactions there are among them, and the more the ecosystem seems apt to resist disturbances (MacArthur and Wilson, 1955). It is true that highly diversified ecosystems such as rainforests and coral reefs give the impression of a great stability of biomass, at least on the scale of human observation. For a long time, ecologists believed that the increasing complexity of an ecosystem led to greater stability (Elton, 1958). According to Colinvaux (1982), the essence of the stability-complexity theory is a simple idea whose only unusual and delicate aspect is that the complex is stable. He adds that naturalists are greatly attracted by this theory because it corresponds closely to the intuitive idea that nature, which is complex, functions harmoniously. Again, the old notion of the economy of nature underlies this discourse.

Thus, in theories of succession and climax, there is an implicit hypothesis that the diversity and stability of an ecosystem increase as it approaches its maturity, the climax, a stage at which a complex structure is capable of regulating itself by means of the diversity of its components. The complexity-stability hypothesis, which has not really been confirmed by observations or experiment, assumes that if the number of relationships increases within a network of interactions, the disappearance of one link will be rapidly compensated for by the establishment of another. In other words, if a species disappears provisionally or definitively from an ecosystem rich in species, another species will take its place (the niche) and fill similar ecological functions. This is in reality a transposition to ecology of principles derived from information theory, developed to analyse the functioning of complex networks. The networks are stable to the extent that their nodes, which can open channels of substitution, are more abundant (Deleage, 1992) (see Chapter 5).

Mathematicians have made many attempts to simulate the behaviour of systems that are more or less rich in species. They have for the most part obtained results similar to those of May, who in 1972 presented his results, entirely opposed to the ideas received till then: when a system becomes more complex in the sense that it contains more species, and the interactions become more diversified, it also becomes more fragile from the dynamic point of view. In other words, according to May, complexity leads to instability. At the time this appeared to be a total challenge to the predominant ideas in the field. May's hypothesis prevailed in the ecological literature for many years and many modellers helped to support the idea that the complexity of ecological systems reduced their capacity to react to disturbances, in such a way that complex systems tend to disappear.

Even though the hypothesis that nature does not like complexity goes counter to the intuition of many ecologists, there were not many objective arguments to oppose the results from models. However, ecologists did not remain inactive, and this hypothesis provoked varying reactions (Polis, 1994). It is now recognized that it does not correspond to the reality. For example, critics emphasized the fact that the communities used in the models of May were put together at random, while that is not the case with natural communities, in which the interactions between species are the result of a coevolution that has probably selected the most stable associations at the cost of unstable associations. Other critics, in a similar way, pointed out the failure to consider a hierarchical organization of the network of interactions, which is an essential characteristic of ecosystems (Frontier and Pichod-Viale, 1998). A system that comprises many distinct elements can exist only if it is made up of a small number of interactive subsystems, which are themselves formed of a few elements, and so on. In these conditions, the subsystems made up of small sets of species that are closely interconnected form highly integrated and highly stable systems and the same would be true for a set of subsystems. This hierarchical organization of relatively autonomous entities explains why highly complex systems are actually persistent.

Recent results have to some extent supported the hypothesis that biological diversity has a positive effect on the functioning of ecosystems. Laboratory and field experiments have shown that greater species richness can increase the primary productivity and retention of nutrients in the ecosystem (Tilman et al., 1996) (see Chapter 12). Besides, the models that earlier disturbed ecologists have more recently come to their aid. Using a series of non-linear mathematical models and incorporating biological hypotheses that are more realistic than those used in earlier models, McCann et al. (1998) ended up with results opposite to those of May. They were able to show that complexity tends to stabilize ecosystems, amortizing the fluctuations of populations and limiting the loss of species. In fact, the many but weak linkages that exist between one species and the community almost always, in different situations, help stabilize the dynamics of that species. These results seem more satisfactory than May's for many ecologists because they confirm their intuition. A debate that began with a clumsy approach but nevertheless received wide publicity in the scientific community seems closed, at least temporarily. We can recall in this context Legay's statement that models cannot be right, cannot be exact, and cannot mislead us; there is no model that is false. However, those who use models may be misled. The user is entirely responsible for the choice of model and the hypotheses that support it. The use of an inappropriate model may mislead the researcher and become an obstacle on the road to knowledge.

The stability-complexity relationship is not neutral in operational terms. In fact, although there is no obvious direct link between the species richness and stability of an ecosystem, we must draw conclusions in terms of the politics of environment protection (Drouin, 1991). On the contrary, to demonstrate that complexity is important for the integrity and stability of natural systems provides arguments to ecologists as to the need to preserve the various species that coexist in systems considered to be ecologically healthy. We can thus better understand the fierceness of the debate.

In reality, there are arguments to show that a certain "stability" of environmental conditions over the long term, or rather a certain permanence of ecosystems, leads to biological diversification. Diverse communities can moreover reflect more than others the phenomena of speciation and coevolution proper to perennial environments. A good example is the ancient lakes and particularly the large lakes of East Africa, where fish and invertebrate communities that are highly specialized in ecological terms have been able to survive for millions of years (Lévêque, 1997). These lakes contain many endemics. Inversely, in environments characterized by a high variability of environmental conditions, only robust communities of species with a vast geographic distribution will survive. This is true for fish communities of the Sahelo-Sudan rivers that are subjected to seasonal cycles characterized by severe low water and to long-term fluctuations that sometimes lead to the near-complete disappearance of hydrosystems.

10.6. THE DYNAMIC EQUILIBRIUM OF ECOSYSTEMS AND DISTURBANCES

Ecology has for a long time involved, for practical and conceptual reasons, the study of homogeneous environments that are more or less autonomous and independent from one another. The reality is quite different. Natural environments are heterogeneous, often fragmented, and they change over time. Ecologists now concur in recognizing that the environmental variability and heterogeneity are essential to the long-term stability of communities. Thus, we transcend the population approach to take into consideration a certain number of parameters that structure the environment, such as soil heterogeneity, fluctuations of habitats linked to flooding of rivers, and the connectivity between subsystems, the synergistic action of which maintains the cohesion of the entire system over the long term.

Long-term studies of ecosystems show that their state depends, at any time, on their history and the dynamics of their environment. They oscillate more or less regularly around an average state, with greater or lesser amplitude. These fluctuations result either from random influences of the environmental framework or auxiliary processes, or from the system's own dynamics. Rather than being in equilibrium, an ecosystem is in reality an interactive system. A change in the environment creates a dynamic response of the system as a whole, with many positive and/or negative feedbacks. Moreover, the ecosystem may exert an influence on its environment, which implies a reciprocal relationship and that the first is not entirely subordinate to the second (Golley, 1993). This dualistic vision of the behaviour of ecosystems contrasts greatly with Tansley's original conception of a deterministic system.

It is interesting to observe that the stability or resilience of ecosystems has primarily been studied from the perspective of biological sciences. Golley (1993) caricatured the attitude of limnologists with respect to this question by recalling that the principal objective of the study of lakes has been to characterize the equilibrium state of these ecosystems. However, since many limnologists were trained as biologists, they have tended to respond to this question by looking only at biological elements of the system. Still, they have pertinently pointed out that the functioning of a lake ecosystem depends on physical and chemical factors such as the hydric balance of inputs and outputs. However, in reality, they believe that the functioning of a whole can be evaluated by the study of the performance of just one of its elements, the biocoenosis. This example clearly illustrates the operational difficulties of a holistic approach to ecosystems.

10.6.1. What is a disturbance?

The recognition of a structuring role of disturbances in the dynamics of communities and ecosystems is one of the most interesting challenges of ecological paradigms in the past two decades. It is also an approach that is different from the idea of equilibrium (Reice, 1994). A general definition of disturbance was proposed by Pickett and White (1985) and modified by Resh et al. (1988): "any relatively discrete event in time that is characterised by a frequency, intensity, and severity outside a predictable range, and that disrupts ecosystems, community, or population structure, and changes resources, availability of substratum, or the physical environment."

According to Townsend (1989), it is "any relatively discrete event in time that removes organisms and opens up space which can be colonized by individuals of the same or different species."

Sousa (1984) considers a disturbance to be "a discrete, punctuated killing, displacement, or damaging of one or more individuals (or colonies) that directly or indirectly creates an opportunity for new individuals (or colonies) to become established."

According to these definitions, disturbance is thus a generally physical phenomenon, and not the biological consequence of this event, which is the response to the disturbance. Predation is in theory not a disturbance because it is intrinsic to the life of a prey species, which must adapt to it. Nevertheless, voluntary introductions of new predatory species in an ecosystem can be considered a disturbance for this ecosystem because new predators and their prey have not manifestly had the time to co-evolve. Moreover, if an event is predictable, the biota is adapted to it. This is true of high water periods in tropical rivers with large flood plains. However, if this predictable event has not taken place, the impact on the system may be highly significant and that is therefore the equivalent of a disturbance.

Disturbances can be characterized by various descriptors: the type (physical, biological), the regime (spatial distribution, frequency, intensity, duration), and the regional context. Depending on their nature and intensity, some disturbances do not evoke any response from the ecosystem, while others can be called catastrophic, for example, when the habitat and its settlement are destroyed. Most often, a disturbance leads to a general restructuring of the ecosystem. In this regard, it is a process of re-initiation in the phenomenon of ecological succession.

10.6.2. Intermediate disturbance hypothesis

Hutchinson, an ardent advocate of theories of competition, observed some years later that the model does not apply to phytoplankton populations in lakes. In a famous publication (Hutchinson, 1961), he spoke of the paradox of plankton, observing that in a relatively simple environment, many more species of phytoplankton coexist than could be predicted by the theory of competition and the limiting effect of resources. He thus logically posed the inverse question: is not the great species richness observed the consequence of temporal fluctuations of the environment that prevent it from reaching a state of equilibrium? That is, the coexistence would be the result of non-equilibrium phenomena rather than equilibrium phenomena. These new ideas did not at first receive enough attention in a context in which theories of equilibrium constituted the dominant paradigm. However, they anticipated later developments about the impact of disturbances on the species richness of ecosystems.

In fact, many ecological studies concluded that disturbances could favour biological diversity by reducing the pressure of dominant species over other species, thus allowing the latter to develop. *The intermediate disturbance hypothesis, proposed by Connell (1978), predicted that species richness would be greater in communities with an average level of disturbance than in communities that are not disturbed or that undergo very significant or very frequent disturbances.* This hypothesis aimed at first to explain the great species richness of tropical forests and coral reefs. When disturbances were less frequent, interspecific competition limited the number of species that could establish themselves, and the more competitive (resident) species occupied the available space. Inversely, when the disturbances were frequent and/or intensive, the dominant competitive species were eliminated and only colonizing species with a short cycle could survive. For a regime of disturbances of intermediate frequency, intensity, and amplitude, the resident species and the pioneer species cohabit, which favours a greater species richness.

Fallen trees, factors that maintain the forest biodiversity

In forest environments, natural regeneration often follows a disturbance of the canopy caused by the accidental fall of a tree, which locally causes a greater insolation of the undergrowth. This fall may result from violent winds or from the death of the tree base. Such events, which have a random spatial distribution and occurrence, favour the establishment of "new" species and help maintain the biological diversity. Recolonization of the new spaces occurs from:

• the seed bank in the soil made up of all the seeds that have fallen near the fallen tree before the fall, and that remain dormant while waiting for favourable light conditions;

• seeds coming from elsewhere after the fall of the tree; and

• shoots from existing trees.

10.6.3. Non-equilibrium dynamics

On the basis of the intermediate disturbance hypothesis, Huston (1994) proposed a model of *dynamic equilibrium* that applied, for a given habitat, to the species richness of competitive species belonging to a single functional type and in competition with one another (Fig. 10.10). The model is based on the relative importance of two opposing forces that influence the species diversity:

— competition that leads to exclusion in situations of high growth rates of populations, with low intensity of disturbances;

— severe or frequent disturbances generating mortalities and limiting reproduction in populations that grow slowly, so that the species do not have time to re-establish themselves.

In the real world, many systems are disturbed frequently enough (by fire, tornadoes, drought, etc.) for the re-establishment of communities to be regularly challenged.

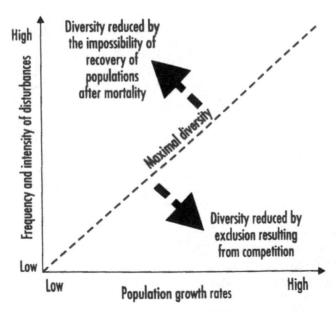

Fig. 10.10. Equilibrium dynamics model of Huston (1979)

The model does not predict a constant level of biological diversity for a certain number of parameters but rather a pattern of fluctuations within limits fixed by the dynamic equilibrium model. Therefore, it is analogous to the model of the island biogeography theory (in which the opposing forces are immigration and extinction) but involves smaller spatial and temporal dimensions.

Actually, the dynamics of non-equilibrium seems best to explain the spatial and temporal heterogeneity observed in most ecological systems. Disturbances and heterogeneity are interdependent factors that create opportunities for recolonization and determine the structure of communities. In such a context of a constantly variable environment, competition and predation and, generally, interspecific relationships seem to have little influence on the structure of communities, except in certain circumstances and during limited periods of relative stability.

Some theories, such as patch dynamics (see Chapter 9), rely on the hypothesis that interspecific competition is an important factor of the structuring of communities. However, in a fragmented and heterogeneous system, competition between species varies in time and in space under the influence of environmental factors. Other theories have a different approach and start from the principle that it is the abiotic factors that determine the dynamics of communities. The significance and speed of changes observed depend directly on the amplitude and frequency of environmental fluctuations. This is true of responses to climatic changes, which confirm the results of palaeoecology. The biological characteristics of species and their capacities of dispersal will, however, have a significant effect on their capacities to respond to environmental changes. Trees with a long life cycle can tolerate locally unfavourable conditions for a long time, while invertebrates with a short life cycle will respond much more rapidly by demographic changes.

10.6.4. Buffering and recovery capacities of ecosystems

The buffer capacity or regulatory capacity of an ecosystem is its ability to implement internal mechanisms that ameliorate the impact of a disturbance. However, another type of question raised by research on disturbed environments concerns the recuperative capacities of ecosystems after disturbance. Many efforts have been devoted to evaluate impacts of various disturbances on the structure or functioning of ecosystems, but studies on the processes of long-term recovery of disturbed ecosystems are much more rare.

This capacity of ecosystems subjected to severe disturbances to recover from them depends on biological characteristics of the populations that constitute them: rates of generation, fertility, dispersal capacity, etc. (or even demographic strategies). However, there are other factors independent of the nature of organisms that are involved, such as modifications of the habitat, residual toxicity, the duration and time of impact with respect to the time of reproduction, and the presence and distance of refuge zones (Niemi et al., 1990). We should add that the recovery capacity of ecological systems depends partly on the existence of refuge zones for the flora and fauna, which serve as reservoirs of recolonizers after the disturbance (Hildrew, 1996).

An experiment on reconstitution of plant cover on fallow land in Ouagadougou (Burkina Faso) showed the capacity of the vegetation to reconstitute itself on soils exhausted by many years of cultivation, stripped by erosion, in which the infiltration of rainwater was reduced by the surface crust. Despite this unfavourable environment, in 15 to 25 years the land left fallow was restored to the vegetation it had before the abandonment of cultivation, for the time that the human and animal pressure was suppressed (Achard, 1994). Another example of recovery of marine ecosystems after a major stress is that of the oil spills from the *Amoco Cadiz* accident.

Oil spills

Oil spills are catastrophic events that punctuate the history of maritime transport of fossil fuels. The *Torrey Canyon* accident of 1967 has been forgotten, as has that of the *Amoco Cadiz* in March 1978, and even the *Erika* accident in December 1999, which polluted the Brittany coastlines. But there is something to be learned from these involuntary true-to-scale experiments that were highly publicized and said to be large-scale ecological disasters.

The study of medium- and long-term consequences of petroleum pollution shows the importance of phenomena of ecological succession and regulation in the restoration of disturbed ecosystems (Laubier, 1991). If we simplify it extremely, after the acute toxicity phase, oils spills have consequences comparable to those that follow the release of organic effluents. With respect to the *Amoco Cadiz*, in stations highly exposed to the pollution, an immediate and total destruction of the macrofauna was observed. After a few months, the fossil fuels lost their toxicity, and anoxic conditions prevailed in the contaminated sediments. A substitution fauna established itself, adapted to anaerobic conditions and the high organic matter content of the sediment. These opportunistic species proliferated, and sometimes species with high biomass and production established themselves. Then, tolerant species established themselves, followed by more sensitive species.

After four years, the biological communities were estimated to be restored and, in the following years, the groups of sensitive species became dominant. At the end of seven years, the communities recovered with a composition and structure very similar to that observed before the disturbance.

The effects of oil spills are extensive at the outset, and it takes about six to seven years before the consequences are erased from the site of the accident. Obviously, that does not justify the lack of efforts to prevent such accidents.

10.7. NON-LINEAR SYSTEM DYNAMICS

The 19th century bequeathed to us the classic laws of nature, such as Newton's laws, which are among the best examples of deterministic law: once the initial conditions are given, we can predict any event, past or future. These laws talk of certainties. Until the discovery of deterministic chaos, it was believed that simple equations, designed to simulate a phenomenon in a simple form, would end in a simple dynamic behaviour. In consequence, a modification of the initial conditions would lead to a more or less proportionate response to the significance of the changes. It was also believed that this simplified model furnished a simplified image of the real phenomenon and that it was possible to improve the image by progressively complicating the model. These hypotheses, which reflect "linear philosophy", were invalidated with the advent of the notion of chaos, which introduced concepts of probability, indeterminism, and irreversibility in the laws of nature (Prigogine, 1994).

10.7.1. Bifurcations and attractors

The trajectory of natural systems in time is controlled by two sets of factors: a set of laws concerning the constituents of the system and their interactions, and a set of constraints from the external world that manifest themselves generally by parameters of control . The geosphere-biosphere relationship is an example of a non-linear dynamic system. In fact, the flow of solar energy, or the concentration of the atmosphere in gases in trace amounts, are examples of parameters of

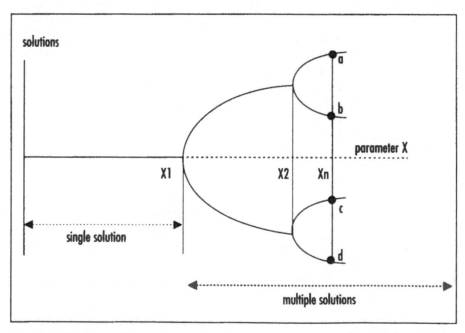

Fig. 10.11. Diagram of bifurcation typical of non-linear dynamic system (Nicolis, 1992)

control. They are not linear functions, notably because of the many feedback processes that exist between the different components of the system (Nicolis, 1992) (see Chapter 5). This is why the fundamental processes that control the dynamics of the system are not constant but vary and become dependent on the state of the system. Non-linearity is the source of complex and unpredictable behaviours in the sense that there is, in theory, more than one possible trajectory.

Figure 10.11 represents a simplified version of the behaviour of a non-linear dynamic system when a control parameter (X) varies. At different values of X (X1, X2, X3), which are called bifurcation points, the preceding solution becomes unstable; in other words, new solutions can be envisaged under the effect of small disturbances. In general, these bifurcations produce a multiplicity of simultaneously possible stages, called attractors (branches a, b, c, d in Fig. 10.11 for the value Xn). The state that will be chosen depends on the initial conditions. The result is that the dynamics of such a system will be an aperiodic succession of intermittent leaps between coexistent attractors (Nicolis, 1992).

Lorenz's butterfly effect

Edward Lorenz addressed the question of predictability in meteorology by constructing a very simple model of atmospheric convection, taking into account three dynamic variables. Wishing to reproduce an evolution that he had already calculated, he decided to start from the middle of the period over which his previous calculation was made. The values of the variables in the middle of this period must be made up of the initial conditions for the new calculation. He decided to retain only three decimal places for these variables instead of six. This approximation seemed to have no consequence since it was of the order of uncertainty of meteorological measurements. Contrary to this expectation, he found that the new calculation led to substantially different results from those he had obtained previously. He thus discovered, by chance, sensitivity to initial conditions. In 1972, Lorenz presented before the American Association for the Advancement of Science a brief communication titled "Predictability: Can the beating of the wings of a butterfly in Brazil trigger a tornado in Texas?" (Lurcat, 1999). This riddle actually expressed a more serious question: is the behaviour of the atmosphere unstable with respect to disturbances of a low amplitude? Today, the chaotic character of the dynamics of the earth's atmosphere seems to be widely acknowledged.

10.7.2. Chaos theory

Chaos theory is largely based on the mathematical theory of dynamic systems, which are systems evolving in an autonomous manner over time. For example, studies on the nature of turbulence by Ruelle and Takens (1981) demonstrate mathematically that a simple system, following simple mathematical laws, does not necessarily have a simple dynamic behaviour and may even have an extremely complicated dynamics. The 1970s were devoted to the experimental demonstration of this mathematical affirmation and the term *chaos* appeared only in 1975. Since then it has caught on.

As remarked by René Thom, the term *chaos* is improper. In common language it signifies the complete absence of order, but it should be reserved for systems with a behaviour that defies description. However, the systems that chaos theory addresses are explicitly described and are, besides, structurally stable. They are thus deterministic, even if they are unpredictable. That is why we speak of *deterministic chaos*.

Variable *x*

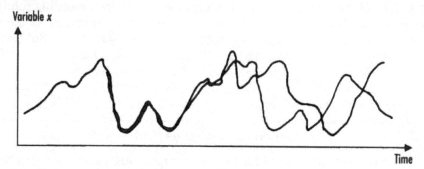

Time

Fig. 10.12. Evolution of a variable from two similar but not identical initial conditions in the case of deterministic chaos (Frontier and Pichod-Viale, 1998)

Chaos is always the consequence of instability. In reality, chaos theory addresses events that are unpredictable because of their sensitivity to initial conditions. Two initial situations that are very similar but distinct lead after a relatively long interval to completely different final situations (Fig. 10.12). A chaotic phenomenon thus has a paradoxical behaviour. It follows deterministic laws but its long-term evolution cannot be predicted because the initial state can never be known with absolute precision. Chaos theory thus opposes Laplacian utopia, according to which at any moment the state of a system can be deduced if we know precisely its initial state (Gleick, 1989). The non-linear models with chaotic behaviour show, in particular, the following (Frontier and Pichod-Viale, 1998):

— that a complicated dynamics may result from a small number of simple and deterministic equations;

— that a simple dynamics can only represent an approximation of a complex dynamics;

— that the dynamics obtained is unpredictable because of its great sensitivity to initial conditions and that it consequently gives the impression of being indeterminate and random.

In the mid-1970s, ecologists interested in the dynamics of plant and animal populations played an important role in the advent of chaos theory. At that time, those who worked on dynamic biological systems, such as predator-prey systems, postulated that such systems tended towards a state of equilibrium. They supposed implicitly that density-dependent regulatory factors tended to control population size. When they observed the presence of disordered, unpredictable, *chaotic* fluctuations, they attributed them to external causes that would be "background noise", such as unpredictable variations of the climate or environment (May, 1991). Nevertheless, some of these ecologists began to say, from the 1970s, that the apparent disorder observed in the dynamics of animal and plant populations could be inherent to these systems. The mathematical models constructed to simulate the behaviour of natural populations, which were of the deterministic type, seemed in effect to give similar results. In closely examining the equations used by fishery experts and entomologists to model the regulatory mechanisms of population size, researchers such as May noticed that as the model was implemented, the final numbers fluctuated indefinitely as if they were random. In other words, for equations based on simple rules that could be used to calculate the value of the number of generation $t + 1$ given that of generation t, the prediction became impossible. It is thus that the concept of *deterministic chaos* emerged in ecology. Later researchers showed that in population biology the mathematical models of population growth and competition led also to chaotic behaviour.

However, it was necessary to ensure that the chaotic conditions simulated by equations corresponded to concrete situations. In a review of the possible role of chaos in ecology, Hastings et al. (1993) concluded that the existence of chaos was highly plausible according to modelling, but that more information was needed to confirm it. The major problem lies in the difficulty of discriminating the chaotic temporal series from those of fluctuations produced by background noise or errors in sampling. Besides, chaotic dynamics appears when a system functions in constant conditions. When these conditions change, the system adapts itself to new conditions and the dynamics becomes different. Nevertheless, in ecosystem ecology, it is quite improbable that the conditions remain unchanged long enough for a chaotic regime to establish itself. Given the widespread variability of environmental conditions, there is a high probability that we observe only transitory regimes of dynamics that could be chaotic in essence, but is incessantly challenged and in a phase of adaptation (Frontier and Pichod-Viale, 1998). In the face of this difficulty, some scientists estimate that chaotic dynamics does not exist in real populations. They add that in creating a risk of gradual extinction of these populations, chaotic fluctuations make it difficult for them to survive in the long term. The question thus remains open.

Nevertheless, certain models of simulation have indicated chaotic behaviour in ecology. For example, according to the results of a simulation model, a simple food chain comprising three species presents a chaotic dynamics in its behaviour in the long term, when biologically pertinent parameters are used (Hastings and Powell, 1991). These results suggest that chaotic behaviour may be more frequent than was thought in natural systems.

The indication of chaotic behaviours has consequences for the strategies of long-term observation of ecosystems. Given the sensitivity to initial conditions, there are intrinsic limits to monitoring and prediction of future states. We must take this into account when using numerical models in order to evaluate the impact of human activities on ecosystems.

10.8. ARE ECOSYSTEMS SELF-ORGANIZED SYSTEMS?

In a deterministic and caricatural manner, we can say that living things are organized systems that pursue a precise objective: to adapt, survive, and reproduce. The different biological functions at all the levels of organization, from genes to organs, must be coordinated and controlled mutually to favour the survival of the organism so that it can reproduce in the best conditions, in spite of the fluctuations of the environment. Otherwise the penalty is simple: the poorly organized biological system, or one incapable of adapting to its environment, permanently disappears. Evolution, in a sequence of trial and error, will thus select the species most apt to survive within certain limits of the characteristics of their environment. We can discuss various aspects of this simplification, but it is a fact that living things are self-regulated systems in which the essential objective, at least as we perceive it, is to ensure the perenniality of the species. It is also clear that, in these systems, the whole is more than the sum of the parts, i.e., that the properties of living things cannot be inferred from the individual properties of the molecules they are made up of.

The essential question posed definitively by ecologists is whether other levels of organization of the living world, especially ecosystems, pursue similar ends to those of living things in the matter of survival and continuity, or whether these levels are only random sets of species in transitory environments. *In other words, are there rules of organization of ecosystems that ensure the cohesion of different parts and their reproducibility in the long term?*

Ecological systems are comparable to self-organized systems to the extent that they are open systems in the thermodynamic sense that maintain themselves far from thermodynamic equilibrium. In a self-organized or self-regulated system, the reciprocal interactions within the system between the structures and the processes help regulate its dynamics and maintain its organization, by means of feedback phenomena. We can call it an *antichaotic* system because, unlike the chaotic systems that are highly sensitive to initial conditions, it channels different initial conditions towards a single final stage (Kauffman, 1993). Such systems will be relatively robust with respect to certain types of disturbances to which their components are adapted, but they will evolve with threshold effects when they go beyond the limits of these adaptations (Perry et al., 1989). This "metastability" (O'Neill et al., 1989) depends not only on internal interactions of the system but also on external forces that could regulate and reinforce the internal factors of cohesion. The result is that the interactions between scales and particularly the limit phenomena are central to our understanding of self-organization in ecology (Perry, 1995).

We can pose various questions about self-regulated systems: Are they frequent in nature? If so, in what spatio-temporal scales? How do they evolve? These questions have so far not received a definitive response because they are particularly difficult.

10.8.1. Homeostasis of ecological systems

Homeostasis is the phenomenon by which a system can spontaneously maintain its internal equilibrium when its environment varies. The principle of homeostasis controlled by feedback is familiar to those who manufacture machines. The study of these processes of control in electronics, mechanics, or biological systems involves cybernetics (see Chapter 5).

Homeostasis is the process of regulation that runs living things. It keeps the general state of the organism constant despite external disturbances. This aptitude has been generalized to ecological systems. The idea that ecosystems have such stability and that it is conferred on them by the diversity of the network was first introduced by MacArthur (1955) to test the hypothesis of E.P. Odum that the great diversity of pathways in an ecosystem and the myriad feedback mechanisms allow us to reconstitute new links of transfer or flows in the event that certain vital links are disturbed. MacArthur proposed to ecologists a quantitative hypothesis that could apparently be tested: the diversity of flows facilitates the homeostasis of ecosystems and ensures the stability of the whole. This hypothesis was received with great enthusiasm in the decades that followed but it faced operational difficulties. In fact, flows are difficult to quantify in ecosystems, so the attention of researchers gradually turns towards the diversity of species. Moreover, it is difficult to quantify homeostasis, so we refer to the notion of stability as it is defined in linear systems. Ecologists thus turned away from the primary idea of MacArthur linked to information theory to end up with a proposition that diversity (of species) generates stability. It is now believed that there is no simple relationship between diversity (species richness) and the stability of ecosystems. There may be ecosystems that are very poor in species but highly stable, and diversified and unstable ecosystems (see Chapter 12). This example shows, however, how an idea, original at first and based on information theory, was progressively turned from its initial direction and even completely changed in essence.

A possible origin of homeostasis of ecological systems lies in the reciprocal interactions between species and their environment. It is now known that organisms not only are constrained by their abiotic environment, they also act in turn on that environment. The Gaia theory, in particular, suggests that organisms contribute to mechanisms of self-regulation that have ensured a certain stability of the environment on the earth's surface for hundreds of millions of years. Ecologists ask the following question: can self-regulation on the planetary level (Gaia theory) come from natural selection at the individual level (Lenton, 1998)? Nothing so far has proved that evolution involving the individual level could be at the origin of the evolution of characteristics that preserve and protect entire systems, as the Gaia theory tends to suggest. This question of the coevolution of organisms and environment remains a subject of debate.

Floreale or the "world of daisies"

Let us imagine a simplified planet in which the principal species are flowering plants. They are subject to radiation from a star, which increases its production of heat as it ages, like our sun. As on earth, the average temperature of this planet is an equilibrium between the heat received from the star and that reflected and dissipated in the form of infrared radiation of a long wavelength. The albedo of bare soil is 0.4, i.e., the soil absorbs 40% of the solar heat it receives. Floreale is a simplified planet in which the environment is reduced to a single variable, the temperature, and a biota composed of seeds of two species of daisies, one with white flowers, the other with dark-coloured flowers. The optimum growth of these two species occurs at 22.5°C and they can survive in a temperature range of 5 to 40°C. Let us imagine that this planet is so cold at the beginning that the daisies could not develop, but that it warmed up gradually. When the temperature reaches 5°C in the equatorial region, the daisy seeds germinate and it can be postulated that the white and dark daisies are equally represented. The white flowers (albedo of 0.7) reflect the light in such a way that they fall below the critical temperature of 5°C and soon wilt. On the other hand, the dark flowers (albedo of 0.2) that absorb the light are favoured in terms of growth and reproduction. During the following season, the dark daisies will be more numerous and help to increase the heat of the soil and the air (positive feedback) while gradually colonizing the planet.
As the temperature rises, the dark flowers always have a selective advantage in terms of growth because they are closer to the optimum temperature. At a particular point, the daisies invade the entire planet and the average temperature is close to the optimum for their growth. But when the temperature goes beyond the optimum, there is a reduction in the selective advantage of being dark, and the white daisies become more competitive. The growth of white daisies is thus favoured by their capacity to reflect the heat, so that they became more abundant than the dark ones. Beyond a certain temperature (but less than 40°C), the black daisies disappear, then the white daisies prevail if the temperature continues to rise.
In simulating this virtual world, and in calculating the flows received in the form of heat as well as the heat lost in the form of radiation, we can show that the competitive growth of black and white daisies results in a regulation of the temperature of the planet at a level close to the optimal temperature. In other words, a property of the global environment, such as temperature, is regulated for a wide range of solar luminosity by the competition between species (Lenton, 1998) (Fig. 10.13).

10.8.2. Adaptive systems theory

The systems theory represents the natural world as a set of reservoirs and flows regulated by various feedback processes. These representations of the world

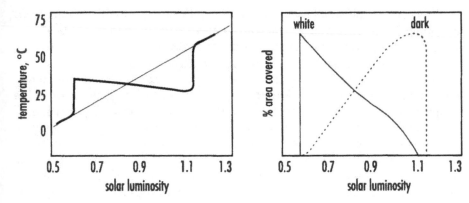

Fig. 10.13. Model of evolution of the planet Floreale. The self-regulation of the planet may be the consequence of natural selection on the individual level of characters that can modify the environment. Left: the average temperature of the planet in the absence of daisies (thin line) and with daisies (thick line). Right: area occupied and respective proportion of white and dark-coloured daisies as a function of the solar luminosity, and thus the temperature of the planet (modified from Lento, 1998).

have been the subject of various mathematical analyses (including many of the deterministic or analytical type) to understand the behaviour of natural systems. However, they are generally not satisfying because they are too simplistic. In particular, mathematical representations of ecological systems have long ignored the potential role of processes of adaptation, whereas organisms perpetually confronted with an uncertain universe have developed a number of mechanisms to minimize the consequences of this uncertainty.

For a long time, the role of evolution in the organization and functioning of ecosystems has been probed. It is known that natural selection favours the genotypes that are more apt to produce an abundant progeny in a given environment. However, do the characteristics selected have a beneficial effect only for the organisms in question or do they also benefit the ecosystem as a whole? In the present state of our understanding, there is nothing to prove that adaptation beneficial at the individual level may also have positive effects (or effects considered positive) for the entire system. There is nothing to support the belief, and nothing to disprove it, that competitive evolution that favours certain species also optimizes the processes at the ecosystem and biosphere level (Levin, 1999). Mutualism and symbiosis among a small number of species also generate feedbacks that affect the ultimate evolution of the partners and help in a way to attenuate the consequences of the unpredictability of the environment.

The theory of complex adaptive systems emerges from systems theory but explicitly takes into account the diversity and heterogeneity of systems instead of representing them only as reservoirs and flows. It explicitly incorporates the role of adaptation in the control of the dynamics and responses of these heterogeneous reservoirs. It allows ecologists to analyse why the processes at the lowest levels of organization (genes, for example) are responsible for patterns at higher levels of organization (ecosystems). Our ability to predict the dynamics of ecosystems on the large scale depends heavily on our understanding of interactions on the small scale, which we can test experimentally. Complex adaptive systems, according to their advocates, can be used to understand how the organization on the large scale is produced, and how it is controlled by processes acting on lower levels of organization.

Although the capacity of species to adapt to changes in environmental conditions is certainly important, the capacity of ecologists to incorporate the adaptation into the models has been limited by the inherent difficulty of including variability and selection in the reservoir-flow sets used in system analysis. Because of conceptual as well as technological advances, it is now possible to increase the complexity of models and to incorporate the variability at the individual level, which allows ecologists to create simple models of adaptive and selective processes in ecological systems (Hartvigsen et al., 1998). This field is yet to be explored.

10.8.3. Self-organized criticality

The great success of theoretical physics has allowed us to hope that simple models can take into account the macroscopic behaviour of complex systems. This is true of models of critical self-regulated systems that, despite their simplicity, can be used to model complex systems. Such systems evolve towards the complex critical state without intervention from external agents, in the course of a self-organization process that covers a long transitory period (Bak, 1999). Specialists in evolutionary science as well as population biologists have somewhat successfully used models of this type, which seem not to have been applied to ecology so far.

Criticality is a term used in physics to describe the transition between qualitatively different states such as solid-liquid or liquid-gas. The critical temperatures for water, for example, are the boiling point and the freezing point. In the physics of solids, the term *criticality* describes a situation in which a small disturbance at one point may virtually have an influence on the entire environment, causing major transformations. Generally, this concept is applied to situations in which a small disturbance rapidly spreads to the entire system, in systems in which the elements are highly interconnected. In water, which is transformed into ice when it is frozen or into vapour when it is boiled, the transitions correspond to critical states during which the crystallization extends rapidly to the entire volume under consideration.

In the example of water, the criticality is controlled by the change in temperature, while with self-organized criticality the system maintains itself by feedback processes that it generates. Self-organized criticality is defined as follows: complex systems of large dimension, made up of many interacting parts and far from equilibrium, evolve spontaneously towards a critical point. We often give the example of a heap of sand to illustrate the phenomenon of critical point. If we take the sand between our fingers and let it fall slowly on a flat surface, at the beginning we have a flat heap and the grains remain where they fall. The grains are independent from one another and their behaviour responds to the forces of gravity and friction. However, if the slope increases, we reach a point beyond which the addition of grains of sand often causes avalanches. A small modification of the configuration can transform an insignificant event into a catastrophe (Bak, 1999). In the critical state, the functional unit is the entire heap and not the individual grain. We cannot comprehend the elements of a critical system in isolation: none of the individual grains can cause the emergent properties of the sand heap. To study the grains individually, by a reductionist approach, would be of no use.

According to Bak (1999), ecosystems are good candidates for theories of self-organized criticality. By analogy with the heap of sand, the addition of new species to an ecosystem will push it towards a critical state at which it will spontaneously become unstable. Bak estimates that ecosystems organize themselves into "critical states" in which continual extinctions act as shock waves in the system, causing other extinctions and reorganizations that bring the system towards a dynamic equilibrium of renewal and replacement. Over long time periods, these events explain, according to Bak, the existence of "punctuated equilibrium", with long periods of "stasis" during which the phenomena of extinction and speciation are limited, followed by periods of rapid changes, as Gould and Eldridge (1977) have shown on fossil series. Over shorter periods, we can draw a parallel with the island biogeography theory of MacArthur and Wilson. However, this applies to non-saturated communities, whereas on the mainland the communities could be compared to the sand heap: cascades of species replacement may be generated by the continuous entry of species into an already saturated environment.

We may consider self-organized criticality the underlying theory of the Gaia hypothesis (Bak, 1999). In the image of self-organized criticality, the entire ecosystem has evolved towards the critical state, so that all the species represent a single coherent organism that follows its own dynamics of evolution. To consider that evolution independent of any species makes no sense. At the critical point, all the species are influenced and act collectively as a single metaorganism, all sharing the same outcome. A single event may cause the collapse of part of the ecological network and finally its replacement by a new, stable ecological network. This point is illustrated by the occurrence of large-scale extinctions. According to Bak (1999), the vigorous opposition to the Gaia hypothesis, which represents an authentically holistic conception of life, is characteristic of the frustration of a science seeking to preserve its reductionist vision of biological evolution.

The ideas of Bak on "self-organized criticality" do not suppose the existence of an unknown force that orients evolution in one direction that will be the best possible for the ecosystem. Developing ecosystems grow in complexity until they reach criticality. Still, the individual fluctuations of species support a cascade of variations and extinctions that are statistically predictable. The generation of innovations, through cascades of extinctions and the continual loss of species, creates a dynamic and self-maintained equilibrium. This is an interesting hypothesis that merits exploration, even though it is difficult to appreciate its relevance at the moment. It is probably too simple because it ignores the role of selection at the different levels at which selection occurs (Levin, 1999).

10.9. CONCLUSION

All other things being equal, it has taken a long time for ecologists to shift from a paradigm of equilibrium to one of non-equilibrium, and it is not evident whether this process has been completed. For cultural, theoretical, and practical reasons, the concept of equilibrium continues to rule over ecological research, especially in its overall aspects. The idea of the balance of nature, deep-rooted in the collective imagination, leads us to believe implicitly that everything that takes an ecosystem far from its point of equilibrium is "bad" for that system. This concept can be understood by administrators and decision-makers, while the paradigm of non-equilibrium suggests to them that it is difficult to make predictions. Thus, the terms *disturbance* or *stress* almost always have a negative connotation in the common language, while research has shown in the last few decades that disturbances have been among the determining elements of the long-term dynamics of ecosystems.

We need to ask the following question: Has the paradigm of equilibrium not been more or less consciously maintained in the discourse of ecologists because it constitutes a means of communication, which is false but convenient, with the public and with administrators? There is solid reason to believe that a temporarily unbalanced ecosystem, like Cartesian diver, will recover its earlier conditions after a few oscillations, provided the disturbance it underwent has not been severe enough to destroy that system irremediably. The concepts of "ecosystem health" and "ecosystem integrity", which intuitively are based on the idea of equilibrium, work to this effect. And we must recognize that many field observations have proved that ecosystems have a strong resilience and can recover quite rapidly after the shock of a disturbance.

Chapter 11

Matter and Energy Flows in Ecosystems

"Size has a remarkably great influence on the organisation of animal communities. A little consideration will show that size is the main reason underlying the existence of ... food chains, and that it explains many of the phenomena connected with the food-cycle. There are very definite limits, both upper and lower, to the size of food which a carnivorous animal can eat."

Elton (1927)

"It is embarrassingly easy to list ways in which existing empirical data on food webs could be improved and depressingly difficult to see how to achieve these high ideals in practice. Food webs studies highlight a crucial problem that confronts all areas of ecology."

Lawton (1989, *Ecological Concepts*)

One of the principal functions of an ecosystem is the production of living organic matter by autotrophic organisms. This organic matter contains energy that will be used and degraded by heterotrophic consumers within the food webs. After organisms die, the organic matter is decomposed: the mineral constituents are recycled and enter a new cycle of production of tissues of autotrophic organisms. According to Evans (1956), *ecosystem ecology studies the transfer, transformation, and accumulation of energy and matter through living things and their activities.*

The pioneering work of Lindeman (1942) on the functioning of Lake Cedar Bog in the United States is the first trophodynamic approach to ecology. In converting the biomass of different species into energy equivalents, i.e., in proposing the calorie as a common unit, Lindeman made it possible to study and compare the productivity of ecosystems while introducing the concepts of matter and energy transfers, and energy yield. In reality, Lindeman's study is the concretization and operational expression of many ideas put forth at that time by different ecologists (see Chapter 2). The trophodynamic approach reached its peak with the International Biological Programme, the major objective of which was to study productivity and compare matter and energy flows in different natural ecosystems in order to derive general laws from their functioning (Golley, 1993). The programme supported an unprecedented mobilization of the scientific community and the accumulation of a large quantity of information, albeit of a descriptive nature.

Ecologists devoted a great deal of effort to studying biological production as well as the pathways of organic matter and energy within the ecosystem.

Many studies focused on food webs in order to deduce general laws about their organization (Pimm et al., 1991; Cohen et al., 1990; Polis, 1994). Still, it has to be recognized that despite a considerable improvement in our understanding, and especially evidence for the major role of microorganisms in matter and energy flows, we still do not have a general theory of the functioning of ecosystems.

11.1. MATTER AND ENERGY FLOWS IN ECOSYSTEMS

The earth is an open thermodynamic system (see Chapter 5): the flow of solar energy (high energy photons) that penetrate the biosphere are gradually transformed into work and heat, which is ultimately dissipated into space in the form of infrared radiation (low energy photons). *One of the objectives of ecosystem ecology is to understand the mechanisms by which this flow of solar energy creates and sustains life.*

11.1.1. Matter cycles and energy flow

The minerals needed for the production of living matter are virtually indestructible. In theory, they are constantly cycled in the ecosystem. This is indispensable for the very existence of ecosystems, given that there is a limited quantity of mineral resources on earth. Nevertheless, it is known that a certain fraction can be stored more or less temporarily in soils or sediments, and that there may be inputs and outputs of matter in an ecosystem. Massive

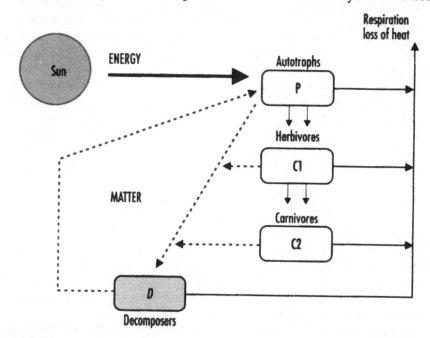

Fig. 11.1. Diagram of flows of matter (broken lines) and energy (solid lines) in food chains

accumulations of fossil fuels as well as chalk accumulations of organic origin are notable exceptions to the dominant ecological dogma that matter is completely recycled in the ecosystem while energy can only pass through it (Fig. 11.1). This last notion is also challenged by Patten et al. (1997), who defend the idea that energy is at least partly recycled with matter, in the form of residual exergy incorporated in the chemical linkages of organic matter.

The discovery of matter and energy cycles in ecosystems has changed the way in which we conceive the dynamics of biological production. The emphasis is on the notion of flux and no longer only on that of mass. The importance of detritic chains and microbial loops is widely recognized.

11.1.2. How matter and energy flows structure ecosystems

Energy and matter are transferred in ecosystems principally through food chains. At each trophic level, a significant part of the energy is dissipated in the form of heat (Fig. 11.1). Dissipation can be defined as a spontaneous change from an organized form to a less organized form, bringing a state far from thermodynamic equilibrium closer to equilibrium, with a diminution of exergy and of information (Straskraba et al., 1999). A continuous flow of exergy of solar origin is needed to keep the system far from thermodynamic equilibrium, in a highly structured form. *The constant dissipation of energy in the food web is thus the driving force of the synthesis of organic substances in heterotrophic organisms from solar energy accumulated by means of primary production.*

Between the fixing of energy by autotrophic organisms and the complete remineralization of organic matter by decomposers, there are a large number of intermediate transformations. This is what Patten et al. (1997) called the energy cascade in the food web. The different stages of this continuum are organized around an overall process, a *charge-discharge cycle* of energy. Each cell or organism has its own charge-discharge cycle: it must acquire energy, which will be stored to be used for maintenance, reproduction, and other functions.

At the physiological level, organic synthesis or *anabolism* is made possible by the input of energy. This is the key to the charge-discharge process in energy: organisms need a continual input of energy to stay alive. The process of dissipation is characterized by the decomposition of proteins in the presence of oxygen to produce CO_2, water, and ammonium, with the release of energy. This is *catabolism*.

At the organism and community level, the dissipation of matter and energy is the result of respiration, excretion, and mortality due to predation or senescence. Excretion, like respiration, is governed by quantitative laws. The dependence of respiration on temperature is expressed by a complex relationship that has often been put into a more simple form, the Q_{10} law. The processes of excretion and exudation lead to a rapid regeneration of nutrients that become available for reuse by autotrophs. These processes accelerate the cycling of matter, which has overall a positive effect on the productivity of the ecosystem, as has been shown by various observations and experimentations.

The intensity of physiological processes is controlled within certain limits by the temperature, which acts on the activity of chemical and biochemical reactions. The well-known Q_{10} law states that physiological activity doubles when the temperature increases by 10°C. However, this linear relationship is valid only within certain limits of temperature that are compatible with the life of the organisms concerned and to which they are adapted. For very low or very high temperatures, the physiological activity diminishes considerably.

At the ecosystem level, the charge phase consists of accumulating the energy in the biomass in the form of energy-rich molecules in successive cycles of primary and secondary production. The energy liberated by catabolic processes (respiration, fermentation, etc.) of all the organisms in the ecosystem is used to keep the organization of the system far from the state of thermodynamic equilibrium. The energy cascade in the food web consists of an anastomosed network of trophic relationships between the different types of organisms present in the ecosystem that have developed specific feeding behaviours over the course of evolution.

11.2. ECOSYSTEM PRODUCTION AND PRODUCTIVITY

Matter and energy are stored in *biomass*, which is the quantity of living organic matter present at a given moment, in relation to a unit of area (m^2 or ha) or of volume (m^3). This biomass or reservoir of living matter is the result of an equilibrium between two flows: one is production and the other is predation and mortality. It can be expressed as fresh weight or dry weight, or it can be converted into caloric value, carbon level, chlorophyll level, or various other units. Biomass is generally evaluated per species, per community, and/or per trophic level.

If the biomass corresponds to storage, *production* has the dimension of a flow: it measures the rate at which organic matter is elaborated, in producers and in consumers. Production ensures the turnover of the biomass that is consumed or dead. In quantitative terms, it is the total quantity of newly synthesized organic matter produced by a given biomass during a given period of time. It thus depends largely on the growth rate of the individuals that constitute the biomass. *Primary production* corresponds to the accumulation of energy of solar origin in green plants with chlorophyll, while *secondary production* is the energy retained by heterotrophic organisms to elaborate their tissues from living matter of plant or animal origin. In general, extrinsic factors that control production (inputs of energy and allochthonous matter) are distinguished from intrinsic factors (rates and efficiency with which organisms within the food chain elaborate their own tissues).

It must be emphasized that biological production does not correspond only to the increase of biomass (ΔB) during the period of time considered. The biological production of an individual is the sum of increase in weight, sexual products, or young that are produced in the period of time considered, and the mucus or other products emitted by the organisms. When a population is being studied, the biological production corresponds to the sum of individual productions, including that of individuals that are dead during the period taken into account.

Productivity is a rate of production that corresponds to the quantity of matter produced (P) per unit of biomass (B) and per unit of time. It can be expressed as the *P/B ratio*, the ratio between the production and the biomass in an environment, for a given unit of time. Primary productivity corresponds thus to the speed with which energy (in the form of carbon hydrates) accumulates in the biomass.

a) Sources of energy: autotrophs and heterotrophs

The chemical reactions that result in the synthesis of organic molecules, whether during the passage of elements of the inert world into the living world or during the transfer of organic matter from one individual to another in food chains, cannot occur without an input of external energy.

For *autotrophic* organisms that extract raw materials from the mineral world and synthesize all the molecules needed for their functioning, the energy source may be light in *phototrophic* organisms or a chemical reaction for *chemotrophic* organisms. For a long time, only phototrophs were known, organisms that transform light energy into chemical energy that can be used for biochemical reactions, by means of *photosynthesis*. However, the existence of other processes was discovered, such as *chemosynthesis*, which uses the energy released by reactions of chemical oxidation that do not require light.

Heterotrophs do not have the capacity of synthesis to elaborate all the molecules needed for their functioning. They need to synthesize their own molecules from organic matter that is already elaborated. Generally, they are chemo-organotrophs because their sources of energy and raw material are organic molecules. Animals, heterotrophic with respect to nitrogen, carbon, and sulphur, as well as fungi are the principal representatives of this category.

b) Notions of trophy

Biological production values vary greatly with the environment and latitude, since they are a function of light and temperature, as well as of nutrient inputs. On the basis of their productivity, aquatic ecosystems can be classified as *oligotrophs* (from the Greek "poorly fed") or *eutrophs* ("well fed"), or *mesotrophs* when we talk of intermediate systems. These degrees of trophy are commonly used to designate the trophic potential of a body of water and the effects of nutrients on water quality.

An aquatic system can evolve from an oligotrophic stage towards a eutrophic stage. At first, the term *eutrophication* referred to a natural process of ageing in a lake under the effect of accumulation of exogenous matter carried by runoff of surface waters. With time, a lake can gradually fill in and evolve into a marsh, then into a terrestrial ecosystem. At the same time, the littoral flora develops and the primary production that increases in the euphotic zone accelerates the process of filling in.

However, in the past 20 years, the term *eutrophication* is increasingly used to describe the enrichment of water bodies by the artificial and undesirable input of nutrients (phosphorous and nitrogenous compounds) favouring the production of aquatic plants. The excessive proliferation of algae and macrophytes leads to changes in the water quality, and notably to phenomena of deoxygenation that have consequences for the dynamics of biological communities.

11.3. PRIMARY PRODUCTION

Primary production, the prerogative of autotrophic organisms, is the first link of life on land and in water. The process of *photosynthesis* consists of transforming solar energy into chemical energy stored temporarily in the form of energy-rich chemical linkages, in specialized organic molecules that could subsequently release them, such as glucides or lipids. Most of the biomass on the earth's surface is made up of plants (99.9% in weight, according to Whittaker, 1975).

11.3.1. Mechanisms of accumulation of biochemical energy

The processes that lead to the accumulation of biochemical energy are involved in the "charge" phase of the ecosystem in the sense of Patten et al. (1997). The two principal mechanisms known are *photosynthesis* and *chemosynthesis*.

a) Photosynthesis

Eukaryotic producers (terrestrial plants and algae) as well as prokaryotic producers (cyanobacteria) capture incident light energy in the form of photons by means of a pigment, *chlorophyll a*. Photosynthesis is the process that leads to the biosynthesis of organic molecules (carbohydrates such as glucose) from inorganic precursors by plants that use light as an energy source. The process is complex (Jupin, 1996) but the overall reaction is written as follows:

$$CO_2 + H_2O + \text{solar energy} \longrightarrow HCHO + O_2$$

where HCHO is the base element of glucides. In the chemical sense of the term, this oxidation-reduction requires the input of 112.3 kcal per mol of assimilated carbon.

Photosynthesis occurs in two stages:

— The *light* phase or photochemical phase marked by photolysis of water in the presence of light. This phase leads to the synthesis of energy molecules present in all organisms, the adenylic nucleotides (ATP, ADP, AMP).

— The *dark* phase (Calvin cycle), which occurs in the chloroplasts of the plant, which capture the carbon dioxide. During this phase, there is synthesis of carbonate molecules, glucides, protides, lipids, carriers of potential energy.

In sum, photosynthesis corresponds to a transformation of light energy into chemical energy stored in the organic molecules that constitute the plant biomass.

Modes of synthesis of glucides and ecological significance

Different modes of glucide synthesis, which has a high ecological significance, have been indicated (Jupin, 1996).

— C3 plants, in which the first stable product formed during photosynthesis is phospho-glyceric acid, which contains three carbon atoms in its molecule. This group comprises all the trees, all the legumes, and many cultivated plants such as wheat, rice, potato and, generally, all the plants of the temperate regions. Ecologically, these plants have a lower yield of photosynthesis than C4 plants and they require more water.

— C4 plants represented by certain cultivated species (sugarcane, maize, sorghum) and many herbaceous species living in dry tropical regions (the percentage of C4 species increases from the temperate regions towards the equator). The first stable product of photosynthesis is oxalo-acetate, which has four carbon atoms. The C4 plants have the advantage of being able to use low concentrations of carbon dioxide of the air more efficiently than C3 plants, but they require more energy and grow poorly below 10°C. The C4 species represent less than 2% of flowering plant species but 10 to 15% of the total production of continental biomass.

— CAM plants (plants with crassulacean acid metabolism), which fix carbon dioxide during the night when the stomata are open. This metabolism, which is similar to the C4 type, was discovered in the family Crassulaceae, for example rhubarb, but involves many families such as Cactaceae, Euphorbiaceae, and the agaves. Its principal advantage is to limit water losses in plants living in arid and desert regions. It is estimated that 5% of flowering plants are involved but the biomass production is limited (Saugier, 1996).

C4 plants appeared about 30 million to 40 million years ago, thus more recently than C3 plants, which are the primitive type present from the Primary Era in marine as well as terrestrial vegetation. Some scientists believe that the C4 mechanism appeared towards the end of the Carboniferous Era, when the oxygen level in the atmosphere was nearly the same as it is today, as a protective mechanism against oxygen (Frontier and Pichod-Viale, 1998). However, this "adaptation" could also have been a response to the decline in the carbon dioxide level in the air during the Tertiary (Saugier, 1996).

---

b) Chemosynthesis

The study of sulphur-reducing bacteria discovered in hydrothermal springs in the deep ocean indicated the role of chemosynthesis as a support of animal life. In these entirely original ecosystems, bacteria are the primary producers of organic matter, drawing their energy from oxidation of sulphurous compounds (sulphur-oxidizing bacteria) and methane. Like plants, these bacteria elaborate their carbon molecules from carbon dioxide. Research has since confirmed the importance of chemosynthesis as a generator of life in the ocean. The bacteria draw their energy either from the oxidation of hydrogen sulphide released in hydrothermal fluids or from the oxidation of methane from the degradation of organic matter and emitted in the form of ooze in marine sediments.

This chemosynthesis is sometimes based on the symbiotic association of an animal with bacteria. For example, in the giant tube worm *Riftia pachyptila*, which has no digestive tube, the blood transports toxic gases produced by hydrothermal springs (carbon dioxide, hydrogen sulphide) to an internal organ in which bacteria (endosymbionts) oxidize the reduced sulphide compounds before returning organic matter to the worm by chemosynthesis.

New biochemical pathways and new biogeochemical cycles are likely to be discovered. Rich and unique bacterial flora have been found in the depths of the water table. A drilling of 2.75 km in northern Virginia led to the discovery of microorganisms in a pocket that were probably completely isolated since the time of the dinosaurs (Kerr, 1997). Some of these microbes, living at temperatures of 75°C, extract their energy from ancient organic matter in the surrounding rock, using minerals such as iron or manganese to oxidize carbon.

11.3.2. Estimation of primary production

Gross primary production (GPP) is the measure of photosynthetic activity. It is the total quantity of organic matter (expressed, for example, in gram dry weight, quantity of carbon, or energy equivalent) produced by autotrophic organisms in a given period of time. It is a function of the available energy and minerals that allow the elaboration of organic molecules. It also varies according to environmental conditions.

However, in reality, photosynthetic activity and respiration of autotrophic organisms are two inverse energy reactions that are produced simultaneously. The result is the quantity of matter and energy available for the first heterotrophic level, the *net primary production* (NPP). The NPP is the difference between the GPP and respiration in an autotrophic organism. *From the evaluation of NPP, we can determine the efficiency with which the energy is used by producers to be transmitted subsequently to other elements of the ecosystem and, in a way, ensure its biological functioning.*

Evaluation of NPP on the global scale is not an easy task. In the oceans, the phytoplankton is largely responsible for photosynthesis, and nearly all the biomass is photosynthetic. In terrestrial environments, the production of plants above the soil surface can be accurately estimated, but such estimations are much more problematic with respect to the underground part. Besides, a large part of the biomass of terrestrial plants is made up of roots and stems that respire but produce little. As a result, the terrestrial biomass is much greater than that of oceans but less productive: while the primary producers of oceans are the source of nearly half the NPP, they represent only 1 gigatonne in biomass, or 0.2% of the global biomass of primary producers. This is to say that the turnover rate of plant organic matter in the oceans (on average one week) is much higher than that of terrestrial environments (on average 19 years) (Falkowski et al., 1998; Field et al., 1998).

Among the most recent figures, Field et al. (1998), on the basis of earlier studies and using models, estimated that the NPP on the global scale was 104.9 gigatonnes (10^{12} kg) of carbon per year, with a contribution of nearly equal parts from oceans and terrestrial systems (Table 11.1).

However, the spatial distribution of NPP is heterogeneous on land as in the oceans, with vast areas in which production is low, and smaller areas with high production. In the oceans, it is the coastal zones and upwelling regions that are the most productive. On land, 25% of the area without permanent ice has an NPP higher than 500 g C/m^2/yr, while in the oceans the corresponding area is only 1.7%. On the global level, the NPP is greatest at the equator, but there is also a peak in the temperate latitudes of the Northern hemisphere resulting essentially from terrestrial production.

The NPP presents seasonal variations, which reflect the influence of physical and chemical factors on the biological processes. In the ocean, these fluctuations are modest: from 10.9 Gt of carbon in spring in the Northern hemisphere to 13.0 Gt of carbon in summer. However, fluctuations of terrestrial NPP are greater: the overall production of the Northern hemisphere is around 60% greater in summer than in winter.

The German chemist von Liegig, who attempted in 1865 to estimate plant production on the continents, used the simplest method available at the time. He weighed the grass produced on the lawn in his garden for one year, about 8 t/ha. Extrapolating to the area of the continents and multiplying by an average level of carbon of 0.5, he concluded that the global production was 60 Gt of

Table 11.1. Contribution of various terrestrial biomes to net primary production (Field et al., 1998)

Terrestrial ecosystems	NPP
Tropical rainforests	17.8
Deciduous forests with large-leaved trees	1.5
Forests of trees with large leaves and needles	3.1
Permanent forests of trees with needles	3.1
Deciduous forests of trees with needles	1.4
Savannahs	16.8
Perennial prairies	2.4
Shrubs with large leaves	1.0
Tundras	0.8
Deserts	0.5
Cultivated land	8.0
Total terrestrial ecosystems	56.4
Marine ecosystems	48.5

The values are in gigatonnes of carbon (10^{12} kg/yr).

carbon per year, which is very close to modern estimates using far more sophisticated methods (Saugier, 1996).

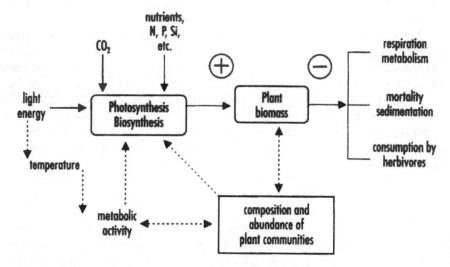

Fig. 11.2. Control mechanisms of phytoplankton growth and development in pelagic environments. Photosynthesis, which is the source of increase in biomass and energy, is controlled by the availability of light and nutrients, as well as by the temperature and species composition of the phytoplankton. The losses correspond to the respiration, natural mortality, and consumption by herbivores.

11.3.3. Control factors of primary production

Primary production depends mainly on four factors: the *availability of nutrients, intensity and duration of insolation, temperature,* and *availability of water* for terrestrial environments. However, the terrestrial primary production is also sensitive to various factors such as soils and plant characteristics. Temperature most often acts as a factor that controls biochemical process and the rate at which plants can synthesize organic matter. This is why in temperate environments there is a seasonal cycle of primary production with a maximal photosynthetic activity in the warm season.

a) Availability of minerals: the notion of limiting factor

The concept of limiting factor was probably introduced by Justus Liebig in 1840. Liebig showed that the bioavailability of nutrient minerals was one of the major control factors of primary production. These observations generated one of the oldest ecological laws, the *law of minimum*, which states that *the growth of a plant is possible only to the extent that each of the minerals essential to the elaboration of the organic matter is present in sufficient quantity in the soils.* The maximum yield of a crop is limited by the rarest essential nutrient in the environment, as a function of the particular needs of the plant in question.

The concept of limiting factor applies to total quantities as well as to the proportions of different nutrients. It can be hypothesized that the rates of absorption and utilization of nutrients by plants reflect the proportion of these elements in their cytoplasm. In marine plankton (Redfield et al., 1963), the proportions of the elementary composition of the plant organic matter (in atoms) are 106 C : 263 H : 110 O : 16 N : 1 P. From this it can be deduced that in order to produce 2.5 g of dry organic matter, 225 mg of N and 31 mg of P are required. In fact, however, the gradual mobilization of nutrients in the biomass leads to the exhaustion of one or several elements in the environment. The concept of limiting resource is thus transferred from organisms to ecosystems. We can thus logically expect that the biomass production will be determined by the nutrient least available with respect to this demand (limiting factor). In the continental surface waters of a temperate environment, the ratio of N/P concentrations is nearly always higher than 7, so that phosphorus is the element that is exhausted first and becomes a limiting factor.

The proportions cited above correspond to an average situation, because different species of algae have particular nutritive needs and do not necessarily assimilate the nutrients with the same rhythm and in equal quantities. The result is that the facility with which a species of alga can compete with another for the available nutrients depends on its capacities to absorb and assimilate the nutrients. This observation is the basis of the theory of competition for resources (Tilman et al., 1982). According to these authors, diatoms are better competitors for phosphorus than other algae. When the phosphorus level is low, the growth of populations of diatoms is thus higher than that of other algae.

In a marine environment, one of the important discoveries of these past few years is the demonstration that the pelagic primary production is probably limited by the availability of iron. This element, which in trace quantities is indispensable for the photosynthesis of proteins, is not cycled like other elements in the pelagic system (see Chapter 4). During large-scale experiments, masses of water of the equatorial Pacific were enriched in iron, which temporarily caused a high algal development, thus demonstrating *in situ* that the availability of iron is the limiting factor for phytoplankton growth in this region.

For a long time, the availability of nutrients was studied without taking cycling processes into account. Therefore, in coastal regions of upwelling, for example, or in zones in which the waters turn over seasonally, the nutrient flows are the source of high productivity. However, in the open waters, the nutrient flows from the deep zone towards the surface are relatively low. In that context, nutrients are cycled very efficiently by the picoplankton by means of the microbial loop (see later). When we speak of the limiting nutrient, we must therefore distinguish the limitation of the biomass by the available stock of that nutrient (notion of capacity) from the limitation of the rate at which the biomass is formed or renewed (notion of intensity), which does not depend on the concentration of the limiting nutrient but on its flows (Capblancq, 1995).

b) Light

On the earth's surface, the intensity of light varies with the daily and seasonal cycles. In aquatic systems, light is a control factor that acts at various levels. On the one hand, it can have an inhibiting effect on the surface of the water when the intensity is too high, and on the other hand it attenuates with the depth. The result is that during a period of maximum illumination in a homogeneous water column, in terms of physicochemical and phytoplanktonic characteristics, the photosynthetic activity is maximum at a depth of a few metres and then diminishes with the depth because of the attenuation of light (Fig. 11.3).

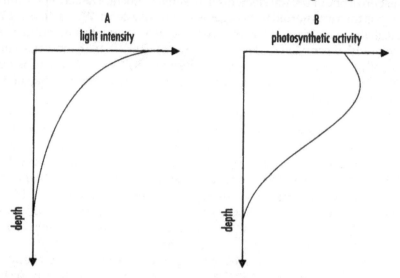

Fig. 11.3. Light attenuation as a function of the depth (A) and photosynthetic activity per unit of phytoplanktonic biomass at different depths (B)

Analysis of results obtained during the International Biological Programme showed that the primary production of lakes and reservoirs was strongly correlated with latitude (Brylinsky, 1980). The latitude factor corresponds to the duration of insolation as well as to the influence of the average temperature. Surprisingly, the visible incident energy is not well correlated to the primary production, but a more precise analysis shows that beyond a certain level high light intensities are responsible for a decrease in production.

All other things being equal, the incident light energy does not seem to play a major role in the control of primary production on the global scale.

c) Temperature and availability of water

The search for the agents responsible for the control of primary production on the global scale established that the NPP of many terrestrial ecosystems is controlled by temperature and availability of water (respectively 31% and 52% of the area concerned), while light energy is important for only 5% of the area (Churkina and Running, 1998).

Plants growing in colder regions are less productive than those growing in warmer environments because the rates of photosynthesis and respiration are lower, as is the rate of litter decomposition. Temperature is thus the principal control factor of the productivity of savannahs with C3 plants, deciduous forests, and evergreen forests with trees that have needles. The deciduous forests and evergreen forests occupy regions with average annual temperatures below or close to the freezing point, and with an evaporation rate close to the precipitation level

On the other hand, forests with large-leaved trees, evergreen or deciduous, are found mostly in regions in which the average temperature is high. The availability of water controls the productivity of savannahs with C4 plants or deciduous plants with large leaves and of desert zones. From the combination of temperature and availability of water, we can thus explain on the global scale the existence of large regions such as savannahs with C3 and C4 plants and forest biomes corresponding to trees with different types of leaves.

11.4. SECONDARY PRODUCTION

While primary production consists of the elaboration of living molecules from minerals, heterotrophic organisms use living matter to elaborate their own molecules. Thus, secondary production is that of heterotrophic organisms that use the energy accumulated by autotrophic organisms directly (herbivores) or indirectly (carnivores, detritivores, decomposers).

Secondary production is more difficult to measure than primary production in the absence of an easily standardized method, and given the complexity of food chains. Besides, a large part of the biomass elaborated by heterotrophic organisms is consumed as it is produced, so that the increase of biomass is nearly nil, while the biological production may have been high.

11.4.1. Turnover rates of biomass

To evaluate the productivity of ecosystems is a cumbersome, time-consuming, and costly task. In theory, the productivity of each of the different species must be studied using methods of population dynamics. This is rarely possible and ecologists have looked for simpler methods of evaluation. One possible approach to simplifying production studies is to use the P/B ratio, which expresses the turnover rate of biomass per unit of time. This ratio could be calculated for a multispecies population or a settlement. It is often calculated over one year, which allows comparisons between species and between environments.

Theoretical studies by Allen (1971) indicated, for populations in which the dynamics are described by different growth models and mortality models of an exponential type, that the P/B was equal to the coefficient of mortality (Z). That is, the P/B ratio diminishes in populations when longevity increases (Lévêque et al., 1977). Lavigne (1982) stated the relation $P/B = 5.78\ W^{-0.266}$ from the study of a large number of animal species.

Generally, there is positive relationship between the turnover rate and the size of individuals: the smaller the species, the shorter its biological cycle, and the more rapid the turnover.

11.4.2. Energy budgets

Biological systems follow the thermodynamic principle of energy conservation: energy that goes through these systems can be converted from one form into another (from light energy into chemical energy, into work, into heat, etc.) but it is never created or destroyed. It is thus possible in theory to draw up a balance of inputs and outputs and to evaluate the energy accumulated in the biomass (Fig. 11.4).

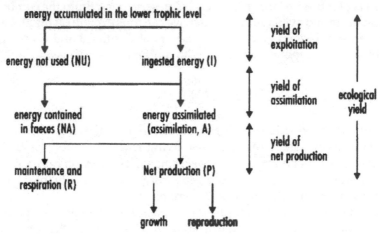

Fig. 11.4. Energy balance: different processes of dissipation of energy during transfer in an individual.

At the level of individuals constituting the biomass, the energy ingested by a consumer will take various routes that can be evaluated conventionally by drawing up an energy budget:

— In relation to the potentially available energy, one part is ingested (I) and another part, often large, is not used (NU).

— Part of the energy ingested is not assimilated (NA) and will be excreted in the form of faeces.

— Part of the assimilated energy (A) will be dissipated by respiration (R) for the maintenance of metabolism, and part will be used for the growth of the individual (P) and stored in the body mass. The biological production (P) of a consumer corresponds to the assimilated energy minus the respiration of the organism.

The energy flows corresponding to the food assimilated are as follows:

$$A = I - NA$$

This assimilated energy flow is then subdivided as follows:

$$A = P + R$$

We can establish an energy budget for an ecosystem by analogy with that of an individual. Ecologists have tried for many decades to quantify the energy and matter flows between different trophic levels of the ecosystem. As we go up the food chain, the quantity of energy accumulated by primary producers dissipates rapidly, because at each change of trophic level a considerable part of the energy is either lost (not consumed) or not assimilated, or it is used by metabolism. This energy loss is estimated at 80-90% during the transfer from one trophic level to another. The ecological efficiency, i.e., the ratio between production of the consumer over the production of the biomass consumed (expressed in energy units), varies according to the species and the periods. It is thus difficult to evaluate globally. Nevertheless, it is estimated that in a lake the light energy assimilated in chemical form by algae represents only 0.1% to 1% of the incident energy. The production of herbivorous zooplankton is on average slightly higher than 10% of the planktonic primary production. This efficiency is higher in zooplanktophagous species and is around 20% for carnivorous zooplankton.

11.4.3. Ecological pyramids

The principle of transfer and dissipation of energy and biomass across successive trophic levels gave common currency to the old concept of *ecological pyramid* popularized by Elton (Fig. 11.5). The base of the pyramid consists of primary producers, and the carnivores are at the top. The pyramid is constructed by the stacking up of rectangles, of area proportionate to the significance of the parameter measured, and it gives a graphic representation of the trophic structure of an ecosystem. Ecologists construct pyramids of numbers (numerical abundance of individuals of a trophic level), biomasses, and/or energies.

The example given in Fig. 11.5 is that of Lake Lanao (the Philippines) (Lewis, 1979). We may question the relevance of these results (and of many others) in

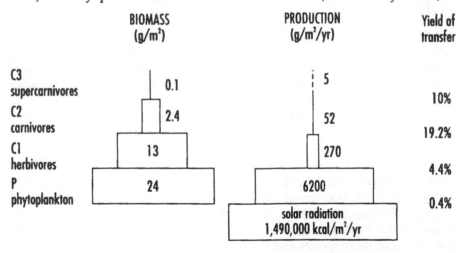

Fig. 11.5. Pyramids of biomass and production in Lake Lanao (from data published by Lewis, 1979)

relation to the ecosystem given that the benthos and microorganisms have not been taken into account, but generally all these budgets, which give somewhat simplistic versions of matter and energy transfers within ecosystems, are only caricatures of reality. Ecological pyramids were popular in the 1960s, but they are now much less so among ecologists.

11.5. FOOD WEB ORGANIZATION

The nature and intensity of trophic relationships established between species living in an ecosystem play a central role in ecological theories. It is through these networks of interactions that matter and energy circulate in the ecosystem.

11.5.1. Trophic levels

In order to live and reproduce, living things need a source of energy. Some, such as *autotrophic* organisms, draw from the abiotic environment the minerals necessary to synthesize their own organic substance by means of light energy from solar radiation. These "primary producers" include all the chlorophyllian plants (algae, macrophytes) as well as bacteria and protists. Autotrophs serve to feed herbivorous consumers, which transform plant organic matter composed of carbohydrates into animal matter that is relatively richer in nitrogenous organic substances.

Herbivores themselves serve as prey for other consumers, the carnivores. All the animals as well as fungi and bacteria that draw their energy by consuming living or detritic organic matter are *heterotrophic* organisms. Finally, *decomposers* are organisms that mineralize dead organic matter to transform it into simple minerals that can again be assimilated by plants.

11.5.2. Food chains and food webs

Matter and energy circulate constantly from primary producers to and then carnivores. In drawing these links of dependence of animals that eat some organisms before being eaten by others, we obtain what is called a *food chain* or *food web*, which diagrammatically describes the flows of matter or energy between different trophic levels, from autotrophic producers to terminal consumers. These are caricatures of nature that describe in a simplified manner who eats whom (Pimm, 1982).

The reality is much more complex. To understand the functioning of an ecosystem, we must evaluate the nature, intensity, and complexity of relationships between the different species that are part of it. The theory of *food webs* is based on the existence of multiple interactions between species, either in a relationship of eater and eaten or in a relationship of competition for the same resources. *Food webs* can be represented by diagrams indicating the different links between these species and their intensity (Fig. 11.6). According to Cohen (1989), if a community is compared to a city, a food web is like a plan of the city that signals what routes are taken by the traffic.

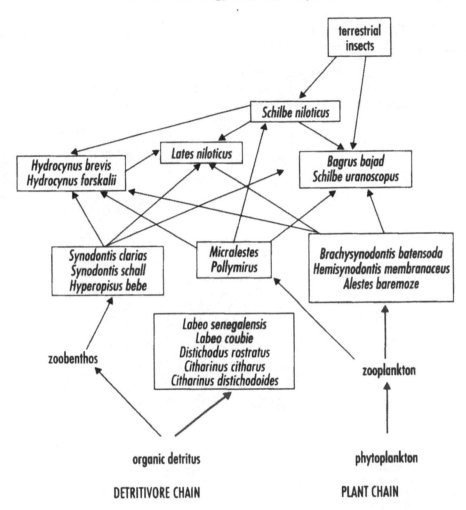

Fig. 11.6. Food web based on feeding habitats of fishes in the free waters of Lake Chad terrestrial insects

The comparative analysis of food webs is of great interest in theoretical ecology because many structural or functional attributes of food webs (e.g., types of interactions and length of food webs, presence of key species, rates of turnover and productivity) play an essential role in transfers of energy and carbon, bio-accumulation of contaminants, or the control of plant biomass. The analysis of ecological mechanisms that control the structure and functioning of food webs thus allows us to elaborate several models, sometimes contradictory, and supports animated debates on the respective role of resources and predators in the control of food webs. Despite an abundant literature, we must recognize that we are still far from being able to formalize and quantify the numerous trophic interactions within ecosystems, more so as recent discoveries on the role of microorganisms have partly thrown open questions that were apparently resolved by earlier studies.

11.5.3. Trophic guilds

Since food webs are often highly complex, it is difficult, even impossible, to take into account all the biological components of the system within a conceptual process. This is why ecologists have developed the notion of *guilds*, or functional groups, which are sets of related species consuming the same type of prey and /

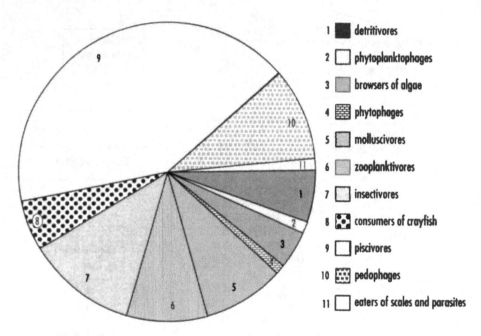

Fig. 11.7. Different trophic guilds and relative proportions (as a percentage of the 240 species surveyed) among the endemic cichlid fishes of Lake Victoria (sources: Witte and van Oijen, 1990)

Table 11.2. Principal trophic guilds in water courses

Trophic guild	Type of food	Mechanisms	Principal groups concerned
Tearers and shredders	Coarse detritic organic matter, fungi	Mastication and digging	Several families of Trichoptera, Crustaceae, Mollusca, Diptera
Filter-collectors	Fine particulate organic matter, bacteria	Filtration apparatus, filets, secretions, etc.	Trichoptera, Simuliids, Diptera, Ephemeroptera
Collector-scrapers	Fine particulate organic matter, superficial film	Scraping surface deposits	Many Ephemeroptera, Chironomids, Ceratopogonids
Browsers	Periphyton (notably diatoms) and organic film	Scratching, scraping, browsing	Several families of Trichoptera and Ephemeroptera
Predators	Animal prey	Biting and piercing	Odonata and some Diptera, Trichoptera, Coleoptera

or feeding in the same manner. In this way we practically reduce the number of components included in a food web.

In lake systems, the trophic guilds are most often defined according to the nature of the prey ingested. For example, there are phytoplanktivores, zooplanktivores, insectivores, piscivores, and so on (Fig. 11.7).

In running-water systems, the guilds are defined by their feeding behaviour (filter-feeder, shredder, etc.) rather than their prey (Table 11.2). Their proportion varies as a function of the size and nature of the river.

11.5.4. Autotrophic and detritic food chains

In freshwater systems, we can schematically distinguish two major types of food chains that are not mutually exclusive but coexist in variable proportions depending on the nature of the aquatic environment (Fig. 11.8).

The first type of food chain starts from the primary producers, i.e., aquatic plants that synthesize organic matter and are consumed alive by herbivores, which themselves serve as prey for predators. Such food chains are most often short (phytoplankton, zooplankton, fish) and generally do not exceed four to six trophic levels.

For the second type, the principal source of energy is detritic matter. Such a food chain is found in small rivers, where allochthonous detritic matter (fallen wood, leaves, or terrestrial insects) plays an important role in the food chain.

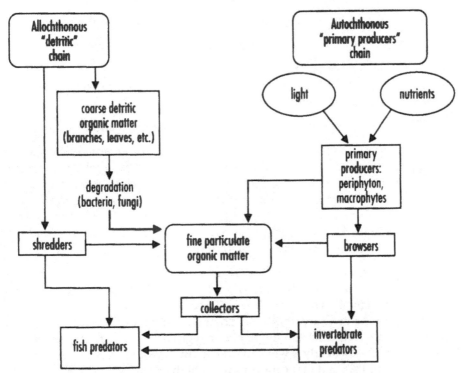

Fig. 11.8. Food chains of autotrophic or detritic origin in an aquatic system

This coarse particulate organic matter (CPOM) is degraded into fine particulate organic matter (FPOM) by microorganisms (bacteria, fungi) and invertebrate shredders. The FPOM is used by invertebrates of the collector guild. Invertebrate shredders and collectors serve as prey for invertebrate predators or for fish.

In the littoral zones of lakes that are rich in macrophytes, as well as in rivers bordered by riparian forests, detritic matter from the banks is often the main base of the food chain. Such systems, based on external inputs, are sometimes called heterotrophic systems. On the contrary, in the pelagic zone, it is the primary producers that are at the base of the chain.

11.5.5. Some generalizations about food webs

Scientists who have attempted to derive general laws about the structure of food webs have all emphasized that the available data are generally of poor quality. Given the often high number of species present in an ecosystem, the relationships between species are often based on occasional observations, rarely take into account the spatial and temporal dynamics, are not quantified, and are not validated by experiments. This generates many biases in the interpretation of the existing data. Nevertheless, some generalizations about food webs have been proposed (Pimm et al., 1991):

— Loops are rare (species A eats species B, which eats C, which eats A), but cannibalism is one form of the cycle known in many species.

— Food chains are generally short, with on average four trophic levels.

— Omnivores are generally rare, but there are several exceptions.

— Contrary to what may be understood intuitively, the length of food chains does not greatly differ in ecosystems having different levels of productivity, except in extreme environments.

— The chains are shortest in zones with frequent disturbances.

— The respective proportions of terminal predators, intermediary species (predatory species that are themselves preyed on), and autotrophic species are more or less constant no matter how many species are involved in the food web. This generalization has also been indicated with results from 50 natural pelagic food webs (Havens, 1992). However, figures that are very different as to the relative proportions have been proposed by other authors: 0.06/0.44/0.50 by Havens (1992), 0.01/0.86/0.13 by Martinez (1991).

The generalizations about food webs drawn from catalogues of studies have, however, been severely criticized by Polis (1993). In brief, Polis says that they are only fantasies of theoreticians based on incomplete and biased data. He finds, on the contrary, that food chains can be long (7 to 10 links), that omnivores are frequent, that loops occur regularly, that the ratio of prey to predator species is nearly always greater than 1, that terminal predators are rare (less than 1%), and that connectivity is high (an individual interacts with tens, even hundreds of species). The points of view are, obviously, radically opposed, and the debate is far from being closed. Beyond the individuals, it is the two worlds that are opposed: the world of modelling theoreticians and that of naturalist ecologists.

As Polis (1993) remarked, food webs, just like biological communities, are complex systems whose structure results from constraints exerted by many factors of endogenous or exogenous origin. The trophic interactions combine predation by consumers with production by prey. For these multifactorial systems, we must stop affirming that this factor or that reason alone explains the structure observed. We need to adopt a pluralistic vision, considering that several factors can control food webs in nature. The role of trophic cascades is certainly important in the dynamics of communities, but neither more nor less so than other processes such as bottom up control, for example. As a function of the spatio-temporal dynamics of the ecosystem, one or the other of these factors will be temporarily predominant.

This more holistic approach to food webs, which are besides considered reticulated and not linear systems, is more difficult to implement. Modelling becomes an indispensable aid to analysis, given the numerous interactions at work. This is not to say that ecology has moved away from a naturalistic understanding of ecosystems. On the contrary, we will need an increasing amount of naturalistic information drawn from observations and experiments to progress in our understanding of complex systems. However, we must also have means of effectively managing this accumulation of information.

11.5.6. Spatial and temporal dynamics of food webs

The initial concept of food web arose from the paradigm of equilibrium of ecosystems (see Chapter 10). It is a representation of ecological systems that involves a double approach: to identify the trophic categories and to determine the nature of interactions between these categories. However, the simplification inherent in the notion of food web must not obscure the fact that the reality is much more complex. With greater knowledge, the simplistic processes that were proposed since the 1950s are increasingly being challenged. One of the concerns of modern ecology is the consideration of spatio-temporal heterogeneity and its consequences for the dynamics of biological communities. Trophic relationships thus cannot remain static representations of relationships between species. Food webs are dynamic systems in time and in space and these characteristics must be taken into account by theory (Polis et al., 1996). However, in the face of a particularly complex set of interactions between organisms and their environments, and between the organisms themselves, levels of inaccuracy are high.

— The recruitment of predators and prey may vary from one year to another, so that food webs are not repeated in an identical way.

— Many species may change their trophic level during their life cycle. A fish may be planktivorous in the larval stage, a consumer of invertebrates in the juvenile stage, and piscivorous in the adult stage.

— A prey may be consumed by predators belonging to different trophic levels, and a single species may consume several types of prey. However, most of the webs described in the literature do not refer to the frequency and intensity with which one species feeds on another.

— The composition of the food changes during the course of the seasons in response to changes in prey availability. Such changes in the nature and quantity of resources have consequences for the structure of food webs. It may be considered more realistic, but also more simple, to describe these food webs over relatively short intervals of time rather than to present aggregated averages.

— There may be changes in nutrients, detritus, or organisms between various types of habitats: aquatic vs. terrestrial, pelagic vs. benthic, and so on. Some species may thus be part of several ecosystems as a function of their life cycle. This is true for many insects that have an aquatic larval stage and an aerial adult stage.

— The transport of organic matter and nutrients by consumers within a system or between systems ensures a rapid redistribution between habitats. A simplified example is that of the hippopotamus, which feeds on the banks and whose defecation is an important source of organic matter in African rivers.

— Many species migrate long or short distances as a function of the seasons and/or to reproduce. For example, salmon hatch in the rivers, move to the sea to grow, and return to the river to reproduce and in some cases to die. Transcontinental migratory birds also participate seasonally in temperate and tropical food webs.

— There has been a tendency to classify species in precise categories, while the omnivorous regime, which is probably the most widespread in nature, is generally underestimated in food webs (Polis, 1994). A planktophagous fish ingests plants, herbivorous animals, and carnivorous animals. Similarly, detritivores ingest detritus as well as the microbes and invertebrates that live in it. There are in reality few species whose food regime is highly specialized. The result is that the omnivores have a large number of trophic linkages including some that may be numerically small with regard to food but overall play a controlling role in the dynamics of systems.

Connectance

The idea that the complexity of webs, linked to the existence of a large number of interacting species, leads to a greater stability of the system is intuitively shared by many ecologists. It is based on the hypothesis that many interactions between species are redundant, so that they can be substituted for one another. *Connectance* is the ratio between the number of trophic linkages that exist between the different species of a food webs and the maximum possible theoretical number. If S is the number of species and L the number of linkages between species, the connectance C is defined as follows: $C = L/S^2$. The number of linkages increases with the number of species, but the mathematical relationship that links the two values is subject to debate, partly because many examples are not available for which all the relationships between species are well known.

11.5.7. The role of microorganisms in pelagic systems

Our understanding of the role of microorganisms in the functioning of the dynamics of ecosystems has long been limited by technological and conceptual constraints. Recent developments show that most elementary transformations at the origin of matter and energy flows in ecosystems are of microbial origin. A major challenge for ecologists is now to identify and quantify the metabolic functions of these microorganisms. In the space of about twenty years, our understanding of the functioning of the pelagic ecosystem and its food webs has considerably evolved.

a) Microbial loops

The existence of the *microbial loop* was proved in the early 1980s (Azam et al., 1983). It can be summarized as follows: if the predation of phytoplankton is conventionally the basis of aquatic food webs, the use of carbon excreted by living organisms and/or products of post-mortem degradation of phytoplankton by bacteria generates a microbial food web called the *microbial loop*. In a pelagic environment, bacteria are abundant and can be consumed by zooplankton, but their principal predators are themselves heterotrophic protists (Ciliates, Flagellates, Amoeba) consumed more easily by the zooplankton. This transfer of matter and energy accumulated in the bacteria via the bacterivorous protists ensures a transfer of the picoplanktonic production to the higher trophic levels and consequently a better cycling of the organic matter that cannot be directly used by the zooplankton (Fig. 11.10).

The microbial loop thus intervenes in predation as well as the regulation of bacterial and autotrophic picoplankton populations, and in the more rapid cycling of nutrients that limits the algal growth. This is a fundamental discovery of the aquatic pelagic systems that has necessitated a questioning of ecological theory. In particular, the trophic status and the functional role of organisms, which were easily identified in the case of linear food chains (Fig. 11.9a), now appear to be simplistic. Rather than going directly to higher trophic levels, a significant part of the primary production, sometimes more than 50%, is diverted to the microbial loop, where nutrients are rapidly remineralized and reintegrated in the stock of dissolved inorganic substances (Fig. 11.9b). The microbial activity also creates a positive feedback at the base of the food chain (Stone and Weisburg, 1992).

By this process of rapid cycling and remineralization, the microbial loop keeps nutrients in the system while making them available almost constantly for phytoplankton. Recent studies have indicated that the phosphorus supply to

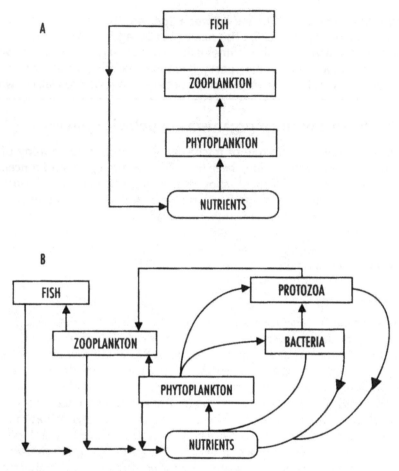

Fig. 11.9. Pathways of nutrients in aquatic ecosystems. (A) In a food chain with a vertical structure. (B) Cycling of nutrients by microbes at the base of the food chain (modified from Stone and Weisburg, 1992).

plankton in the temperate lakes comes essentially from cycling within the planktonic community and not from external sources (Hudson et al., 1999). Within the range of 4 to 80 µg P/l, there is a linear relationship between total phosphorus in the water and the rates of regeneration. In other words, the cycling is equally efficient whether in a eutrophic or oligotrophic environment.

The role of microorganisms, largely underestimated at first, proves to be at least as important for the functioning of ecosystems as the chain based on autotrophic plants. The bacterial production in various lakes is between 20 and 600 mg C/m²/day, which is of the same order of magnitude as the phytoplanktonic production. As an example, the bacterial production in Lake Leman is about 75 g C/m²/yr, which corresponds to a quarter of the net primary production of the lake. Nevertheless, there remain many uncertainties as to the efficiency of the carbon and energy transfer across the microbial loop (Azam, 1998). Some results suggest that the microbial loop acts as a sink for carbon and energy, and that its ecological importance lies in remineralization rather than in biomass production (Karl, 1999).

b) The role of viruses

Dissolved organic matter is produced by various metabolic and ecological processes. *However, recent studies have indicated that the lysis of cells by viral infection may be an important process in the production of dissolved organic matter in pelagic habitats.* In fact, viruses are among the most abundant groups in the oceans (Fuhrman and Suttle, 1993) in any geographic location, around 5 to 25 times as abundant as bacteria.

This lysis, which involves 30% of the mortality of bacterioplankton and 10% of that of phytoplankton, releases debris in the form of dissolved molecules, colloids, and cell fragments (Fuhrman, 1999). Most of these products become immediately available to bacteria, thus creating a viral loop. Recent studies show that the diversity of viruses is much greater than was earlier thought. Advances in methodology, especially in molecular biology, make it possible to better understand the role of viroplankton in the structure and functioning of food webs and aquatic ecosystems (Amblard et al., 1998).

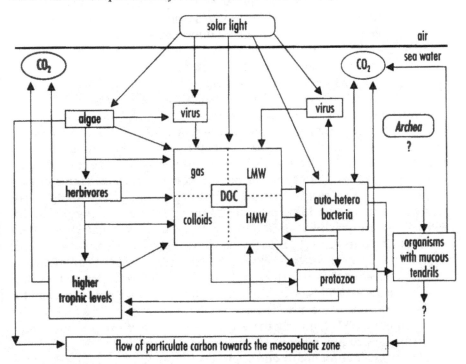

Fig. 11.10. Diagram of an oceanic food chain. At left, the classic pathway of carbon and energy flows from algae to higher trophic levels. At right, the microbial loop, which uses the energy stored in dead detritic matter to produce biomass in microorganisms that can again enter the classic carbon and energy cycle. At present, the role of *Archea* is not clearly understood. DOC, dissolved organic carbon; HMW, dissolved organic matter of high molecular weight; LMW, dissolved organic matter of low molecular weight.

11.6. THEORIES ABOUT THE CONTROL OF ECOSYSTEM FUNCTIONING THROUGH FOOD WEBS

11.6.1. The top down and bottom up theories

Conventionally, food webs have long been considered linear chains in which nutrients flow from the primary producers to higher trophic levels. Referring to this vertical structure of the food chain, and strongly influenced by the hypothesis of interspecies competition, ecologists have considered that competition between primary producers for the use of nutrients plays a major role in the regulation of populations. *This is the theory of control of communities by the resources, or bottom up control. The available resources, regulated by the physicochemical environment, control the food chains from the producers to the predators.* In other words, a given trophic level is strongly constrained and limited by the abundance of the lower trophic level. The role of phosphorus, for example, as a limiting element for primary production and as a stimulant for eutrophication, conforms to this hypothesis.

Many ecologists subsequently helped to demonstrate that there is also an inverse effect and that *the functioning of an ecosystem is strongly constrained by the predation exerted by the higher levels on the lower trophic levels (top down control).* One of the consequences of the many linkages within food webs is the principle of cascade interactions: predation exerts a direct effect on the prey and indirect effects on the other lower trophic levels (Carpenter et al., 1985). In an aquatic environment, the hypothesis of top down control (Northcote, 1988) postulates that predation by fish modifies the structure and functions of an ecosystem, and that the effects are transmitted in a cascade along the food chain, definitively controlling the state of the entire ecosystem.

Ecologists disagreed on the relative importance of top down and bottom up controls, and each camp presented arguments to support one hypothesis or the other. However, this pointless debate has given way to a more consensual approach: that these two types of control coexist and are nearly always complementary in a single ecosystem, but dependent on the spatial scale considered and on overriding parameters such as climatic changes (Schultz, 1995). They are generally assimilated with regulatory negative feedback processes. The question is thus to find out how they interact and under what conditions one provisionally overtakes the other (Polis, 1994). Studies on lake environments have shown that top down and bottom up factors each explain about 50% of the variability of the biomass, i.e., that the biomass of phytoplankton and its production are controlled equally by nutrients and by food chains (Carpenter et al., 1991). Similarly, these factors seem to have an equal importance for algal benthic communities in a fluvial environment (Rosemond et al., 1993). In highly productive (or highly eutrophic) lakes, bottom up control seems to be dominant (Person et al., 1992). Nevertheless, bottom up factors generally seem to be more important, for the following reasons (Polis, 1994):

— Consumers cannot exist without autotrophs, whereas autotrophs can exist without consumers.

— Energy follows a single direction in food chains and must necessarily, in one form or another, control the production and biomass of higher trophic levels.

a) Top down control: impact of predation by fish in freshwater systems

The influence of fishes on the structure and dynamics of aquatic ecosystems has long been underestimated. However, since the 1980s, limnologists have realized that fishes are an integral part of aquatic ecosystems and can have an important influence on the other trophic levels as well as on the cycling of nutrients (see Chapter 12). In particular, the predation of fish on invertebrates has a considerable impact on size structure and on the species composition of zooplankton and zoobenthos.

Zooplanktophagous fish prey either by passive filtration of zooplankton on the gills, which selectively eliminates large planktonic organisms, or by visual selection of prey (Lazzaro, 1987). In both cases, the result is a reduction of the average size of organisms composing the zooplankton, i.e., the disappearance of large species in favour of smaller species.

Studies conducted in Lake Paul (Michigan, USA) (Post et al., 1997) are a good example of dynamic interactions between predators and prey. In this lake, the black bass (*Micropterus salmoides*) is a predatory fish that in the adult stage consumes planktivorous fish and eventually practices cannibalism on its own juveniles. However, in the course of its development the black bass feeds mostly on zooplankton and then on invertebrates before it becomes piscivorous. To the extent that there is high variability of recruitment of juveniles of *M. salmoides* from one year to another, in the years in which recruitment is very good we can in theory observe the impact of the cohort on the planktonic settlement with, notably, high predation on the largest species to the advantage of the smaller species. The results obtained in Lake Paul allowed us to verify these predictions: the young in an exceptionally abundant cohort hatched in 1993 eliminated, large cladocerans (*Daphnia*) during the summer, which favoured the appearance of small species of cladocerans (*Bosmina*).

The impact of zooplanktophagous fish on the structure of settlements of pelagic crustaceans was indicated following introductions of new species in aquatic environments. For example, the disappearance of large Cladocera and Copepoda was recorded in Lake Kivu after the introduction of a planktivorous clupeid, *Limnothrissa miodon* (Dumont, 1986). In the United States, the introduction of *Alosa* in the Great Lakes also led to the disappearance of large calanoid copepods (Brooks and Dodson, 1965). There are many similar examples throughout the world.

Predation by fish may also have a significant impact on the demographic dynamics of zoobenthos when prey of certain sizes are selectively preyed upon. The malacophagous fish of Lake Chad, such as *Synodontis clarias* or *Hyperopisus bebe*, consume essentially the young individuals of benthic molluscs (*Cleopatra, Bellamya, Melania*). This high predation explains why, despite nearly continuous reproduction, the size structure of mollusc populations exhibits a constantly low proportion of small individuals (Lévêque, 1997).

b) Bottom up and top down controls in marine systems

Eutrophication and fishing are presently the principal factors of disturbance in marine pelagic systems. The first corresponds to a bottom up disturbance that modifies the availability of nutrients to phytoplankton, while the other is a top down disturbance that brings about the reduction of predators with consequences in terms of trophic cascade on the herbivores (mesozooplankton) via the carnivores (zooplanktivorous fish). There is good reason to investigate the consequences of these disturbances on the function of marine systems.

An analysis of several experiments in mesocosms as well as long-term observations made in marine ecosystems indicate that there is no strong link between the primary producers and the herbivores (Micheli, 1999). Predation by zooplanktivores is manifested in a reduction of the abundance of mesozooplankton, but the latter does not seem to have an effect on the phytoplankton. Inversely, the increase in nitrogen availability favours the development of primary producers but without having discernible effects on the mesozooplankton. Generally, the consequences of resource-consumer interactions do not seem to be transmitted in a cascade in the pelagic food chains. It is thus unlikely that the pelagic fishery directly affects the phytoplankton biomass, but it may have an influence on the zooplankton biomass and the availability of food for other carnivores.

The weak relationships between primary producers and herbivores may be the result of the following:

— Interactions between the species of zooplankton that attenuate their top down effect on the phytoplankton. The simplification inherent in the food chain concept ignores the complexity of interactions between species and does not describe them adequately.

— The efficiency of transfer of primary producers to the higher trophic levels and the impact of herbivores on the primary producers, which depend on the quality of the food and the proportion of edible and inedible algae in the phytoplankton. An increase in the proportion of inedible algae generally goes with the productivity increase due to higher levels of nutrients of anthropic origin.

— Phenomena of advection or losses of nutrients or individuals in the open pelagic systems can attenuate the local biological interactions.

11.6.2. Theory of trophic cascades

The concept of trophic cascade arises from a known principle of fishery management in a lake environment: in a system with four trophic levels, for example (piscivorous fish, zooplanktophagous fish, herbivorous zooplankton, phytoplankton), an increase in the biomass of piscivores will have consequences at all the lower levels of the food chain. The increased predation by piscivores leads to a decline in the biomass of zooplanktophagous fish. The predation pressure on the zooplankton is relaxed, and the result is an increase in the zooplankton biomass and thus a reduction in the biomass of the phytoplankton, which is subjected to greater predation.

In theory, if the biomass of predators can be controlled, we can control the cascade of trophic interactions that regulate the algal dynamics. By adding or removing some predators we can thus influence the overall dynamics of the system (Carpenter et al., 1985). From this arises the idea of bio-manipulation of aquatic environments, which consists of modifying the algal dynamics by planned introduction of species or selective fishing (Fig. 11.11).

a) Trophic cascades in aquatic ecosystems

Some *in situ* observations are available that serve to corroborate the hypothesis of trophic cascades. From 1978 to 1986, some regularity was reported in the planktonic successions of Lake Mendota (Vanne et al., 1990). In early spring, cyclopoid copepods were abundant, then gave way to the cladoceran *Daphnia*

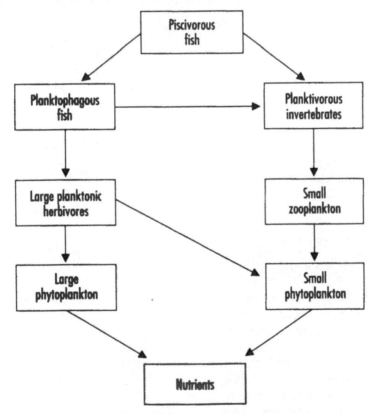

Fig. 11.11. Diagram of the principal interactions between the organisms in a lake leading to a trophic cascade. Large species of herbivorous zooplankton are consumed selectively by planktivorous fish. The carnivorous zooplankton feeds principally on small zooplankton. When planktivorous fish are abundant the zooplankton is composed of small individuals. On the other hand, when planktivorous fish are rare, the large species of zooplankton are dominant because of predation by planktivorous invertebrates. The nutrients (mostly organic P and N) come from inputs and cycling by animals.

galatea mendotae at the end of spring. The calanoid and cyclopid copepods again became dominant during the summer. Two zooplanktivorous fishes, the coregonid *Coregonus artedii* and the perch *Perca flavescens*, were also abundant in late spring and summer. They thus selectively consumed the large cladocerans such as *Daphnia*, With respect to the phytoplankton, there was a bloom of diatoms after the snow melted, followed by a period of clear water (coinciding with the development of *Daphnia*), then a development of cyanobacteria in summer.

An unusually warm summer in 1987 caused a massive mortality of planktivorous fish in Lake Mendota and the near disappearance of coregonids. The population of large cladocerans *Daphnia pulicaria*, very rare in the lake over the 10 preceding years, grew in that summer to supplant the smaller cladocerans such as *D. galatea mendotae*. The greater biomass and size of *Daphnia* led to an increase of 2 to 3 times in the rate of browsing on phytoplankton. The result was a significant reduction of the phytoplankton. This series of events provided an indirect large-scale demonstration of the influence exerted by predators on lower trophic levels.

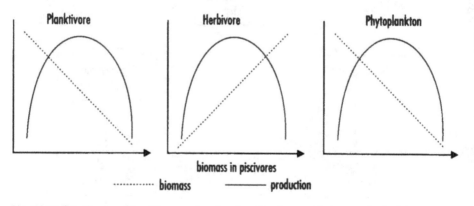

Fig. 11.12. The theory of trophic cascades showing the changes that occur in the biomass and production of zooplanktivores (zooplanktophagous fish), herbivores (zooplankton) and phytoplankton as a function of the biomass in piscivorous fishes (Carpenter et al., 1985)

These trophic cascades, from the fishes to the quality of the lake water, are good examples of feedback between populations and ecological processes. A change in the composition and size distribution of fish will modify the composition and size structure of herbivorous zooplankton. The impact of the latter on the phytoplankton is thus a function of the relative abundance of certain species through their size, their rate of browsing, the nature of the prey they select, their demography, and other factors. Ultimately, these cascade effects are determined in relation to the cycles of nutrients in the pelagic ecosystem (Fig. 11.12).

b) The green world hypothesis

In a system with three trophic levels (plant, herbivore, predator), where predators are abundant, the herbivores will be rarer while the plants, consumed less, will be abundant, from which we get the expression *green world*. This hypothesis (Hairston and Hairston, 1993) suggests that trophic cascades are the key factor of the structuration of living communities and of the dynamics of ecosystems. We note, however, that the hypothesis applies to systems with an odd number of trophic levels (three or five), while with four levels (see above) we end up with control of the plant biomass.

An interesting example of trophic cascade in the terrestrial environment illustrating the green world hypothesis was proposed recently (Post et al., 1999). In a national park in the United States, scientists examining 40 years' worth of data observed that, in response to more snowy winters resulting from the North Atlantic Oscillation, wolves (*Canis lupus*) hunt in larger bands and kill three times as many moose (*Alces alces*) as during less snowy winters, during which they hunt in smaller bands. The reduction of the number of moose diminishes their consumption of balsam fir (*Abies balsamea*), which is their principal food, and the growth rate of balsam firs, which are attacked less because of the fewer herbivores, increases considerably after snowy winters. This study shows that a local trophic cascade can be the result of behavioural change in a terminal predator resulting from a large-scale climatic variability. The environment can thus significantly modify the behaviour of species that have a complex social organization, with important ecological consequences for the functioning of the ecosystem. This study also shows that primary production and the dynamics of nutrients are strongly influenced by herbivores such as moose.

c) Limits of trophic cascades

In practice, interactions between trophic levels are much more complex than might be imagined from the simplified diagram of trophic cascades. There are thus several reasons why the principle of trophic cascades cannot always be verified (Polis, 1994):

— The classic food chain mobilizes only a small part of the energy stored by autotrophic organisms. It has been observed that herbivores consume on average only 10% (between 1 and 50%) of the plant production. The rest passes into the detritic system, via dead organic matter. It is eaten by micro- and macro-detritivores or reenters the microbial loop. Eventually, some detritivores are consumed by generalist predators, so that part of the energy reenters the classic chain (Fig. 11.13). In reality, this link between the two chains circumvents the herbivores. In this context, where a large part of the energy passes through the detritic chain, it seems difficult for herbivores to be able to control effectively the primary production, or for predators to make the system evolve towards a green world.

— A predator may consume prey using the same type of food resources as it does. This is called intraguild predation (Polis and Holt, 1992), and it seems more widespread than earlier thought. For example, an omnivorous filter-feeding fish may consume herbivorous zooplankton and the phytoplankton browsed by the herbivores. However, some species may also change their food regime depending on the resources available. Moreover, many species of fish change their food regime during their life cycle. Thus, we distinguish ecophases corresponding for example to zooplanktophagous fry, juveniles that eat crayfish, and ichthyophagous adults. The existence of this intraguild predation and the plasticity of food regimes also explains the difficulties sometimes encountered in verifying the predictions of trophic cascades.

— The difference in the vulnerability of prey is another compensatory mechanism of the dynamics of complex communities. All plants are not consumed in the same way because they are edible to varying degrees. It is thus unlikely that the pressure exerted by herbivores would uniformly affect the autotroph level as a whole. We must add that spatial and temporal heterogeneity may curb the transmission of trophic cascades.

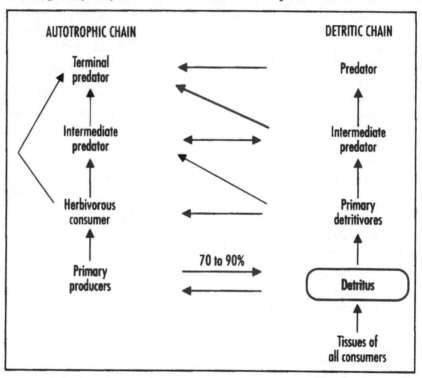

Fig. 11.13. Relationships between the autotrophic food chain and the detritic chain showing the role of detritus (Polis, 1994). The line between the detritus and the primary producers represents the use of nutrients by plants.

According to Strong (1992), trophic cascades exist only in relatively species-poor environments, where one or more of species play a clearly more important role than the others. In a way, trophic cascades are observed in situations in which the disappearance (or introduction) of key species has marked consequences on the system. Such a situation is apparently not the norm in natural ecosystems and even represents a quite unusual category of mechanisms of trophic interactions. Because of this, most trophic interactions do not give rise to cascades. In systems with a high species richness, the effects of predation are buffered by the diversification of food webs and the many adaptations developed by the prey. Most known examples of trophic cascades come from relatively limited and isolated aquatic systems (mesotrophic lakes, in particular), in which algae constitute the basis of the trophic system. Examples in the terrestrial environment are rarer.

11.6.3. Energy flow through size class in a community

In the wake of Lindeman (1942), classic ecology gave an important place to species in the food web to the point of making it a dogma. However, other authors (Cousins, 1985; Frontier and Pichod-Viale, 1998) propose a different analysis of ecosystems based on the size of organisms. The functional importance of the body size of organisms is an easily measured descriptive variable that allows us to discriminate the individuals of different species or belonging to a single species and to take into account a large number of physiological, ecological, and demographic traits of animal populations.

Without entering into detail, we can find arguments to say that the size of an organism is an important ecological parameter that determines largely what it does in the ecosystem.

— The organism changes its food regime as it grows.

— For each predator, there is a limited interval of possible prey (neither too large nor too small).

— An allometric relationship has long been indicated between sizes of organisms and their metabolism. The smaller the organism, the greater the matter and energy flows needed to maintain and reproduce a unit of biomass, and therefore the more oxygen the organism consumes in proportion to its weight than a larger organism. There is thus overall an allometric relationship between the size of individuals and the respiration (R), production (P), and the specific P/B ratio per unit of area of a community. With respect to respiration, there is a relationship of the following type between the weight of organisms and their respiration: $R = aW^b$. When we are interested in a range of weight from bacteria to mammals, the value of b is close to 0.25 (Straskraba et al., 1999). Generally, the smaller an organism, the faster its turnover rate. The result is that matter and energy flows and the turnover rates are higher in the communities made up of small organisms than in those with large organisms.

— The density of large individuals is lower than that of small individuals. In fact, large animals use a great deal of energy per capita per unit of time, so that a determined quantity of resources may maintain more small individuals than large individuals. Some studies (Damuth, 1987) have indicated an inverse relationship between the density of organisms in the habitats where they are studied and their size. The relationship is of the following form: density = body weight$^{-0.75}$. This overall pattern is not always verified, however, and the slopes of regression curves vary with the groups taken into account.

In other words, the size of organisms largely determines their place in the ecosystem, whether terrestrial or aquatic. This is why some ecologists have suggested the study of energy flows in ecosystems, based on the spectrum of size of biomass (Sheldon et al., 1972) and the idea of an "ecology of size" has become common. The distribution of the abundance of organisms as a function of body size has great ecological significance in that it allows us to better understand the functional bases of organization of communities (Blackburn et al., 1993).

The hypothesis of the "cascade model" starts from the principle that species can be classified a priori in a hierarchic system (or in a cascade) so that a species can consume only species of a lower level and can only be consumed by species of a higher level. This hypothesis finds application in the concept of a "trophic continuum" introduced by Cousins (1987), according to which size plays a crucial role in food webs in which a food particle is sought by a predator, which is a larger particle. In theory, along a food chain we observe:

— an increase in size of living particles;

— a reduction in the number of particles as a function of size; and

— a concentration of the biomass in increasingly large units.

Taking these observations into account, the level of biomass present in an aquatic ecosystem can be expressed in the form of an allometric function of the size of organisms. This relationship represented by a log-log graph is called the biomass spectrum. This standardized representation can be used to compare different ecosystems for which we have detailed data on the density of the biomass as a function of the size of organisms belonging to different trophic levels (phytoplankton, zooplankton, benthos, fish). For a series of freshwater systems, the slope of the regression curve is parallel, but the points of interception are a function of the trophic state of ecosystems (Boudreau and Dickie, 1992).

In marine organisms, the size spectrum is nearly continuous, from bacteria to Cetaceae. The size leap between a prey and a predator is about 10 in a linear dimension, and 1000 in volume and weight (Frontier and Pichod-Viale, 1998). In different marine ecosystems, it has been observed that the form of the size spectrum, i.e., the frequency distribution of biomass or abundance, is relatively stable. Duplisea et al. (1997) observe that the biomass spectrum of the Canadian coast was highly stable between 1970 and 1991 despite wide variations in the species composition. In the ecosystem of Georges Bank, there were significant changes between 1963 and 1991, with, for example, an increase in the elasmobranches, which represent 65% of the biomass against 20% of the total biomass in the early 1970s. However, there were only minor changes as to the size structure of the ecosystem (Bianchi et al., 2000). This relative stability of the size spectrum suggests that there may be internal regulations that tend to distribute the biomass of a community over a wide range of sizes, thus favouring a better use of resources.

Ecology of size and functioning of marine food chains

Ecology of size belongs in the context of hierarchical structures and scales. In a pelagic marine environment, two systems of production coexist and function at very different spatio-temporal scales (Frontier et al., 1992). The existence of these two systems is linked to the Stokes law, that in water an organism is quicker to sink passively when it is larger. Each system corresponds to different sizes of organisms, different metabolism, and different spatial extents.

One of the systems corresponds to microorganisms. Because of their small size, organisms of the picoplankton (bacteria, protozoa) are sensitive to the viscosity of the water and remain confined to the surface layers instead of sinking to the depth. The picoplankton serves to feed ciliates and flagellates of 5 to 10 μm (the ultraplankton). These organisms are consumed by appendicularia that are larger but have a rapid metabolism that quickly restores the nutritive salts, which are immediately taken up by the picoplankton. They follow the horizontal or vertical displacements of the water masses in which they are found and this system of production functions almost entirely in a closed circuit.

Another category of producers is represented by larger cells such as diatoms and dinoflagellates, which grow in waters that are well supplied with nutritive salts. These large cells are consumed by crustaceans and other zooplanktonic herbivorous organisms, themselves consumed by larger predators (fish larvae, cephalopods, etc.). These chains lead to large living particles that passively sink to the deep layers, where the organic matter is gradually mineralized. The speed with which they sink is a function of their size. The result is the build-up of large reserves of nutrient salts in the deep layers, reserves that are brought to the surface only during vertical movements. Some large organisms are capable of active displacements (vertical migrations) that can put them in contact with different water masses.

This coexistence of two systems of production in a marine environment is now well established. In the *microbial loop*, the relatively closed and rapid cycle of the matter initiated by organisms of the picoplankton is independent of the large-scale cycling. The physical processes involved are often on the scale of the second and the millimetre. Biologist-oceanographers have perceived, moreover, that the world of microbes (flagellates, autotrophic bacteria, cyanobacteria, etc.) constituting the picoplankton and ultraplankton plays a much more important role than was ever imagined in terms of productivity. On the other hand, for the cycle starting with larger organisms, the spatial scale is that of the oceanic circulation with its upwellings and turbulence due to wind, and time scales from the order of seasonal cycles to that of several years.

This ecology of size, which is mostly applied to the aquatic pelagic environment, is not really verified in the terrestrial environment, where the size distribution of organisms is rather linked to the structure of the physical habitat divided into many distinct niches (Holling, 1992). The structure of terrestrial communities that exploit these heterogeneous spaces is more influenced by the characteristics proper to each of the niches than by all the other species present in the ecosystem. Plants are consumed by many herbivores of various sizes, including some smaller than the plant (parasites, for example). We thus speak of the substrate food, which is larger than the consumer that lives in it but that, nevertheless, initiates a food chain of the "particle" type.

Chapter 12

Biological Diversity and Ecosystem Functioning

"In order to explore further the relation between a system and its components, an analogy of the functioning of a car may be utilized. Everyone has experienced the breakdown of a car. Opening the hood will not enable one to recognize the function of most components. One needs to know the function of the components in detail for any repair. There are very simple components, which are absolutely necessary for the function of the total automobile, such as the gasoline line that connects the gasoline tank with the motor. Other components improve the function but are not essential to the use of a car, such as the exhaust, but its malfunction will result in increased cost, noise, and pollution. There are parts which are not essential for immediate function, such as the bumper, but it is this part which may save lives under extreme conditions. Brakes are used intermittently and for emergencies.... Last but not least, there are parts that make the car more attractive, such as chromium parts, which have nothing to do with function, but which may become important when selling the car. Even if all components of the car are present and are intact, the car may still not run properly, if it is not well tuned, i.e., if the assembly of individual components is not acting together."

Schultze and Mooney (1993)

For a long time, population biology and the ecology of communities were focused on the study of the distribution and dynamics of species and communities as a function of physical, chemical, and biological environmental constraints. In a way, it was a search for the rules that govern the distribution of living communities. Ecosystem ecology was interested in the overall flows of matter and energy at different temporal and spatial scales, greatly emphasizing the processes rather than the species involved. *The new challenge facing ecology is to know to what extent each species present in an environment acts individually or collectively on the general functioning of the ecosystem, whether in the context of biogeochemical flows or in that of biological productivity* (Solbrig, 1991).

In fact, we can pose the hypothesis that species are agents characterized by functions that they assume within the ecosystem (Barbault, 1997). We speak in this context of *ecological niche*. The functioning of the ecosystem is thus the reflection of the collective activity of plants, animals, and microbes, and the consequences of this activity on the physicochemical environment. But do all the species present in an ecosystem participate in an equivalent way in the functioning of the system? For example, is it necessary to have a few hundreds of phytoplankton species in a lake when only a few contribute significantly to the phytoplankton production? Do some species serve redundant functions and/or are there species that serve more important functions than others (key species)? More generally, we need to probe the relationships that may exist between biological diversity and the "stability" of ecosystems and/or their capacity to respond to disturbances.

Ecological niche

The term *ecological niche* has evolved semantically as a function of the maturation of ecological concepts. According to Elton (1927), it designates what the species does in the community to which it belongs (what it eats and by what it is eaten). This functional definition was revived by Hutchinson (1957), who defined niche as the set of conditions in which the population lives and perpetuates itself. This was actually a generalization of the notion of habitat, that is, the domain of tolerance vis-à-vis the major environmental factors, which are symbolized by a hyperspace with *n* dimensions. Subsequently, Odum defined the niche of a species as its role in the ecosystem: the habitat of a species is its address, the niche is its profession. The niche thus corresponds not only to its place in the food web, but also to its role in nutrient cycling, its effect on the biophysical environment, and other characteristics. At present, the ecological niche is generally characterized in relation to three principal axes that combine most of the relevant variables of the environment: a habitat axis (climatic and physicochemical variables), a trophic axis, and a temporal axis (mode of use of food resources and of land cover as a function of time).

The answers to these questions are important for the understanding of ecosystems and their functioning. However, decision-makers, more concerned with taking action, sometimes pose the problem in the following terms: Are all the species indispensable to the proper functioning of ecological systems, and will the disappearance of some of them have marginal or major consequences? In other words, what species are helpful in rehabilitating certain types of ecosystems? We must own that scientists most often feel poorly equipped to provide precise and practical answers to these questions. They are still far from being able to propose general laws, if that is at all possible, from the information available to them. Still, there is a growing interest in working on this subject.

12.1. WHAT IS BIOLOGICAL DIVERSITY?

The term *biodiversity*, a contraction of biological diversity, was introduced in the mid-1980s by naturalists who were uneasy at the rapid destruction of natural environments and their species and demanded that society take measures to protect this heritage. The term was then popularized in discussions that were held from the time the convention on biological diversity was signed during the conference on sustainable development at Rio de Janeiro in 1992 (Lévêque, 1997).

Biodiversity or biological diversity?

The question of biodiversity arises because human societies have acted with an unprecedented amplitude on the diversity of the living world. In this context, there is an environmental problem in the strict sense. The origin of the problem and the responses that can be made lie in the social behaviours or choices of societies with respect to economic development. It thus seems reasonable to reserve the use of the term *biodiversity* to activities linked to the conservation of natural environments in the context of sustainable development. There is a term, *biological diversity*, which is perfectly adapted to the framework of ecological studies such as those we are considering here. This term will therefore be used here in preference to *biodiversity*.

The biological diversity in an ecosystem is expressed at three levels of integration in the living world (Fig. 12.1).

- *Intraspecific diversity*, which concerns the genetic variability of populations belonging to a single species. It is because of this genetic diversity that the species can respond to changes in the environment, selecting the genotypes best adapted to the prevailing conditions at a given time.

- The *diversity of species*, from the perspective of their ecological functions within the ecosystem. There is a huge variety of forms, sizes, and biological characteristics among species. Each of them, acting individually or in groups within food webs, influences the nature and extent of matter and energy flows within the ecosystem. The interactions between species, not only

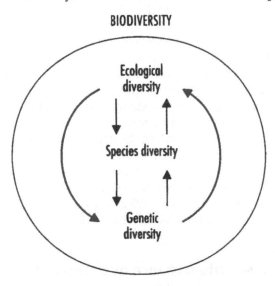

Fig. 12.1. The concept of biodiversity involves all the interactions between the diversity of species, their genetic diversity, and the diversity of ecological systems (di Castri and Younes, 1995)

competition but also mutualism and symbioses, are another aspect of the role of biological diversity in the dynamics of ecosystems.

- *Ecosystem diversity* from the perspective of the variety and temporal variability of habitats. Species richness is generally considered a function of the diversity of habitats and the number of ecological niches that can potentially be used. Large ecosystems, because of their biological diversity, help regulate geochemical cycles (fixation, storage, transfer, and cycling of carbon and nutrients) and the water cycle and influence the gas composition of the air.

Biological diversity, in the ecological sense, is thus a dynamic phenomenon of interactions within each of the levels of integration of the living world, and between these different levels. The functioning of the ecosystem, including its biogeochemical flows, is controlled by these interactions. This integrated approach is a virtual epistemological breakthrough given that specialists so far have tended to be interested in just one level of integration. *However, it is also a recognition that the living world acts on its physicochemical environment by modifying it.* The functional processes of ecosystems such as matter and energy flows are thus controlled jointly by physicochemical and biological processes.

12.2. GENETIC DIVERSITY AND ADAPTATION OF BIOLOGICAL SYSTEMS TO ENVIRONMENTAL CHANGES

The diversity of life begins at the molecular level because each individual belonging to a species is slightly different from the others in genetic terms. Because of this genetic polymorphism, the individuals making up a population may respond in slightly different ways to environmental constraints. As summarized so well by Mayr (1982), organisms are doomed to extinction, unless they change continually in order to survive in a constantly changing physical and biotic environment. These changes are omnipresent: climates vary, concurrent species invade a domain, predators extend their range, food sources fluctuate. In fact, few environmental components remain constant. Hutchinson put it more succinctly when he said that, in the ecological theatre, an evolutionary play is being played out. In other words, variations in the structure and functioning of ecosystems are simultaneously the cause and the effect of natural selection. In this context we can speak of a *biological diversity loop*, the biophysical environment being a source of variability and biological diversity the product of adaptation and selection that in turn interacts and modifies the environment.

12.2.1. Natural selection: chance or necessity?

Chance generates variations (mutations) that then become part of the process of selection. The principle of *natural selection* involves two complementary processes:

the existence of a hereditary genetic variability, and selection of the best-performing individuals in reproductive terms in a given type of environment. When conditions change, the genotypes that produce the phenotypes most apt to respond to new constraints have an adaptive advantage and are selected over the course of successive generations. Selection essentially affects the frequency of genes: any gene controlling the adaptations that reinforce the chances of successful reproduction will have an advantage and its frequency can therefore increase over generations.

The fundamental source of innovation in evolution lies in the mutations that are manifested by a modification at the gene level. Some of these mutations are not expressed, others are lethal for the host, and still others bring about an adaptive advantage in the organisms that carry them. Generally, the biological traits that have an ecological importance often necessitate the interaction of several genes; the favourable mutations thus have a fairly low probability of appearance, but over the long term this probability is not nil.

A fundamental question that biologists ask is whether chance is the only master of the long history of evolution or whether there is an invisible hand, *an unidentified force that gives evolution a direction?* Here we find the ancient paradigm of the economy of nature (see Chapter 10). According to Jacques Monod and many other authors, chance is the only possible solution. In this game of evolution, the result of which is to select the organisms that have the largest number of descendants (principle of fitness), the only real prize for success is the opportunity to continue to play (Levin, 1999). There does not seem to be a global perspective over the long term in this phenomenon of evolution. However, this position is not unanimous, and the question of determinism of evolution will probably give rise to further debate in the future.

12.2.2. Adaptation and phenotypic plasticity

Organisms can adapt to variations within certain limits in their environment. In the changing world in which they evolve, this capacity of living things is essential to their survival. *Adaptation, which is possible because of genetic polymorphism, is thus a functional mechanism that allows species to face the variability of environments.*

There are many facets to adaptation, ranging from a behavioural response to the physical environment to protection against enemies or to adjustment of physiological processes and the internal metabolism. An adaptation is characterized by a modification of the phenotype in response to a specific signal from its environment, a modification that has a close functional relationship with that signal and leads to an improvement of biological functions such as growth, reproduction, and/or survival (Stearns, 1989). Adaptation at the local level can thus be defined as an increase, within the population considered, of the frequency of biological traits that ensures better survival or more successful reproduction in the particular conditions offered by the environment.

A preliminary response lies in the *plasticity* of organisms. The biological characteristics of the phenotype, which could be morphological, physiological, behavioural, and so on, are in fact the result of an interaction between the genes and the environment. Plasticity corresponds to the variability induced by the environment in the expression of the phenotype. The flexibility of the species is tested, especially in the use of resources it needs to survive. In an article on *phenotypic plasticity*, West-Eberhard (1989) recalls a basic rule: the phenotype is the product of the genotype and the environment but it is also a target for selection.

Phenotypic plasticity can thus be considered the production by a single genome of a diversity of adaptive responses that will eventually be subjected to natural selection. Consequently, phenotypic variability merits closer attention as a factor influencing the speed and direction of evolution. Behaviour is a characteristic of the phenotype that presents great plasticity. Some authors feel that the expression of phenotypes is manifested primarily by behavioural change, particularly when a population must face a new habitat or niche. Behavioural adaptability is important because the activity of individuals can, in theory, reduce or exacerbate the effect of the heterogeneity of the environment (Wcislo, 1989). The role of behaviour in evolution has already been suggested by Lamarck, and according to Mayr (1988), changes in behaviour are nearly always the first step toward evolutionary changes. Among the hypotheses that emphasize interactions between evolution and changes in animal behaviour, we must mention the capacity of apprenticeship that allows animals to exploit new situations and to gain access to new resources. Imitation and apprenticeship thus facilitate the acquisition of "evolutionary novelties" and their transmission within the populations concerned.

Microbes, champions of adaptation

Microbes have the peculiarity of adapting rapidly to environmental variations. In any microbial population, there are individuals, the mutators, that constantly maintain a high variability in their progeny. This variability is most often useless, but it takes on an adaptive value when there is an abrupt change in the environment: among the variants, some individuals may have advantageous mutations and thus be better adapted than the initial colony, which they will ultimately replace. However, adaptation of bacterial colonies is acquired by gene transfer between bacteria rather than by mutation. Many cases are known of this natural genetic engineering by which bacteria rapidly adapt to modifications in their life medium (Balandreau, 2000). For example, soils that are planted with maize adapt to the use of the herbicide atrazine in a few years: at that point, all those soils contain bacteria that can degrade atrazine. Another example is that of soybean, a Chinese legume introduced in North America without its symbiotic nitrogen-fixing bacteria. In a few decades, American bacteria became symbiotic with soybean, with the same efficiency found in the continent of origin.

12.2.3. Genetic diversity and functioning of biological communities

Processes of adaptation that are observed at the infraspecific level have direct consequences on ecological processes. In particular, they can modify the nature and intensity of biotic interactions between individuals, whether within the population or between different species. At a macroscopic scale, spatial and temporal heterogeneity can promote genetic polymorphism and the existence of genotypes adapted to conditions of a specific habitat. Besides, adaptations can considerably modify the role of species in processes such as nutrient cycling or in food chains. *As a result, through the phenomenon of adaptation, which is simultaneously the consequence and the cause of genetic diversity, we can see the emergence and maintenance of a biological order in ecosystems.*

Two examples illustrate the interactions between genetic diversity and the functioning of biological communities, as well as the consequences for the functioning of the ecosystem. One shows short-term adaptive responses to seasonal variations of the biophysical environment, and the other shows a long-term response in the context of an evolution between environment and species that is expressed in a better use of all the resources offered by the environment.

a) Genetic diversity of planktonic crustaceans and migratory behaviour

The existence of a great genetic diversity within populations is well known in small freshwater planktonic crustaceans of the genus *Daphnia*. These Cladocera are peculiar in that they are parthenogenetic, at least during the summer, and at the same time they have the ability to hybridize. The result is a coexistence in a single environment of many clones that correspond to either different genotypes of a single species or different genotypes of hybrids. These clones can reproduce sexually on certain occasions and thus can cross. There is also a significant phenotypic variability in Cladocera, which needs to be studied in the context of the vulnerability of individuals to predators (Mort, 1991).

In a small lake in Germany, the existence of at least 54 different clones of the species *Daphnia galatea* and the hybrid *D. galatea* × *D. cucullata* was demonstrated. There were differences in the daily migratory behaviour of different genotypes of *D. galatea*: while the adults tended generally to rise to the surface during the night, there were genotypes that did not migrate and one of them, on the contrary, tended to migrate to the depths. For the hybrid population the general trend was rather to migrate to the depth. These results, which have been corroborated in other lakes (De Meester, 1996), show that there is a genetic component in the migratory behaviour of *Daphnia* clones.

Other observations have also indicated that in the copepod *Diaptomus kenai* Wilson there are populations with genotypes selected for the presence of certain predators. The vertical migrations for each copepod stage (period, amplitude) are adapted so as to minimize the impact of predators that normally attack them at that stage. The migratory behaviour may thus change during development according to the variability of the predators.

In the present state of our understanding, the genetic diversity of populations is thought to be a way of minimizing the risks while simultaneously maintaining a large variety of clones and genotypes that can adapt to different environmental conditions of an abiotic and biotic nature (predation, parasitism). It is a kind of all-risk insurance.

b) Adaptive radiations

The evolutionary process known as *adaptive radiation* is the colonization of several ecological niches of a single ecosystem by populations or species descending from a single ancestor. There is a coevolution between biological species and ecological functions that leads, over a sufficiently long time, to the constitution of "species flocks", which are monophyletic groups of endemic species that are morphologically very similar. The existence of these species flocks is interpreted as the evolutionary response of an ancestral species whose populations were gradually differentiated to specialize in the use of different resources (food, habitat, areas of reproduction) offered by the environment. In a way, it is an optimization of the use of available resources that is expressed by changes in the general functioning of the system, with increased complexity in the pathways of matter and energy flows.

Adaptive radiation in lakes

There are many examples of adaptive radiation in lake systems. Among them are the cichlid fish of the large East African lakes. The "haplochromines" of Lake Victoria descend from a single species of fluvial origin that colonized the lake and gave rise to some 300 living species that presently occupy a wide variety of ecological niches. The specialization of behaviour is probably an important element in the success of adaptive radiation. In particular, the fish developed virtual trophic specializations and all the available resources seem to be used by the cichlids, some being used by them exclusively. Many trophic categories can be distinguished (Witte and Oijen, 1990): the detritivore-phytoplanktivores consuming benthic debris as well as a combination of planktonic and benthic elements; phytoplanktivores; browsers of algae growing on rocks (epilithic) or on plants (epiphytic); phytophages consuming plants; molluscivores, consuming bivalves and gastropods, some species having a highly developed pharyngeal bone that allows them to crush shells; zooplanktivores; insectivores, consumers of crayfish and crabs; piscivores; pedophages feeding on embryos of larvae of other species; lepidophages essentially consuming the scales of other species; and consumers of external parasites.

Beyond the trophic specialization, there is reproductive behaviour, including displays and sexual and specific colorations, and territorial and parental behaviour (Lévêque and Paugy, 1999). There is thus a combination of all the conditions favourable to a sympatric speciation.

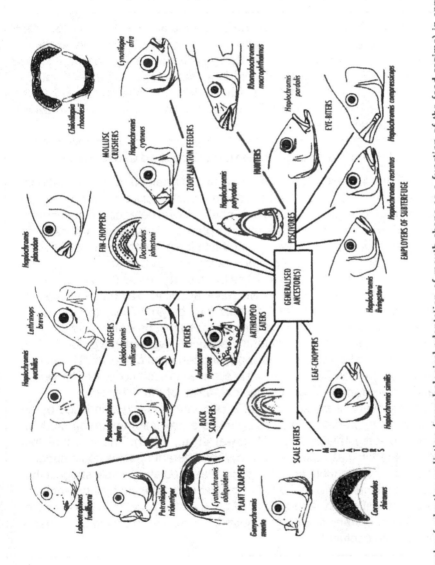

Fig. 12.2. Example of adaptive radiation (morphological adaptation of mouth shape as a function of the food regime) in some species of Cichlidae in Lake Malawi (Lévêque and Paugy, 1999).

Similar phenomena have been observed in other African lakes, such as Lake Malawi (Fig. 12.2) and Lake Tanganyika, as well as in various "ancient" lakes (older than 100,000 years) such as Lake Titicaca and Lake Baikal. There are also cases of adaptive radiation in aquatic groups other than fish: e.g., ostracods, molluscs, crustaceans (Martens et al., 1994).

A universally known example of adaptive radiation is that of the chaffinches of the Galapagos Islands, studied by Darwin. The 13 species known are descended from a common ancestor that reached the Galapagos some 5 million years ago. Each species is established in an island and in a type of habitat in which it is differentiated from others. The group resulting from this diversification is more efficient in the use of the different local resources than was the ancestral species.

Is there genetic transmission of acquired experience?

Predators are important agents of natural selection of populations. Experimental studies showed that the introduction of a predator can trigger a rapid evolution of morphological or behavioural defence mechanisms in animals and chemical defences in plants. These induced defence mechanisms are considered a manifestation of genotypic plasticity, i.e., the possibility for a genotype of producing various phenotypes according to the environmental characteristics in which it evolves (Haukioja, 1999). However, recent experiments have drawn attention to a new phenomenon: the experience of parents in this field may be genetically transmitted to the descendants. In effect, parents exposed to a predator have progeny that are better prepared to defend themselves against that predator than the progeny of parents that have never been in its presence. This is true for the cladoceran *Daphnia cucullata* exposed to kairomones of *Chaoborus*, and for wild radish exposed to the caterpillar of *Pieris rapae*.

The discovery of transmission of a capacity of defence induced by an exposure to predators to the progeny is a new element in the study of predator-prey relationships (Agrawal et al., 1999). The fact that this phenomenon was observed in two very different species could suggest that it may be frequent and, with a slight extrapolation, could suggest that certain effects of the environment could be transmitted by inheritance to later generations in various types of organisms. Could this be a revival of the theory of transmission of acquired characters proposed by Lamarck in the 19th century?

12.3. THE ROLE OF SPECIES IN ECOSYSTEMS

Ecologists are interested in the functions of species in ecosystems, seeking especially to identify those that play a dominant role in food chains. The most abundant species in terms of biomass are nearly always those that ensure the highest productivity and nutrient cycling. We can thus expect that their disappearance from the community will have a greater impact on certain functional processes of the ecosystem than the disappearance of other species. Therefore, those that have the greatest probability of modifying the ecosystem processes are those that modify the following (Chapin et al., 1997):

— The *resource dynamics*. The introduction or loss of a species may alter the resource availability for other species.

— The *trophic structures*. Several significant changes in ecosystem functioning come from the introduction or disappearance of predators or diseases that have greater effects than expected, given the biomass of the species. There are many examples that prove that the introduction of allochthonous fishes may profoundly modify the composition of piscicolous communities, plankton, and benthic invertebrates. Inversely, in the tropical lagoons, overexploitation of fish favours the development of Diadema sea urchins, which causes a gradual replacement of corals by algae in the absence of eutrophication (Jackson, 1994).

— The *disturbance regime*. Organisms modifying the disturbance regime favour non-equilibrium processes such as colonization over equilibrium processes such as competition. This is true for engineer organisms such as the beaver (see section 12.3.1).

12.3.1. Key species

The concept of key species postulates that some species are more important than others in the network of interactions within an ecosystem. It suggests that we can focus on these "structuring species" and ignore the others, which is an attractive idea because it seems to be an easy route to conservation and management of ecological systems. However, there have many critics who have questioned the reality and practicality of this concept. Because of this, although the concept of key species is relatively easy to highlight in the case of large predators, it has often been used in a vague and excessively broad sense. As pointed out by Mills et al. (1993), the concept must be more clearly and objectively defined before it can be used practically. A key species can be defined as a species belonging to a functional group without redundancy, so that the loss of that species causes massive changes in the structure and processes of the ecosystem and may precipitate other extinctions. This concept has some heuristic value in ecology.

The concept of key species was introduced by Paine (1966). On the rocky banks of the intertidal zone of the American coasts, Paine noted a constant association of mussels, acorn-shell (*Balanus glandula*), and a starfish (*Pisaster ochraceus*), which served as predator. In an experiment, Paine eliminated the starfish in some zones. A few months later he observed that the acorn-shells increased in number and occupied 60 to 80% of the area available in the zones that had no starfish. A year later, it was the mussel *Mytilus californianus* that dominated nearly the entire area. The conclusion was that the predator allowed the coexistence of many species of invertebrates competing to occupy the rocky belt located in the intertidal zone. The disappearance of the predator, which Paine called a key species, resulted in the impoverishment of the intertidal community (the composition of which fell from 15 to 8 species), allowing a dominant species to monopolize the resource, in this case the substrate.

Among the key species, there are some major groups (Mills et al., 1993).

- *Key predators* are species whose presence severely limits that of other species. For example, planktivorous fish limit the abundance, even the presence, of large zooplankton in lakes. The piscivorous predatory fish, which contribute to the structuration of ecological systems by means of trophic cascades (see Chapter 10), are also considered key species. Similarly, the disappearance of large terrestrial predators in Europe led to the proliferation of certain crop pests. Most often, the existence of predators allows the coexistence of a large number of species by limiting the development of invading species.

- *Engineer organisms* are those that directly or indirectly modify the availability of resources for other species by causing changes in the physical state of their environment. Acting in this way, they modify, maintain, or create habitats.

  Autogenous engineers modify the environment by their own physical structure, as for example trees and corals. The production of branches, fruits, or coral tissue is not in itself an engineering activity, but these organisms constitute physical structures that modify the environment and create habitats for the other resources. Trees have an impact on the temperature, light, and nutrient cycle and help to structure the physical environment of terrestrial ecosystems. Corals play a similar role and modify the rate of currents as well as the rates of sedimentation.

  Allogenous engineers modify the environment by transforming living or non-living matter from one physical state to another (Lawton, 1994). A classic example is the beaver (Naiman et al., 1988). By cutting down trees to construct dams on rivers, it modifies the morphology of rivers and the hydrology of the system. Beaver dams, which create a small reservoir, slow down the current, retain sediments and organic matter, and modify the structure of the riparian zone and wetlands. Beavers thus indirectly modify the composition of benthic communities by favouring the development of species of calm water at the expense of species of running water. They also have an impact on the structure and productivity of the riparian zone by cutting down trees and shrubs. Other examples of allogenous engineers are termites and ants, as well as most of the burrowing organisms (benthic fauna) or drilling organisms. Earthworms, which dig up and turn over soils, modify their organic and mineral composition, as well as the nutrient cycle and the drainage.

Earthworms—soil engineers

Earthworms are essentially saprophagous animals that feed on plant debris from roots or from fallen leaves and twigs, as well as the organic matter already present in the soil. Many earthworms are virtual "biological farm workers" that dig large networks of horizontal or vertical galleries in the soil, galleries that may be a few millimetres wide. These galleries favour the penetration of water, air, and roots into the soil, aerating and draining the soil. The direct mechanical action is also manifested in the transport and turnover of soil components. In the temperate regions, it is estimated that earthworms ingest on average 300 t of soil per hectare per year, as opposed to 850 to 1150 t/ha/yr in tropical regions. Their faeces form clods composed of organic matter and fine mineral matter, consolidated by iron oxide and manganese as well as by bacterial mucus. Earthworms thus help maintain a lumpy soil structure and increase the microporosity, offering favourable conditions for microbial life and roots (Abdul Rida, 1994). Earthworms also serve as decomposers, digesting the litter and releasing nutrients that can be assimilated by chlorophyllian plants. Finally, they are an important reserve of nitrogen that can be mobilized when they die and serve to feed many other vertebrate or invertebrate animals. Earthworms can effectively improve the properties of soil and protect it from degradation.

Engineer organisms probably exist in most ecosystems and play an important role in their functioning. However, till now, apart from spectacular examples such as the beaver, we have given only limited attention to the role of organisms in structuring the habitat (Jones et al., 1994).

Key mutualists are organisms that are directly or indirectly necessary to the maintenance of other associated populations. This is true of pollinators, which ensure the fertilization of many wild and cultivated plants. The harvest of fruits, vegetables, and oilseeds depends on the presence of an abundant and varied fauna of pollinators, mostly insects. Insects also require a wide diversity of environments to complete their biological cycle.

Another role of biological diversity within ecosystems is to ensure the dispersal of plants by a number of seed dispersal systems using animals as vectors. This is called *zoochory*. Birds, bats, and mammals also disseminate seeds, a process essential to the regeneration of plants in forest systems. Birds and mammals also disperse animal organisms. For example, temporary aquatic environments are colonized by organisms that are transported by sticking to the hairs or feathers of terrestrial vertebrates.

Seed dissemination by vertebrates in the tropical forest

Active transport by an animal (zoochory) is the most widespread mode of seed dispersal in the tropical rainforests. In Guyana, 80% of plant species produce zoochorous fruits that involve 72 species of birds out of 575, and

36 species of mammals (including 23 chiropters) out of 157 (Ministry of Environment, 1998). A seed attracts an animal by its nutritive value (pulp, aril) and by structures that form an attractive signal (colours, movements, odours, etc.), often adapted to the sensory capacities of disseminating vertebrates. For each type of zoochorous fruit, here is a more or less diversified group of frugivorous animals whose size and mode of life are compatible with the characteristics of the fruits (size, shape, chemical composition of the pulp, position on the branches, etc.). For each plant species, the dissemination thus depends on the feeding habits of the vector animals, their daily rounds, their resting areas, and their social behaviours. The result is a spatial redistribution of plants that is always different from what existed before. Many plants with large seeds are dependent on large animals such as monkeys or large birds (toucans, agamis) for their dissemination. The disappearance of these animals, which are often hunted, can reduce the plant diversity.

12.3.2. Rare species

The term *rare* refers to the abundance and distribution of a species. There are several possible interpretations of rarity. A species may have a limited distribution but abundant populations, a limited distribution and low numbers, or a wide distribution but low numbers. The causes of rarity vary and include the need to find highly specialized habitats, a low dispersal capacity, and the trophic position. The species may have a particular behaviour that makes it less detectable by the usual sampling methods. It may also be considered a relic species, on the way to extinction.

Species with a limited distribution are generally endemic species. But what about species with a wide distribution but low numbers? What role do they play in the functioning of ecological systems, given that the common species ensure the basic functions? According to some ecologists, even if these rare species do not serve major ecological functions at present, they represent a form of assurance or guarantee of the stability of ecosystems to the extent that they can replace the presently abundant species if ecological conditions happen to change.

Walker et al. (1999) explored the hypothesis of the functional similarity between some groups of dominant and rare species in Australian savannahs, postulating that this similarity may buffer the effects of disturbances or environmental variability in an ecosystem. According to their hypothesis, the dominant species serve certain functions in a given environmental context but the rare species will serve the same functions in different environmental conditions. In other words, they will substitute for the dominant species if the context changes in order to continue to serve similar functions within the ecosystem. Having established a typology of guilds encompassing rare and abundant species serving the same functions, Walker and his colleagues were able to test and confirm their hypothesis. The resilience of Australian savannahs, for example, is ensured in each functional guild by a set of species having different capacities to respond to environmental changes. This resilience of communities is the result of a diversity of responses at various scales of disturbance as well as of the diversity of responses possible at a given scale. These results raise the question of the origin of such functional complementarity. It could be linked to the history of environments that have fluctuated greatly over the long term, which favoured the selection of species fulfilling the same functions but in different contexts.

12.3.3. Guilds and functional groups: complementary and compensatory effect

In an ecosystem, each of the processes involved in the circulation of matter and energy, the spatio-temporal structuration, and the maintenance and regeneration of the biological diversity is implemented by a variable number of species. In many cases, the species richness observed seems greater than that required to ensure the continuity of the function considered, which suggests that there is some functional redundancy between species ensuring the same function. The Spanish ecologist Ramon Margalef emphasized the existence in nature of a diversity of species much greater than what was apparently required to ensure the proper functioning of the ecosystem. He called this the "baroque in nature", a profusion of species and processes that are able to coexist but do not seem indispensable to the overall functioning of the ecosystem.

In fact, however, it is not always possible to determine precisely the relative contribution of species to ecological processes and we often talk of *functional groups*, sets of species exerting a comparable action on a determined process or responding in a similar manner to changes in external constraints. For example, it is the set of species that exploit the same category of food resources, or the set of species involved in certain major biogeochemical cycles (nitrogen, carbon). In other words, the functional group exists only in relation to a previously defined function, and a single species many belong to various functional groups during its life. The guild is actually the result of a process of simplification by ecologists that consists of isolating functional entities in often complicated food webs.

We also talk of functional complementarity when the rates of ecosystem processes remain perceptibly constant despite disturbances causing modifications in the structure of populations that govern those processes (Frost et al., 1995). A reduction of the biomass of a species may thus be compensated for by an increase in the biomass of another species serving the same ecological functions (Lawton and Brown, 1994). In fact, we observe repeatedly that processes such as primary production can be maintained at relatively constant levels despite disturbances that generate substantial changes in the composition of species controlling these processes.

The relatively unvarying structure of food webs, despite a great variability of species composition, is also an argument in favour of functional redundancy. This is probably not independent of the observation site. A function may be ·

served by a single species or a few species in one ecosystem, while it is served by a large number of species in another ecosystem. The degree of functional redundancy depends on the function considered as well as the constraints proper to the organization of the ecosystem, its ancient or recent history, and the evolutionary dynamics of species (Van der Hammen, 1992).

Results of *in situ* experiments confirm this hypothesis of functional complementarity. In an experiment of acidification of Little Rock Lake in the United States, the biomass of cladocerans, copepods, and rotifers remained high despite the loss of some species in each group (Brezonik et al., 1993). For each of these groups, at least one rare species was identified in the biomass before the experiment that became more numerous as the acidification progressed. The increase seemed to be a compensatory response due to the reduction of interspecies competition for resources, resulting from the disappearance of another species. More generally, in disturbed lakes, modifications are observed in the composition of communities at moderate levels of disturbance, while changes in the ecosystemic processes appear only when the disturbances are severe (Frost et al., 1995). Toxic substances, for example, generally cause changes in the species composition but few changes in the functional processes (Howarth, 1991). These phenomena of complementarity and compensation could constitute a sort of insurance for ecosystems, in that the essential functions are maintained despite the loss of certain elements.

In the framework of the complexity/stability debate in ecosystem ecology, we might ask whether greater diversity generates greater complexity, which in turn generates greater stability. It may be more relevant and less difficult to look at questions of functional redundancy of species as an element ensuring a more stable dynamic behaviour in ecosystems (Naem, 1998).

12.3.4. Mutualism, symbiosis, parasitism

Since Elton (1927) proposed the concept of the food chain, studies on population ecology and the organization of communities emphasized principally the relations of predation and interspecies competition. Competition was defined by Darwin (1859) as the demand of at least two organisms for a single environmental resource (food, habitat, etc.), the availability of which is less than the demand. The principle of *competitive exclusion* describes the phenomenon by which a population of a sympatric species, which has a competitive advantage in the appropriation of a resource, ensures the control of this resource and eliminates populations of other ecologically similar species belonging to the same settlement by depriving them of access to this resource. However, this slightly monolithic approach is no longer widely received, and ecologists have rediscovered the importance of relationship other than competition in the organization of communities.

Commensalism, mutualism, and symbiosis are interactions between species with reciprocal benefits. In a *commensal* interaction, the host does not benefit (in principle) from the organism to which it offers shelter and cover. Epiphytic plants such as lichens or orchids are examples of commensal organisms. *Mutualism* is a more evolved form of commensalism since it is obligatory for the organisms involved, with reciprocal benefits, but the species may still live an independent life. *Symbiosis* involves an obligatory and indissoluble association between two species. In reality, the distinctions are not always easy because they depend largely on the knowledge we have on the biology of the species involved.

a) Mutualism

Ecologists have challenged the idea that interspecies competition alone explains the great biological diversity presently observed in some communities. They suggest in fact that other types of relationships such as *mutualism* and commensalism have been neglected in studies on the organization of communities (Kawanabe and Iwasaki, 1993). In mutualist relationships, the two partners draw a reciprocal benefit from their association. One partner plays a role that may be compared to a service for its associate and, in return, receives a "compensation", i.e., it finds an advantage in the association. Such relationships could favour the coexistence of many species having different ecological niches. Mutualism is widespread in nature under extremely diverse forms and comprises relationships between species that live freely and/or species that live in close association throughout their lives. Over the course of evolution there are many examples of interactions that change, so that some species progress from a situation of antagonism to commensalism and even to mutualism.

b) Symbiosis

When the mutualist relationship is strong and involves the entire life cycle of the partners, it is called *symbiosis*. At present, each of the major types of ecosystems has a cortege of symbiotic associations, which is the expression of adaptive solutions to various environments. Examples are also found in alimentary gut bacteria of vertebrates as well as in coral reefs, where madrepores enclose in their tissues unicellular algae, zooxanthella, which live in symbiosis with the polyps, providing them with various substances that they produce by photosynthesis.

On virgin substrates and in cold environments (lava, tundra, mountains), it is the lichens that are predominant. Lichens are a symbiotic association of an alga and a fungus. The photosynthetic partner is a cyanobacterium (in 10% of the cases) or a green alga (in 85% of the cases). The fungal partner is most often an ascomycete. While the photosynthetic partner ensures the carbonate supply, and sometimes the nitrogen supply for the association, the fungi optimize the volume explored, which is an efficient adaptation to the exploitation of soil resources (Le Tacon and Selosse, 1994). They can develop very large areas of contact by proliferation of hyphae and the hyphae have the capacity to excrete into the external environment large quantities of protons, which disorganize the crystalline networks (Lapeyrie et al., 1991).

The importance of mycorrhizal symbioses has been proven experimentally (van der Heijden et al., 1998). It was shown that the composition and species richness of mycorrhizae controlled the species composition, productivity, and biological diversity of plants in artificial microcosms and macrocosms. Microbial interactions thus seem to have an influence on the functioning of terrestrial ecosystems, a phenomenon that has till now been largely underestimated.

Origin and role of mycorrhizae

Many recent studies show that symbiotic microorganisms play an important role in the dynamics of plants. A plant is surrounded by a cortege of parasitic or symbiotic microorganisms and, in nature, 95% of plants are associated with fungi to form a mixed symbiotic organ, the *mycorrhiza*. The appearance of terrestrial plants was made possible by means of Glomales, lower fungi forming vesicular and arbuscular (VA) mycorrhizae, made up of species that could not live without being associated with plant roots and that have lost all their sexual reproductive capacity. The diversification of endomycorrhizal fungi is traced back to the time plants emerged from water (Simon et al., 1993). These primitive mycorrhizae were well adapted to warm climates, which were most common in the late Primary and the Secondary Era, and to terrestrial ecosystems with rapid mineralization of organic matter. These mycorrhizae are still dominant in tropical ecosystems (Le Tacon and Selosse, 1997).

When the global temperature fell, terrestrial plants, especially trees, associated with different partners. In a cold temperate climate, many trees are associated with higher fungi, Ascomycetes and Basidiomycetes, to form ectomycorrhizae (Smith and Read, 1997). The oldest ectomycorrhizae known date from 50 million years ago and were diversified in the temperate regions, following a major drop in temperature about 40 million years ago, at the beginning of the Oligocene. It was the higher fungi that conserved their sexuality and could grow in the absence of plant roots. In the temperate forests, ectomycorrhizae (ceps, chanterelles, boletus, etc.) colonize the exterior of the tree roots and insinuate themselves between the root cells without penetrating the cells. These ectomycorrhizae play a major role in the nitrogen and phosphorus supply to trees and help protect them from pathogens. If fungi are experimentally suppressed, the plants suffer nutrient deficiencies and wither (Delosse and Le Tacon, 1999). It has been shown in fact that ectomycorrhizae can exploit resources to which isolated plants do not have access (plant debris and organic matter, insoluble minerals) and that they allow trees access to nutrients.

---

c) Parasitism

Parasitic organisms, or pathogens, often play a role equivalent to that of predators in controlling populations and structuring communities. Plant pathogens can have direct or indirect effects on the physiological state, fertility, or viability of the hosts. The most spectacular, and those that have the most obvious impact on the composition of ecosystems, are the "killers", which cause the slow or rapid death of the host (Dobson and Crawley, 1994). An example is elm disease, which in the 1980s decimated *Ulmus campestris* in western Europe and was caused by a pathogenic fungus. Parasitic diseases of plants are sometimes correlated to variations in climatic conditions. For example, experts suspect that the mildness of recent winters was responsible for the virulence of the parasite *Alder phytophthora*, which caused severe mortality in alder, as well as the apparent resurgence of canker diseases in oak and beech.

Parasitism may also have effects on the physiology of hosts, without killing them immediately. In populations of red grouse (*Lagopus lagopus scoticus*), it has been found that a parasitic nematode, *Trichostrongylus tenuis*, which invades the intestinal caeca of the bird, affects its fertility and life span. It is the source of cyclic fluctuations, with periodic collapse of the population, as has been observed in the past: when the nematodes become abundant, the fertility of female partridges may be reduced by half and lead to a decline in the population. Experimentally, the preventive treatment of red grouse populations against the parasite prevented their decline in numbers (Hudson et al., 1998). The impact of this single parasite on red grouse populations is thus considerable.

Pathogens can also indirectly modify the structure of plant communities by acting on herbivores. In the national park of Lake Manyara in northern Tanzania, two successive anthrax epidemics greatly affected the populations of impalas in 1977 and 1983 (Prins and Jeugd, 1993). Because of the considerable reduction of grazing after the epidemics, young plants of *Acacia tortilis* established themselves and significant cohorts corresponding to the epidemic years are now present. In this example, the parasite actually functions as a terminal predator, so that there is a trophic cascade (Chapter 11).

Long-term interactions

In host-parasite systems, two organisms with different genetic information live together, often one inside the other. The genetic information of each partner is thus expressed side by side and in the long term. This is called *long-term interaction*, as opposed to the *short-term interaction* found in prey-predator relationships.

The information of the parasite may be expressed in the phenotype of the host and reciprocally. Besides, DNA may be exchanged between genomes of the parasite and of the host. The result is that the parasite can, in theory, manipulate the physiology and behaviour of its hosts in a direction that favours its transmission or survival, i.e., its *fitness*. The hosts respond by elaborating varied defences.

Manipulation and exchange of information gradually transform some systems (not all) from a parasitic state towards a mutualist state. A mutualist association functions as a single entity with respect to evolution, but each partner remains governed by the advantage it finds in the association.

Any host-parasite system exerts selective pressures on its biotic environment and receives pressures from it. In this sense, parasites participate in the equilibrium of the biosphere as a whole (Combes, 1995).

12.4. HYPOTHESIS ON THE ROLE OF BIOLOGICAL DIVERSITY IN ECOSYSTEM FUNCTIONING

Living organisms constantly transform minerals into organic molecules, and vice versa, Because of this, they participate in the regulation of the physicochemical quality of the water, soils, and atmosphere. However, the species differ from one another in the way they use and transform the resources, in their impact on the physical environment, and in their interactions with other species. Research on biological diversity theoretically emphasizes the singularity of each species and its actual contribution to the functioning of the ecosystem, rather than the averages resulting from the aggregation of several species that are thought to play a similar ecological role.

We can thus state the hypothesis that species richness is important for two reasons: first of all, the number of species is a measure of the probability that species having important biological traits will be present (e.g., key species, engineer organisms). Subsequently, the greater richness theoretically allows the coexistence of more diversified biological traits in the ecosystem, a factor

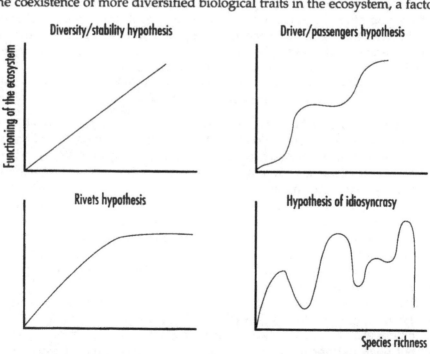

Fig. 12.3. Hypothesis of relationships between biological diversity and ecosystem functioning (biogeochemical processes of ecosystems)

favourable to better use of resources in a variable environment (Chapin et al., 1997). Many hypotheses have been proposed to explain the relationship between the richness and nature of the species present in an ecosystem and their role in its functioning (Fig. 12.3).

- The *diversity-stability hypothesis* proposed by MacArthur (1955) predicts that the productivity of ecological systems and their capacity to react to disturbances increases when the number of species in the system increases. The underlying idea is that the energy flows in the food webs are maintained better when there are a large number of interspecies linkages. In these conditions, there is a greater probability of activating alternative pathways when certain linkages are broken following the disappearance of one or several species. In other words, a greater number of interacting species increases the collective capacity of members of a community to maintain the functions of the ecosystem after a disturbance.

- The *rivets hypothesis* introduced by Ehrlich and Ehrlich (1981) proceeds by analogy. On an aeroplane wing there are more rivets than needed to ensure the integrity of the wing, but the progressive removal of rivets beyond a certain threshold may lead to a sudden collapse of the wing. Similarly, the capacity of an ecosystem to absorb modifications in the species richness decreases to the extent that some species disappear, even if the performance of the ecosystem remains apparently unchanged. This hypothesis recognizes explicitly that each species plays a role in the ecosystem, given that each disappearance progressively weakens the integrity of the system. However, it also implies that the ecological functions of different species overlap so that if one species disappears, the ecological function it serves could be compensated for by other species that serve the same functions. In practice, an ecological function does not disappear till all the species that serve it have been eliminated from the ecosystem. From this point of view, the rivets hypothesis is similar to the *redundancy hypothesis*, which postulates that many species have a similar effect on an ecosystem and can extend their functions in the ecosystem in order to compensate for the drop in abundance or the disappearance of similar species. The hypothesis of compensatory mechanisms (Frost et al., 1995) makes a similar case. An important corollary of the *redundancy hypothesis* is that ecosystems comprising the same number of species may have different functional structures. The result is that it is difficult to compare different multispecies assemblages and attempt to establish a relationship between species richness and stability of ecosystems.

- The *drivers and passengers hypothesis* (Walker, 1992) is an alternative to the preceding hypothesis. In relation to the rivets hypothesis, Walker suggests that species do not play an equivalent role and that many species are superfluous (passengers), only a few of them (the drivers) playing a role essential to maintaining the integrity of the wing. These species, with a more important ecological function than the others, structure the ecosystems.

They may be ecological engineers or key species. It is their presence or absence that determines the stability of an ecological function. The rivets hypothesis and redundancy hypothesis consider the loss of species a random factor: the order and nature of species lost has no importance. On the contrary, the driver and passenger hypothesis takes into account the identity of species: removal of species in an A, B, C sequence may have different results from removal in a C, B, A sequence.

- According to the *hypothesis of idiosyncrasy*, the functions in an ecosystem change when the species diversity changes, but the amplitude and direction of these changes are unpredictable because the role of each species is complex and may vary from one environment to another (Naem et al., 1995). The functions of the ecosystem depend largely on the ecological history and evolutionary history of interacting species. However, several studies have also shown that ecosystems having different biological communities nevertheless possess important similarities in terms of ecological function.

12.5. THE ROLE OF BIOLOGICAL DIVERSITY IN NUTRIENT CYCLES

The productivity of ecosystems depends closely on the availability of nutrients that control the primary production at the base of the food chains. The flow of nutrients in ecosystems is under the control of both chemical processes and biological components of the system.

The role of living organisms in nutrient cycles is complex and we are still far from knowing all its implications. Still, the recognition of this role is an important theoretical advance. Nutrient cycling is not a simple restoration of minerals needed for primary production. It involves numerous biotic components of the ecosystem that will determine, for example, the nature and complexity of circuits and regulate the importance of flows, as well as the rate at which the nutrients will be cycled. Many functions such as nitrification and denitrification, nitrogen fixation, methanogenesis, and decontamination are served only by microorganisms, the diversity of which in the ecosystem is still poorly understood. The bacterium *Nitrobacter* on its own ensures the function of nitrification in soils. It is not the number of species that is important, but the function of species taken individually.

The following examples are meant to illustrate the role of living organisms in the biogeochemical cycles of nutrients.

12.5.1. Microorganisms and functioning of aquatic systems

Despite the significant methodological progress made during the last decade, nearly 90% of microorganisms present in the environment have not been described. Still, the development of methods to quantify correctly the biomass and activity of bacteria has allowed us to take into account the importance of the bacterial compartment in the ecological functioning of aquatic environments (Amblard et al., 1998). While the role of bacteria and protozoa was considered negligible in the traditional concept of the food chain, we now know that they can significantly control the main energy and nutrient flows (see Chapter 11).

Biologists now recognize three major phyla in microscopic organisms: the eukaryotes (*Eukarya*) and two groups of prokaryotes: *Bacteria* and *Archea* (Pace, 1997). What biologists for a long time called "blue algae" are actually bacteria (cyanobacteria) containing chlorophyll *a* and thus capable of photosynthesizing their own organic molecules. It was long thought that the photoautotrophic plankton was dominated by eukaryotic organisms such as flagellates and diatoms, and that it was consumed by mesoplankton larger than 200 im. This traditional view of a classic phytoplankton-herbivore food chain has radically changed after the 1980s. In the northern Pacific, photoautotrophic bacteria discovered in the late 1980s were found actually to be responsible for most of primary production as a whole. The prokaryotes, especially *Bacteria*, are predominant in the biomass. It was also shown that production by photoautotrophic picoplankton (0.2 to 2 μm) occurred constantly, while the eukaryote-browser chain was more seasonal.

With methodological advances (sequencing of ribosomal RNA), we have also proved the existence in the northern Pacific of representatives of *Archea* that were confused till now with phototrophic bacteria. However, the ecological role of these organisms is still not well understood. New discoveries are likely to be made in this field. In any case, it now seems established that the phylogeny and physiological diversity of prokaryotic communities are essential for the maintenance of bioelement cycles (Karl, 1999).

12.5.2. Plant species and transport of nutrients by animals

Vegetation exerts a strong influence on soil properties. In particular, the nature of plant species may determine the soil fertility in natural ecosystems. Plants create positive feedback, directly intervening in the nutrient cycle and indirectly affecting microbial activity and consumption by herbivores. In nutrient-poor environments, for example, plants grow slowly and use nutrients more efficiently. On the contrary, plants in nutrient-rich environments grow rapidly and take up nutrients less efficiently, thus favouring a short cycle.

In nutrient-poor ecosystems, such as tundras and boreal regions or highly eroded tropical systems, the rates of mineralization of organic matter are low, so that the nutrient availability is also low. In these environments, most of the resources are destined for the underground parts to the detriment of the above-ground growth, unlike in plants living in nutrient-rich environments. Besides, plants invest in carbonate metabolites such as terpenes and polyphenols, which are defences against herbivores as well as antimicrobial agents. Because of the existence of these compounds, the litter is of poor quality and decomposes slowly, not favouring nutrient cycling by microorganisms. On the contrary, in ecosystems in which the soils are rich in nutrients, plants favour a more rapid growth of above-ground parts and the roots assimilate nutrients more rapidly than in species of poor environments (Hobbie, 1992). They produce litter that is easily degraded and the high rates of consumption by herbivores further accelerates the nutrient cycle.

Experimental studies have also shown that nitrogen availability in soils can influence the structure of plant communities. Nitrogen input may modify the composition of prairie ecosystems, with the replacement of dominant species by other species (Tilman, 1996). Inversely, other studies have shown that the composition and diversity of plants have an effect on nitrogen levels in the soil (Hooper and Vitousek, 1997).

Biological fixation of nitrogen

Nitrogen is the most important constituent of a plant after carbon. The concentration of nitrogenous forms that can be assimilated by plants (ammonium, nitrates, simple organic compounds) in the soil and water is often insufficient to ensure plant growth. As for molecular nitrogen (N_2), a major constituent of the atmosphere, it is chemically inert and can be used only by certain prokaryotic microorganisms called nitrogen-fixing organisms, which can be free or symbiotic. It has been shown that biological fixation of molecular nitrogen is actually a very important microbial activity for the maintenance of life on earth. It is estimated that around 175 million tonnes of atmospheric nitrogen is reintroduced annually into the life cycle by biological fixation, in comparison to say the quantity of nitrogenous fertilizers used in agriculture, which is about 40 million tonnes per year. There are three groups of nitrogen-fixing bacteria associated with the higher plants:

— the vast group of *Rhizobium* associated with the legumes (families Papilionaceae, Mimosaceae, Cesalpiniaceae);

— the *Frankia*, sporulant filamentous bacteria (Actinomycetes) associated with trees and bushes (Ganry and Dommergues, 1995), e.g., *Alnus, Casuarina*; and

— cyanobacteria associated with various hosts including *Azolla*, a small aquatic fern used for centuries as a green manure in rice fields in the temperate regions of China and Vietnam.

12.5.3. Nutrient cycling and transport by consumers

Many conceptual models note that predators directly influence the lower trophic levels through consumption of prey. They can also modify the species composition of biological communities and thus of the overall functioning of the system. In aquatic systems, there are many examples that show that higher trophic levels (fish, for example) act on the abundance and dynamics of primary producers (Vanni, 1996). However, ecologists have also shown that predators could act by other means on the dynamics of prey, modifying the quality and quantity of nutrient inputs to primary producers. Thus, the effects of a predator on a lower trophic level are not simply direct top down effects but also indirect bottom up effects (Andersson et al., 1988).

Among the roles organisms play in nutrient cycles, the following may be important:

— Heterotrophic consumers cycle nutrients by means of excretion and defecation. Cycling by herbivorous zooplankton and zooplanktivorous fish is an important source of nutrients for the phytoplankton in marine and continental waters (Braband et al., 1990; Schindler et al., 1993). For some algal species, the excretion of zooplankton may stimulate growth, to the point of counterbalancing the mortality by predation. The excretion of zooplanktivorous fish is also a source of nutrients for the phytoplankton. To the extent that nutrients, once used, are cycled, there is a positive feedback.

— Heterotrophic organisms constitute a reservoir that stores nutrients for short or long periods depending on the life span of the organisms. Just like trees in terrestrial systems, fish confer a greater stability to nutrient cycles in aquatic environments, in that they are reservoirs that are not subjected to the same annual fluctuations as other components of the system. For example, phosphorus contained in the biomass of fish is remineralized slowly, and this permanent source serves to maintain a biomass of algal plankton during periods in which less phosphorus is available from other sources.

— Consumers, by modifying the qualitative and demographic composition of prey communities, can act on the dynamics of nutrients. Predation by fish that modifies the size composition of zooplankton (see Chapter 11) can act indirectly on the rates of cycling of nutrients that can be used by the phytoplankton. We have seen that small zooplanktonic species have a better rate of excretion than the large species, per unit of biomass.

— Because of their capacities of displacement, consumers can transport nutrients to various places in the system considered. Zooplankton, which makes vertical migrations that follow a daily rhythm, can transport nutrients between the bottom and the surface. Many species of zooplankton feed on and excrete nutrients on the surface but excrete nutrients only in deep waters. The result is likely to be a reduction of nutrients available for phytoplankton confined to the surface, at least in the stratified lakes. Fish can also transport nutrients between the littoral zones and the pelagic environment by moving between these habitats. They feed in one zone and excrete some of their nutrients in another zone (Fig. 12.4).

The nutrient flows depend on the behaviour of the species with respect to the places where they search for food. Fish that feed on the bottom can also transport nutrients from the benthos towards the pelagic environment, transforming the particular forms stored in the benthos into dissolved inorganic forms (NH_4, PO_4), which thus become available for the phytoplankton (Braband et al., 1990; Carpenter et al., 1992). We often cite the example of Pacific salmon, which come to lay eggs

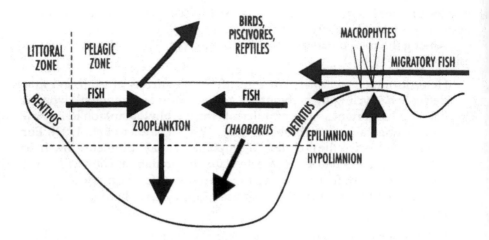

Fig. 12.4. Diagram of nutrient transport by consumers that is the source of exchanges between the various ecological zones of a lake. Fish transport nutrients from the littoral zone towards the pelagic zone. Zooplankton and larvae of *Chaoborus* feed usually in the epilimnion and egest their excreta in the hypolimnion. Migrating fish import large quantities of nutrients into the lake, while macrophytes extract the nutrients stored in the sediment and produce detritus that is used by the detritivores. We must not forget the loss represented, for the ecosystem, by human predation and predation by terrestrial animals on aquatic flora and fauna.

and die in the upper courses of streams after having grown in the sea. They thus provide a considerable quantity of nutrients to the streams.

Salmon and nutrient transport

Coho salmon of the Pacific (*Oncorhynchus kisutch*) pass a large part of their life in the sea, where they grow, and then return to reproduce in the river, where they subsequently die. The massive migration of salmon into the less productive rivers of the western coast of North America is a capital input of food and nutrients. The living salmon serve to feed species of terrestrial vertebrates such as brown bears, eagles, and otters. The eggs and juveniles are consumed by aquatic organisms. The carcasses of dead salmon are rapidly invaded by necrophagous invertebrates and large numbers of insect larvae. The juveniles of coho salmon feed in turn on these carcasses and their necrophages, and up to 95% of the stomach contents of young fish may come from carcasses (Levy, 1997). Isotope analysis of nitrogen has also shown that salmon carcasses are a source of nutrients for vegetation along the banks and in the flooded areas, and that we can find traces of nitrogen coming from salmon in the herbivorous vertebrates (Moore, 1998). Salmon, which has perhaps adapted to poor trophic conditions in water courses by migrating to the sea, is thus an essential element in the functioning of the western American rivers. The significant quantities of nutrients imported each year during the reproductive migration help maintain the functioning of the system and the growth of juvenile coho salmon.

Recent studies have thus considerably modified our ideas about the nutrient cycle. *It is now clear that animals play an important role in the dynamics of nutrients, a role that was long ignored by geochemists.* This role is apparently complex and we have only begun to understand it in a few specific situations. Figure 12.5 presents an example of integration of various levels of organization of biological diversity and predictable consequences for the dynamics of nutrients.

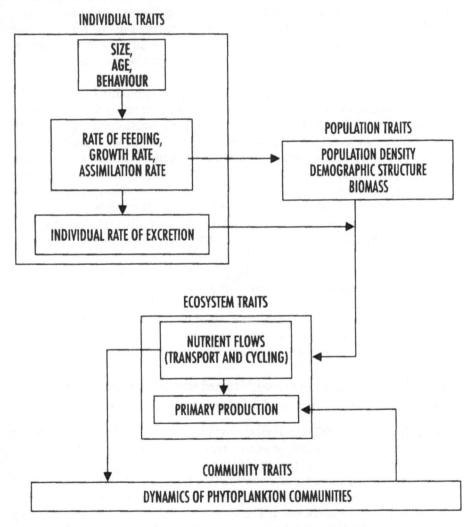

Fig. 12.5. Example showing integration of levels of organization of biological diversity from the individual to the ecosystem: nutrient transport and cycling by detritivorous fish in a lake. Individual traits such as age, size, and behaviour determine the choice of habitat, searching and feeding habits, and rates of food assimilation and growth. At the population level, characteristics such as demographic structure and abundance interact with individual characteristics to influence the functions of the ecosystem such as nutrient flows and primary production. The nutrient flows in turn control the structure of planktonic communities. By feedback, the dynamics of planktonic communities influences the primary production through species composition and variations in the rates of production (Vanni, 1996).

12.6. SPECIES DIVERSITY AND BIOLOGICAL PRODUCTION

Is there a relationship between the biological diversity and productivity of an ecosystem? This question could be an example of a "faulty good question", to which ecology sometimes holds the secret. It needs to be stated more precisely in every way. Are we talking of relationships between primary production and species richness, or the functional richness of autotrophs, or relationships between primary production and the biological diversity of heterotrophs, or relationships between biological diversity and the biological productivity of heterotrophs?

In reality, ecological theories concerning relationships between the productivity of ecosystems and biological diversity are rare, and the points of view are far from unanimity. Studies of the energetic functioning of ecosystems have so far focussed on the biomass present at each trophic level rather than on the diversity of organisms present at each level. It is true that some observations seem to indicate that energy flows in ecosystems are less sensitive to the number of species present. Species-poor environments such as deserts and tundras are also less productive systems in comparison to tropical rainforests, which are rich in species. On the other hand, wetlands or agricultural systems are examples of high biological productivity with a small number of species. Similarly, marshes along the sea coast, in which the benthic fauna is less diversified, are highly productive environments, while the deep ocean is less productive despite a high species richness (Boucher, 1997). Observations in the natural environment suggest that high productivity is not necessarily associated with a great species richness (Holdgate, 1996). It is probable in our present state of knowledge that biological diversity is not an essential element of the productivity of ecosystems. Still, it may play a determinant role.

12.6.1. Experimental approach to relationships between diversity of autotrophs and biological production

In the early 1990s, scientists could not answer questions about the role of biological diversity in the functioning of ecosystems with any degree of precision. Experiments were thus planned to investigate this relationship. In the United States, for example, Tilman and Downing (1994) studied the effects of a severe drought on 207 experimental prairie fields in which the floristic composition was controlled and manipulated. The experiment demonstrated that primary production was better maintained and established itself more rapidly after a period of severe drought in diversified communities than in communities with a low species richness. The authors concluded that reduction of the biological diversity alters the capacity of ecosystems to react to disturbances. They also showed that beyond 10 species there was no longer a relationship between stability and species richness, probably because the principal functions were being served. Simultaneously, Naem et al. (1995), working on model ecosystems of Ecotron (see Chapter 4), showed that diversified systems are also the more productive.

These two studies gave rise to many debates and criticisms. In fact, beyond species richness we must distinguish the role that each species plays in functional processes such as the remineralization or fixation of nitrogen. This is what some call *functional diversity*. According to Steele (1991), functional diversity is "the variety of different responses to environmental change, especially the diverse space and time scales with which organisms react to each other and to the environment."

Thus, to better understand the role of biological diversity in the nitrogen cycle, we must know which species fix atmospheric nitrogen or associate themselves with nitrogen-fixing prokaryotes. There is not always a correlation between the taxonomic criteria and the ecological functions. Experiments have also shown that the stability of ecological processes was linked to the diversity of functional groups rather than to species richness. In a study varying the number of functional groups in 342 fields measuring 13 m × 13 m from 1 group to 5 groups, it was observed that the productivity of American prairies is greater when the number of functional groups is larger (Tilman et al., 1997). In other words, a community composed of N congeneric species is functionally less diversified than one composed of N species belonging to different genera or families.

BIODEPTH

The European project BIODEPTH (Biodiversity and Ecological Processes in Terrestrial Herbaceous Ecosystems) was launched in 1996 with the objective of confirming or disproving the existence of a relationship between species richness and the productivity of a plant ecosystem such as a prairie. Out of 8 sites distributed from Sweden to Greece, 480 parcels of 4 m² each were sterilized in order to control the number of species present. These parcels were reseeded with different levels of species richness, from the monoculture of a leguminous, grassy, or other herbaceous plant to more or less complex combinations of species, with a maximum of 32 species in a single parcel. In total, 200 different combinations were tested. The first results demonstrated a general effect of plant diversity on the production of biomass regardless of the type of prairie and its geographic position. Among the principal mechanisms responsible for this positive effect of species diversity, we must mention particularly the functional complementarity between species that improves the collective performance. In a combination of grassy and leguminous plants, for example, the former exploit the soil nitrogen and the latter fix the atmospheric nitrogen, which overall increases access to the nitrogen resource and thus the productivity of the combination (Hector, 1999).

However, the role of microorganisms and animals has been shown to be important in some studies. Microbial communities react when the species richness of plants within a given functional group varies. Other studies have shown that the number of mycorrhizal fungal species present can alter the production and diversity of plants (van der Heijden, 1998). In reality, most of the experiments conducted so far have not taken into account the variations in the number and identity of herbivores, carnivores, parasites, decomposers, and other organisms. It is thus possible that there exist uncontrolled biases in these experiments due to the activity of other organisms.

In an experimental study of microcosms, Naem et al. (2000) indicated that the productivity of an aquatic ecosystem depends on the diversity of producers and that of decomposers. Producers such as algae draw their nutrients from inorganic sources controlled largely by decomposers, while decomposers such as bacteria and fungi extract their carbon from organic sources that are provided by the producers. This codependence of producers and decomposers is an important factor in the functioning of ecosystems. Simultaneously manipulating the producers (green algae) and decomposers (heterotrophic bacteria), the authors showed that there is no simple relationship between diversity and productivity and that productivity is a complex function of the diversity of algae as well as bacteria. They also proved that in productive experimental systems the bacteria were capable of using a greater variety of carbon sources than in less productive systems.

In other words, the capacity of bacteria to use different carbon sources not only depends on the diversity of bacterial species present but also increases when the algal diversity increases. In this case also, the positive feedback must be pointed out. We can draw the lesson from this experiment that studies of productivity that address only the producers evidently underestimate the complexity and importance of the processes of interdependence.

The experimental approach tends to reconstitute simplified ecosystems that have a larger or smaller number of functional groups. However, the approaches are highly varied, some considering only one species per functional group and others varying the diversity of species within a functional group. In consequence, the results of various experiments are sometimes difficult to compare, and even contradictory. Besides, it is not always easy to identify the functional groups. What conclusions can we draw from these experiments? On the basis of the information presently available, we can say the following:

— *A greater species diversity is a form of insurance for the long-term functioning of ecosystems.* Ecosystems in which there is a redundancy of functions seem better prepared to respond to disturbances than those in which each species fulfils a single and unique function. The consequences of the removal or addition of a species depend thus on its degree of functional similarity with other species of the community. If this species belongs to a functional group that is already represented, its addition or removal will have less important consequences than if it belonged to an unrepresented functional group. In other words, if several species exploit the same resources, as is true for generalist herbivores, the gain or loss of a species has effects on the composition of communities but less effect on the ecosystem processes to the extent that there is a compensatory reaction of other species. Such communities have a behaviour that is relatively predictable (see rivets hypothesis and redundancy hypothesis).

— *As soon as the species richness diminishes, the response of the system depends on the species composition of the community and the identity of the dominant species becomes determinant.* In this context, the behaviour of the system is not easily predicted (see driver and passenger hypothesis).

— The variability of processes in an ecosystem is reduced if the functional groups are dependent on one another (as with functional groups in the food chain) and if the number of species per functional group is high.

— The variation of diversity in a functional group in the food chain has marked consequences on the production of other functional groups.

We cannot help but feel that all this sophistication in the approach has only confirmed what ecologists have suspected for a long time: a greater biological diversity is favourable to the production and stability of ecosystems and ensures the perpetuation of biogeochemical cycles. We must add that since these ecological experiments were conducted in more or less controlled conditions and on a small scale, the question naturally arises of their representativity in relation to environments that are natural or that are artificial but much more complex than the microcosm communities.

12.6.2. Relationships between heterotroph species richness and biological production

We can state the hypothesis that in ecosystems having short food chains, all other things being equal, the terminal production is greater than in ecosystems having long food chains and thus there is a significant energy loss at each change in trophic level. In other words, for an equivalent input of energy, a system composed of only phytophagous species will be more productive in theory than a system that includes many carnivorous species. This hypothesis is verified empirically in livestock farms—but is it true for natural systems?

The theory of intermediate production

In the framework of complex relationships between productivity and species richness, several observations made on various animal groups seem to show that this relationship is unimodal: the species diversity increases with the production up to a maximum value and then decreases when the productivity (or an index of productivity) continues to rise (Rozenzweig and Abramsky, 1993).

The ascending phase is linked to the hypothesis of Wright, who indicated that a positive relationship between the quantity of energy and the local or regional biological diversity on a set of data (Wright et al., 1993). However, the causes of the decrease are not clear. After having tested several hypotheses, Rozenzweig and Abramsky (1993) came to the conclusion, slightly deceptive according to them, that high productivity reduces the heterogeneity of the environment. Therefore, in a regime of high production, it is sometimes observed that a reduction in the species richness is associated with an increase in the density and/or biomass of a smaller number of species, which are the only ones to benefit from this increase in energy.

The data available to test this hypothesis are rare. Nevertheless, quantitative data about piscicultural production estimated by the production of fish or catches by birds were compared in four shallow lakes of intertropical Africa having very different piscicultural communities (Lévêque, 1995) (Table 12.1). These data, if we understand them correctly, are to be considered with a great deal of caution given the many sources of uncertainty. However, the fish production of Lake Nakuru, a saline lake with a single species of introduced tilapia, was observed to be clearly higher than that of other lakes. This tilapia, the production of which was estimated from consumption by piscivorous birds, feeds on the cyanobacterium *Spirulina platensis*, which has very high production in this type of environment. The short food chain could partly explain the high productivity. In the three other lakes (Lakes Chad, George, and Chilwa), the terminal production estimated from fish catches seems to be equivalent (100 to 200 kg/ha/yr). The food regimes are mostly *planktivorous* in Lake George, *detritivorous* and *zooplanktivorous* in Lake Chilwa, and highly diversified with many carnivores in Lake Chad. Thus, the length of the food chains or the diversity of species present does not seem to have an overall influence on the ichthyologic production of these environments.

Similar conclusions could be drawn from research on the American Great Lakes. The Great Lakes were subjected to pollution for more than a century, the fishing was intensive, they lost many indigenous species following introduction of exotic species, and their present functioning is not in any way comparable to their earlier functioning. In spite of all that, the secondary production of these ecosystems has changed relatively little throughout this period (Pimm, 1993).

Table 12.1. Species richness and dominant trophic groups in fish

Lakes	Nakuru	Chilwa	George	Chad
No. of species	1: *Oreochromis*	3: *Clarias, Barbus, Oreochromis*	30, including 21 Cichlids	100 varied species and families
Food	Phytoplanktivore	Detritivores, zooplanktivores	Biomass 64% phytoplanktivores, 20% piscivores	All types
Fishing production (kg/ha/yr)	625-2436	80-160	100-200	100-150

Fish production was estimated from captures in four shallow lakes in tropical Africa (Lakes Chad, Chilwa, George) or consumption of fish by piscivorous birds (Lake Nakuru) (Lévêque, 1995).

12.6.3. Relationships between primary production and biological diversity of heterotrophs: degrees of trophy in waters and composition of fish communities

What are the reactions of ecosystems and biological diversity to an increased level of nutrients? This question has been extensively studied in aquatic systems, where the term *eutrophication* is commonly used to designate an intense proliferation of algae and macrophytes that accumulate excessively in the water bodies (see Chapter 11). This accumulation of plant matter can lead to alterations in the quality of water and biological communities of the water body. In other words, the eutrophication encourages higher productivity of waters by the input of nutrients that usually limit the primary production. It is a "doping" of the autotrophic production.

Whether it is of natural or anthropic origin, one of the principal manifestations of eutrophication is the reduction of oxygenation of water consecutive to a high

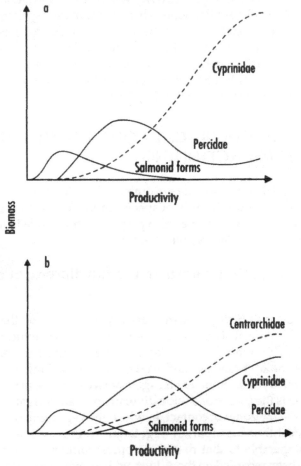

Fig. 12.6. Composition of ichthyologic settlements along productivity gradients of lake systems (Persson, 1991)

oxygen demand by bacteria that decompose the organic matter produced in abundance. The increase in plant biomass and the ecological consequences that result generally occur at the cost of species richness of the system. Thus, one of the effects of eutrophication on fish communities is an increase in the biomass accompanied generally by a modification in the species composition (Lee and Jones, 1991). This is a situation in which a close but complex relationship can be found between production and biological diversity (Carpenter et al., 1996).

In European lakes, there are substantial modifications (biomass and species composition) of the ichthyologic fauna when we consider a gradient of primary production controlled by phosphorus inputs (Persson et al., 1991). In less productive systems, it is the salmonids that dominate, while the Percidae are abundant in the moderately productive systems. The abundance of Cyprinidae increases regularly with the productivity of the system but they are predominant only in highly eutrophized systems. This pattern of succession has been indicated by a comparison of variously eutrophized systems and a study of the consequences of increase in productivity within a single system over time.

In detail, the decline of Percidae in highly productive systems corresponds to the reduction of perch, while ruffe (*Gymnocephalus cernuus*) and zander (*Stizostedion lucioperca*) increase with productivity. Similarly, in the Cyprinidae, the abundance of certain species, such as rotangle (*Scardinius erythrophthalmus*) diminishes with eutrophication, while bream (*Abramis brama*) seems to replace roach in highly productive systems (Fig. 12.6).

12.7. ROLE OF BIOLOGICAL COMMUNITIES IN ECOSYSTEM FUNCTIONING

In the preceding sections, we have examined the role of genetic diversity and the individual and collective role of species in the functioning of ecosystems. However, biological communities also play a functional role in the ecosystem, as we will see in the examples given below.

12.7.1. Role of gallery forests in the functioning of river systems

Water courses are dynamic systems strongly influenced by the surrounding terrestrial environment. They erode their banks and periodically flood their major bed. This modifies their physiognomy by creating and destroying dead branches, secondary branches, and wetlands over the long term. The spatial and temporal dynamics of exchanges between terrestrial and aquatic environments is the source of a wide diversity of plant formations and animal communities in the zones of contact, in which the vegetation plays a dominant role The autochthonous riparian vegetation in Europe has a high species richness, comparable to that of some tropical rainforests. Some 1400 plant species have been recorded on the Adour, with an average of 314 species over a stretch of 500 m (Tabacchi and Tabacchi, 1994).

These riparian tree-like plant formations (or riparian forests) play several roles in ecological functioning (Fig. 12.7) (Maridet and Colin-Huet, 1995).

- *Stabilization of banks.* The roots of many species of trees (willows, alder) and shrubs create a biological web that retains the sediments and retards the erosion of the banks.

- *Flood prevention.* The vegetation influences the water flow. The above-ground parts of grasses, shrubs, and bushes reduce the speed of the current and the chance of flooding.

- *Creation and diversification of habitats.* Fallen trunks and branches from the riparian forest have long been considered a cause of reduced flow and a potential risk for riparian activities. For these reasons, river management focused on the clearing of the river bed. However, it is now understood that drifting woody debris plays a role in the ecological equilibrium of water

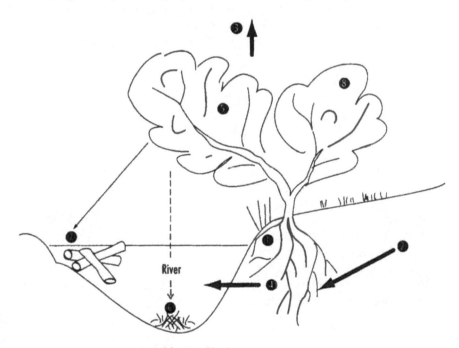

1. Stabilization of banks
2. Filtration
3. Evaporation
4. Denitrification
5. Biological assimilation of nitrogen
6. Organic matter
7. Fallen trunks: shelter for aquatic
   organisms and regulation of water flow
8. Habitat for terrestrial organisms

Fig. 12.7. Influence of riparian forests on ecological processes of water courses (Figure by Maridet)

courses, favouring the creation and diversification of habitats. The succession of waterfalls and calm zones created by fallen trunks and branches provides micro-environments that favour the establishment of many species, as well as a heterogeneity that allows their cohabitation. The dams also help retain organic matter. Besides, the riparian vegetation serves as a temporary habitat for reproduction, feeding or refuge for many terrestrial species (amphibians, birds, mammals). The natural corridor of the Ain valley has 180 species of birds (including teal, egret, heron, and sand martin), of which 100 are nesting species (Michelot, 1990). In general, this wide diversity observed in less degraded systems contrasts with the poverty of highly artificial systems such as poplar or conifer plantations.

- *Regulation of light and temperature.* In small streams, a dense canopy may affect the light intensity at the water surface and regulate the growth of periphyton and macrophytes. In the same way, the riparian vegetation controls the temperature of the water. In temperate water courses, a shaded sector may reduce the water temperature by 3°C. This relative coolness is favourable to trout and salmon, which are sensitive to thermal variations.

- *Source of organic matter.* Riparian woods are a source of allochthonous organic matter (leaves, stems, animals) for the water course. These inputs are degraded by microorganisms present in the water (fungi, bacteria). The quality of inputs varies with the nature of the species constituting the vegetation: for example, the needles of conifers are much less degradable than the more tender leaves of alder or willow.

- *Denitrification and decontamination.* By means of their root system, the riparian woods also act on the nutrient cycle and help eliminate diffuse pollution of agricultural origin. The vegetation absorbs the nitrates and stores them provisionally, but we have also seen that it serves as a filter. Besides, the wetlands and woods offer conditions favourable to denitrification of microorganisms, releasing gaseous nitrogen. Ground water passing through the riparian woods is naturally purified of nitrates from adjacent agricultural areas. According to Pinay and Decamps (1988), a riparian wood of 30 m width eliminates most nitrates.

- *Social and aesthetic role.* The riparian vegetation shapes the landscape and makes the banks of water courses attractive, encouraging recreational activities.

12.7.2. Role of benthic communities in the functioning of marine ecosystems

The knowledge of pelagic environments in aquatic systems has attracted much more interest than that of sediment habitats. It is true that sediments are more difficult to sample than the water column. In any case, these environments cover vast extents in the ocean and in continental waters and harbour an abundant fauna of organisms of all sizes. Snelgrove et al. (1997) attempted to synthesize the role of the benthic compartment in the control of biogeochemical processes in the stability of the sediment, and in the productivity of environments (Fig. 12.8). These different processes involve a close interaction between microorganisms, meiofauna, and macrofauna.

- *Benthos interacts with the water column above*. The benthic food chain functions principally from sedimentary detritic inputs from the pelagic environment and to a lesser extent from the primary production of benthic algae in the shallower environments. The filter-feeding and sediment-feeding organisms use a part of this particulate organic matter (POM) and thus help to improve the transparency of the water. Part of this organic matter will be used by microorganisms, but some part of the detritic matter will also be stored in the sediment and disappear from the system. The microorganisms decompose the POM into dissolved organic matter (DOM) and release nutrients that partly return to the water column. The benthic organisms are also consumed by predators, some of which are benthic and others of which live in the water column.

- *The benthos controls the carbon, nitrogen, and sulphur cycles*. This role in the biogeochemical cycles is partly linked to the trophic cycles mentioned above.

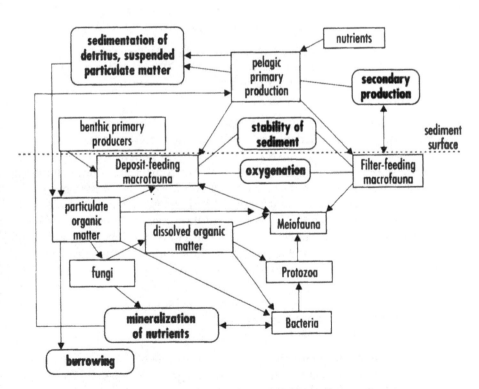

Fig. 12.8. Very simple diagram of relationships between major categories of organisms in sedimentary marine environments, and influence of these groups on some functions of the ecosystem (modified from Snelgrove, 1997).

However, the benthos is also involved in the irrigation of the sediment and particularly in the penetration of oxygen to the depths by means of burrowing organisms. The resulting oxygenation of sediments has a significant impact on the distribution of microorganisms, their activity, and the biogeochemical processes that flow from them.

- *The benthos acts on the stability and transport of sediment.* The mucus secreted by the macrofauna, in particular, aggregates the particles and stabilizes the sediment.

12.7.3. Importance of viruses in the structure and the functioning of aquatic food webs

After the discovery of the microbial loop in pelagic aquatic ecosystems in the early 1980s, studies in the last few years showed the importance of the role of viroplankton in the structure and functioning of aquatic food webs (Sime-Ngando, 1997). Viruses, which do not have an autonomous metabolism, need an organisms or a host cell that is metabolically active for any process requiring energy, including their reproduction. We still have only a sketchy understanding of the species and genetic diversity of viruses, and much remains to be studied, but their density in aquatic systems is relatively high (between 10^4 and 10^8 m/l). In functional terms, viruses play three principal roles:

— They are agents of microbial mortality in the aquatic environment. It is presently estimated that viruses cause at least 30% of the bacterioplankton mortality and 10% of the phytoplankton mortality in the marine environment.

— They regulate microbial diversity in the aquatic environment. The functional impact of viruses is particularly significant in processes that are random and difficult to quantify, such as the increase in genetic exchange or the maintenance of species diversity within the microbial communities. Indeed, the viral pressure results in a constant modification in the genetic make-up of host organisms given that their genetic diversity constitutes a barrier of resistance to viral infection. Some studies show also that viruses may have a greater impact on the genetic structure and thus the species composition of pelagic algae, rather than on their abundance. These observations ultimately support the hypothesis that viral infections, in conjunction with other factors such as spatial heterogeneity or selective predation, are important mechanisms in the processes of exchange and recombination of genetic material and consequently in the maintenance of microbial diversity in the aquatic environment.

— They recycle organic matter in a pelagic environment. The lysis of cells due to virus results in an increase in DOM that can greatly increase the metabolic activity of non-infested planktonic bacteria. Actually, there is a functional loop linked to bacterial lysis through viruses (bacteria to bacteriophages to DOM to bacteria) that contributes to nutrient cycling in microbial food webs, reducing the contribution of bacterial production to the matter and energy flows towards the higher trophic levels.

12.8. CHANGE IN COMMUNITY STRUCTURE AND CONSEQUENCES FOR ECOSYSTEM FUNCTIONING

Species differ in the ways in which they use resources, in their effect on their environment, and in their interactions with other species. Consequently, we can expect that changes in the species composition, whether of natural or artificial origin, will lead to modifications in the nature and intensity of ecological processes. For example, there may be changes in the use of nutrients by the vegetation, which modifies in turn the processes of competition between species and the predator-prey relationships. By retroaction, other changes intervene in the composition of settlements. The modifications of processes resulting from changes in the local species composition can also interact with processes on the regional scale, such as the transfer of nutrients in aquatic ecosystems, and alter the quality of drinking water.

It is still difficult to quantify these modifications in the context of a systemic approach in light of the complexity of the processes at work. We can nevertheless draw some conclusions from modifications induced by human activities and/or the introduction of exotic species into an environment, through their consequences on the structure and functioning of ecosystems.

12.8.1. Consequences of species introductions

Introduction of new species in ecological systems is an *in situ* experiment from which we can study the reactions of a system to a disturbance. Beyond the obituary notices of indigenous species that often disappear as a result of the simultaneous action of various synergic factors, in a certain number of cases impacts have been observed on the structure and functioning of ecosystems.

a) Introduction of Lates in Lake Victoria

In the late 1950s, a large predator fish (*Lates niloticus*) was introduced to develop sport fishing in Lake Victoria. The lake harboured some hundreds of endemic fish species of the family Cichlidae (haplochromines), which constituted a highly complex food web (Lévêque, 1997). In the 1980s, the growth of the *Lates* population resulted in the near disappearance of tens of species of pelagic Cichlidae that consumed phyto- and zooplankton. Subsequently, most of the other species of fish, Cichlidae and others, also became rare as a result of predation by *Lates*. From the perspective of the functioning of the ecosystem and its food webs, the many autochthonous haplochromines were replaced by just two indigenous species: the detritivorous crayfish *Caridina nilotica*, and a small pelagic fish that feeds on zooplankton, the cyprinid *Rastrineobola argentea*. These two species were abundant and served as the main food source for larvae and juveniles of *Lates* after the disappearance of the cichlids. The introduction of *Lates* thus led to an extreme simplification of the food chains (Fig. 12.9), and cannibalism now plays an important role. The large *Lates* feed on their own young, which serve the function of zooplanktivore that the haplochromines used to serve. Simultaneously, the

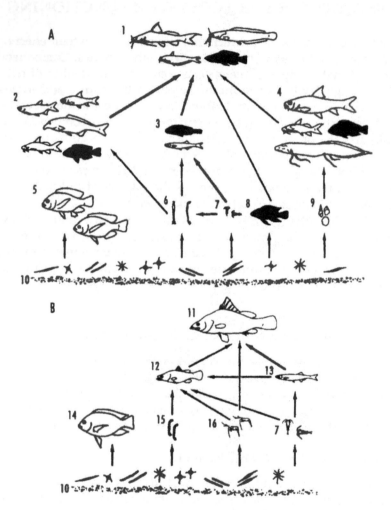

Fig. 12.9. Diagram of food chains in populations of demersal fish of Lake Victoria before and after introduction of *Lates* (Witte et al., 1992). Only organisms that are a significant part of the food of the principal fish species are mentioned. In 1970, *Haplochromis*, shown here in black, were dominant in number and biomass in all the trophic groups, with the exception of piscivores. (1) *Bagrus docmac, Clarias gariepinus, Schilbe mystus,* piscivorous *Haplochromis.* (2) *Brycinus* spp., *Barbus* spp., Mormyridae, *Synodontis afrofischeri,* insectivorous *Haplochromis.* (3) Zooplanktivorous *Haplochromis, Rastrineobola argentea.* (4) *Barbus altianalis, Synodontis victoriae,* molluscivorous *Haplochromis, O. esculentus.* (5) *Oreochromis variabilis, O. esculentus.* (6) *Chaoborus* spp., chironomids. (7) Zooplankton. (8) Detritivorous or phytoplanktivorous *Haplochromis.* (9) Molluscs. (10) Detritus, phytoplankton. (11) *Lates niloticus.* (12) *Lates* juveniles. (13) *Rastrineobola argentea.* (14) *Oreochromis niloticus.* (15) *Caridina nilotica* (Ligvoet and Witte, 1991; Witte et al., 1992).

food regime of the pied kingfisher *Ceryle rudis* has changed: while it once consumed the haplochromines, it now feeds essentially on the pelagic cyprinid *Rastineobola* (Wanink and Goudswaard, 1994).

Although the *Lates* has been accused of being responsible for the disappearance of the endemic Cichlidae, we must remember, to be objective, that the ecological conditions in Lake Victoria have also been modified for many years by the development of agricultural activities in the watershed and urbanization near the lake. The leaching of fertilizers has initiated a process of eutrophication that led to greater algal development, reduced transparency of the water, and deoxygenation at the depths of the lake: factors that could have profoundly disturbed the reproduction of Cichlidae. The introduction of new fishing gear in the 1970s (bottom trawlers) also endangered the populations of cichlids that reproduce mostly on the lake bed. Finally, for some years Lake Victoria has been invaded by water hyacinth, which also modifies the ecological conditions. The impact of the introduction of a large predator such as *Lates* on the lake fauna must therefore be studied in the more general context of environmental change. The impact of the predator was greater because the fish populations of Lake Victoria were already made fragile by other anthropogenic disturbances.

b) Bison in the North American prairies

A controlled experiment of reintroduction of bison (*Bos bison*) in the great plains of North America was an opportunity to observe the impact of this herbivore on the prairie ecosystem, which for many years was susceptible to bush fires (Knapp et al., 1999). Like many large herbivores, the bison does not browse at random. It prefers zones that have been burned. Subsequently, it consumes mostly the dominant grasses (C4 plants) but avoids the bushes and woody species. The grazing pressure on the dominant grasses reduces the competitive pressure on the subdominant species. The result is the development of C3 grasses and forbs and a significant increase in the species richness of plants.

The grazing is not spatially uniform in the prairies, and a mosaic of grazed and ungrazed zones develops. It has been observed that bison tends to select species-poor zones in which grasses are abundant. Under the effect of grazing, these zones become more diversified. However, another factor that can influence the selection of grazing zones is the quality of plants. It has been observed that the leaves of plants are richer in nitrogen in zones where the bison has urinated and thus are more nutritive. The bison has a tendency to frequent those zones that are already grazed. The feeding behaviour of the bison thus results in an increased heterogeneity of the ecosystem, a factor that is in itself favourable to greater species richness. Indeed, the reintroduced bison is a key species in this ecosystem.

12.8.2. Consequences of species removal

The disappearance of species is generally much less documented than the introduction of new species. Two examples illustrate the modifications that can be expected from elimination of species.

a) Consequences of fishing on marine systems

The exploitation of the living resources of the oceans has greatly depleted the stocks of marine fish, particularly demersal fish, catches of which have fallen by 20 to 60% in many areas. The total fishing production approaches 100 million tonnes a year, while nearly a quarter of the exploitable biomass is overexploited and nearly half is exploited at its maximum level (Botsford et al., 1997). Fishing has a large number of direct and indirect effects:

— Reduction or even disappearance of stocks of target fish or reduction in the age and size of populations, especially in species that have a long life span and a low rate of reproduction. Fishing statistics over the past 45 years show that catches of small planktivorous pelagic fish with a short life cycle are increasing, while catches of large piscivorous benthic fish with a long life cycle are declining, particularly in the Northern hemisphere. Intensive fishing thus results in a reduction of the biomass of higher trophic levels, which must have implications for the marine food chains and the functioning of the system overall (Pauly et al., 1998). A similar tendency is observed in freshwater environments.

— The elimination of predators is compensated for by the development of other predators or species of other trophic levels. For example, the elimination of nearly 90% of the Antarctic population of the blue whale, which consumes mainly krill (*Euphausia superba*), has led to the development of other predators on krill, such as the Minke whale (*Balaenoptera acutorostrata*), the seal *Lobodon carcinophagus*, and some species of penguins. Studies conducted in the northwestern Mediterranean support the hypothesis that fishing can alter the structure and organization of marine communities (Sala et al., 1998). They show that fishing has decimated predators of the herbivorous sea urchin *Paracentropus lividus*, which has therefore proliferated. Intensive browsing by this sea urchin has transformed the plant communities of the coastal zones, otherwise made up of erect and fleshy macroalgae, into stagnant communities that are less edible (Steneck, 1998). However, the relationships between the fishing and predation are complex. Sala et al. (1998) showed, for example, that the principal predators of sea urchin are fish and that the predation is a function of the architecture of the habitat. Where shelter is abundant, the sea urchins coexist with the predators, while in habitats that offer little shelter sea urchins are rare. In fact, two guilds of fish consume sea urchins: one consists of small Labridae and Gobiidae, which consume the small urchins, and the other consists of large-mouth Sparidae, which eat the large urchins. While fishing affects the large predatory species more than the smaller species that have no commercial value, the impact of predation is greater on the small sea urchin, whose survival depends greatly on the presence of shelter.

— Captures of non-target species (around one-third of the total captures), often juvenile forms, resulting from the use of non-selective fishing gear.

— The use of gear (dragnets, trawlers) that destroy benthic habitats and the sessile fauna. In many invertebrates that release their gametes in the ambient environment, successful reproduction depends on the density. The disturbance of bottoms probably brought the California abalone (*Haliotis sorenseni*) to extinction (Tegner et al., 1996).

What marine communities existed before fishing?

In the North Atlantic, it is estimated that the biomass of demersal fish, such as the cod and the haddock, was around 10 times as high before exploitation as it was in the second half of the 20th century. What kind of structure of the ecosystem allowed the existence of such stocks (Steele and Schmacher, 2000)? The causal factors, with regard to the present situation, could have been the following:

— A slow rise in the numbers of demersal fish whose populations were made up of larger and older individuals.

— Reduced stocks of pelagic fish. There is proof that herring and mackerel were less abundant before the increase of demersal fishing. The predation of juveniles of pelagic species by the large demersal fish was probably also greater.

— Negligible biomass of predatory invertebrates (jellyfish), which now play a major role in the energy flows. Their present abundance may be the result of a decline in the stock of demersal fish.

— A more effective transfer of the primary production towards the fish, characterized by the small quantity of detritus.

In the context of a systemic approach, the consequences of fishing on the functioning of the ocean system, including the fixation of carbon and other biogeochemical cycles, has not really been studied. Such a project is ambitious and difficult, but not more so than, probably, the study of the impact of greenhouse gases on climate change.

b) Elimination of species introduced in the Antarctic Territories

The damage caused to autochthonous ecosystems by mammals introduced into the French Southern and Antarctic Territories is significant. A programme for the eradication of herbivores was undertaken in some islands in order to restore those islands. This was an *in situ* experiment that provided information on the functioning of these rather undiversified ecosystems with respect to the autochthonous species. However, there are many species introduced from temperate North European regions, and the climatic changes observed recently at Kerguelen interfere with the evolution of communities (Chapuis, 2000).

Among the significant results, we must emphasize that the relationships of competition between native and introduced species in plants, as in invertebrates, have an important role to play in the evolution of communities after the eradication of herbivores. In Amsterdam island, which has a relatively temperate climate, most of the native species that had disappeared from pastures in the southern part of the island reappeared about 20 years after the eradication of herbivores. Besides, the number and abundance of introduced plant species favoured by the trampling of bovines diminished on all the sites. At Kerguelen, on the other hand, the climatic conditions are more severe and there has even been an increase in the species diversity of plant communities after elimination of rabbits, but recolonization by autochthonous species has been slow because of their slow growth and late reproduction. The environment was moreover invaded by two introduced plants, dandelion and groundsel. Dandelion seems to have benefited from the disappearance of the rabbit, which preferentially feeds on it. This is another illustration of the hypothesis that predators limit the proliferation of invasive species.

Chapter 13

The Biogeochemical Cycles

Living matter is made up of minerals: it is estimated that there are around 40 chemical elements in animals, of which 26 are essential to life on earth. The most important biogeochemical cycles involve a simple body, water, and some major elements: oxygen, carbon, nitrogen, and hydrogen, which make up nearly 95% of the atomic composition of living organisms. Phosphorus, sulphur, chlorine, sodium, potassium, calcium, and magnesium represent about 5%. The fact that they are quantitatively less abundant does not signify that they play a less important role in the elaboration of living matter. Phosphorus and/ or nitrogen are often considered limiting factors of primary production. There are also trace elements (sometimes called micronutrients) that are necessary to living organisms. Examples are aluminium, molybdenum, fluoride, strontium, zinc, and cobalt.

Minerals circulate in ecosystems through the food chains, incorporated into organic molecules or in mineral form. During the elaboration of living matter by autotrophic plants containing carbon, various minerals are incorporated into organic matter so that, to put it simply, the resources of the environment are exhausted to the extent that the mass of living things increases. Fortunately, organic matter is remineralized when organisms die, and these constituents return to the mineral state to be ultimately reused. The matter flows thus describe cycles between autotrophic organisms, heterotrophs, decomposers, and the physicochemical environment.

Some essential principles need to be underlined in the study of biogeochemical cycles.

- The various elements (e.g., water) or chemical elements (carbon, nitrogen, sulphur) are stored in what are called reservoirs (e.g., soil, oceans, atmosphere, lithosphere, biomass). There may be exchanges in the form of flows between these reservoirs (Fig. 13.1). For a long time, *geochemistry*, the study of the behaviour of chemical elements on the earth's surface, was devoted to the study and quantification of these exchanges. A good illustration of interactions between these various reservoirs is the oxygen cycle (Fig. 13.2), closely linked to that of carbon and water, since it is the essential source of free molecular oxygen resulting from photosynthesis. It is because of autotrophic plants that the air contains nearly 21% oxygen. A small part of this oxygen is transformed into ozone in the stratosphere (see Chapter 16). Another part serves to oxidize different mineral compounds and accumulates in the lithosphere. Another part is involved, most often through the aquatic ecosystems, in the formation of biogenic structures such as carbonaceous formations that sediment.

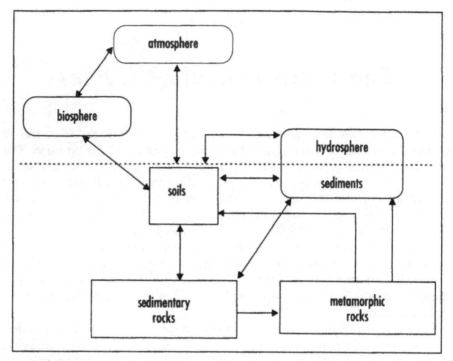

Fig. 13.1. Biogeochemical cycles. Diagram of reservoirs of minerals and exogenous and endogenous cycles of exchanges between them.

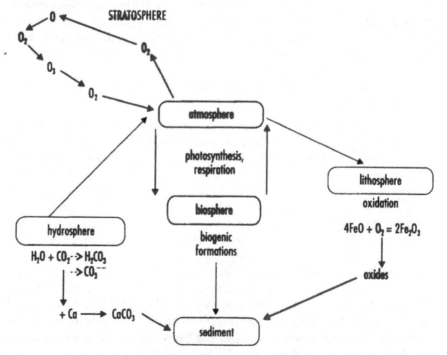

Fig. 13.2. Oxygen cycle: diagram of exchanges between reservoirs

- Over the past few decades, it has been proven that the cycle of elements involves not only physicochemical processes. Living organisms are the source of an entire set of reactions that have long been unknown. *Biogeochemistry* is the study of organic and inorganic chemistry as well as the role of living things in the often complex pathway of chemical elements (Gall, 1995). For example, since the beginning of the Primary Era, the transfer of chemical elements into the lithosphere was considerably modified by the rapid increase in biomineralization. Many organisms capture Ca, Si, or P ions in the external environment to build skeletons, shells, and carapaces, which after the death of these organisms accumulate and contribute to the genesis of sedimentary rocks, considerably affecting the rates of transit of these elements. The *biogeochemical cycle* of an element thus corresponds to its pathway and to the transformations it undergoes in various reservoirs of the planet, including the biosphere. The interactions between living organisms and cycles of elements have assumed particular importance since it was discovered that they may have a significant impact on climatic cycles.

Until the 1960s, the nutrient budgets in lake environments were determined essentially by chemists on the sole basis of water and sediment analysis. It is symptomatic that in the synthesis of the International Biological Programme (Golterman and Kouwe, 1980) the biological processes were nearly entirely ignored in the study of the phosphorus cycle in freshwater environments. In the 1970s, limnologists began to look at the role of consumers in nutrient cycling, particularly through grazing and its consequences on the structure of plankton populations. At present, it is widely recognized that animals—predators and herbivores—control the biomass of primary producers and the phosphorus cycle. More generally, herbivores have a direct impact on the general functioning of systems by their control of the dynamics of prey and by their capacity to transport nutrients from one place to another in the system (see Chapter 12).

- The cycle of each mineral element is not independent even though, for convenient representation, they are treated separately in studies. In fact, all the essential elements must be present simultaneously and in sufficient quantity for the synthesis of organic molecules.

Table 13.1. Principal sources of nutrients available in soil for plants (percentage of total) (Chapin, 1991)

	Atmosphere	Weathering of rocks	Cycling
Temperate forest			
N	7	0	93
P	1	<10?	>89
K	2	10	88
Ca	4	31	65
Arctic tundra			
N	4	0	96
P	4	<1	96

- The needs and modes of nutrient use by organisms are the result of a long coevolution. The result is that in undisturbed terrestrial ecosystems the losses are generally minor. Most of the nutrients used by plants, for example, come from the cycling of elements in the ecosystem (Table 13.1).

13.1. THE CARBON CYCLE

Carbon constitutes less than 1% of the mass of the earth's crust, but it is one of the most important of the chemical elements characterizing the living world, with oxygen.

13.1.1. Carbon and oxygen cycles: two strongly linked cycles

At the origin of the earth, carbon dioxide (CO_2) was the essential constituent of the atmosphere. Very quickly, significant quantities of CO_2 were taken up in the atmosphere and in the ocean because of the development of autotrophic plants containing carbon and chlorophyllian assimilation. The carbon and oxygen cycles are intimately linked in the processes of photosynthesis: the combinations of carbon with oxygen and hydrogen during chlorophyllian assimilation produce glucides while releasing oxygen, according to the following simplified formula:

$$6\,CO_2 + 6\,H_2O + energy = C_6H_{12}O_6 + 6\,O_2$$

This reaction requires the energy derived from the transformation of light energy into chemical energy. Glucides are basic constituents of other complex organic molecules, such as proteins or cellulose.

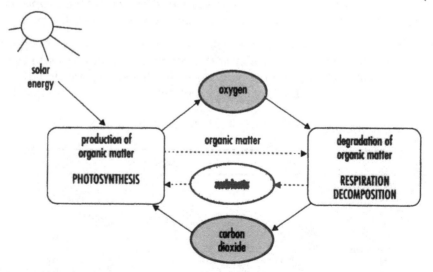

Fig. 13.3. Principal biological processes (photosynthetic production of organic matter and degradation) and relationships between the cycle of dissolved inorganic elements

The transformation of mineral CO_2 into the organic world is the key reaction of life on earth. When animals consume plants, the stock of glucides produced by photosynthesis is degraded by the inverse reaction, which releases water and carbon dioxide and provides energy (Fig. 13.3). This reaction is called *respiration*. Such a reaction of oxidation is also produced during the combustion of carbon, or each time organic matter is burned (burning of coal or petroleum, brush fires).

13.1.2. Carbon reservoirs

The different modes of storage involved in the carbon cycle are quite well known. Carbon is present in a certain number of reservoirs in different forms, such as free carbon dioxide (atmosphere) or dissolved carbon dioxide (aquatic systems), living or dead biomass, carbonate sediments, or fossil fuels. The stocks are estimated as follows: 750 Gt (gigatonnes) of carbon in the atmosphere; around 1000 Gt for the surface waters of the ocean and 34,000 to 38,000 Gt in the depths; 500 to 600 Gt in the terrestrial vegetation; and 1500 to 1600 Gt in the soils (the figures indicated here are of orders of magnitude because they differ in the calculations of various authors). The ocean thus represents a stock of inorganic carbon 50 times as large as that in the atmosphere. The atmosphere is quantitatively the smallest reservoir of carbon but it is the largest in terms of exchanges with the biosphere.

13.1.3. Atmosphere-biosphere exchanges

The dimensions of reservoirs have no direct relationship with the flows (Fig. 13.4). The quantity of carbon present in the biomass is smaller than that contained in the atmosphere, while the vegetation plays a determinant role in carbon flows between the atmosphere and the biosphere. On the earth's surface, there is a constant storage and release of carbon due to photosynthesis and respiration. Exchanges with the atmosphere, the ocean, and plants account for around 200 Gt/yr (Fig. 13.5), which means that more than a quarter of the atmospheric CO_2 is cycled each year. These exchanges follow seasonal cycles. Each year, 8 to 10 Gt of carbon is fixed by plants in spring and summer in the Northern hemisphere, from atmospheric carbon. An equivalent quantity returns into the atmosphere in autumn.

The forests, which are immense reservoirs of carbon, are important agents of the evolution of the earth's atmosphere because the carbon dioxide level depends on the equilibrium between photosynthesis and respiration. The dynamics of carbon in forest ecosystems involves the following processes:

— assimilation of CO_2 by photosynthesis;

— release of carbon by respiration of plants;

— transfer of carbon into soils in the form of woody debris and leaves, and a multitude of organic compounds;

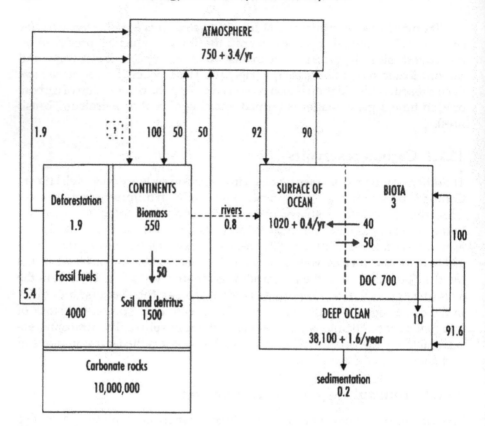

Fig. 13.4. Carbon reserves (gigatonnes) at the biosphere level and flows (Gt/yr) between reservoirs. Values were estimated for the period 1980-1989 (modified from Siegenthaler and Sarmiento, 1993).

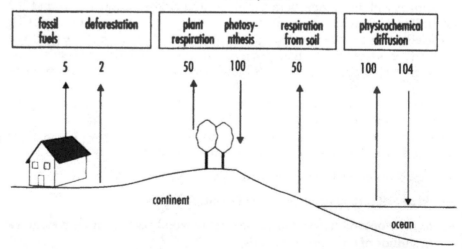

Fig. 13.5. Budget of carbon exchanges between the atmosphere and biosphere (Gt/yr)

— exudation of organic compounds into the rhizosphere;

— eventual release of carbon stored in the soils into the atmosphere.

The terms of this budget have been quantified for an Amazonian tropical forest. The gross production is 30.4 t C/ha/yr, of which around half serves for respiration and the other half (15.6 t C/ha/yr) corresponds to net production, i.e., the quantity of carbon fixed in the tissues. The total quantity of carbon stored is around 180 t C/ha in the epigeal biomass and 64 t C/ha in the hypogeal biomass. In the soil, organic matter represents 162 t/ha. The time of stocking of carbon is 16 years in the biomass, after which it is transferred into the soil, where it is stored for 13 years (Malhi and Grace, 2000).

These studies also indicate that net production in a forest has often been underestimated because of the lack of accounting for root growth and exudation. The result for the tropical forests, which represent 50% of the world's forest area, is an annual production of around 25 Gt C, a value higher than that of 18 Gt calculated till now. With respect to the global carbon budget, it is estimated that tropical forests sequester around 2 Gt C/yr, while deforestation and natural processes are a source of 2.4 Gt/yr, which corresponds to a net source of 0.4 Gt C/yr.

Studies conducted in the Amazon region show that over the period 1989-1998 it functioned as a source of around 0.2 Gt C/yr because of the deforestation and abandonment of agricultural land in the Brazilian Amazon region. This source is clearly equivalent to the storage of carbon by the natural forest. The net flows of carbon are thus nil in this region (Houghton et al., 2000). Another quantitative budget was drawn by European researchers in 15 forests from Italy to Iceland. Depending on the sites considered, the budget varied from a fixation of 6.6 t C/ha/yr to a release of around 1 t C/ha/yr. The CO_2 budget is mainly controlled by respiration, which does not seem to be influenced by the average annual temperature (Valentini et al., 2000). Another study conducted on forest soils in five continents made the same observation with respect to the CO_2 emitted by the decomposition of organic matter (Giardina and Ryan, 2000). These results challenge the idea that global warming leads to an increase of CO_2 emissions by forests.

The role of freshwater ecosystems in the global CO_2 budget is a subject of debate. It has long been thought that many freshwater ecosystems function as sources of CO_2, while marine ecosystems act as sinks. However, recent studies (Duarte and Agusti, 1998) based on the metabolism of communities seem to show that aquatic ecosystems in general act as carbon sources when the respiration rate (R) is greater than the gross primary production (P) in aquatic communities, and as sinks when the inverse is true. Generally, unproductive aquatic systems tend to be heterotrophic (R > P), while productive systems (P > R) are more autotrophic. On the global scale, aquatic environments are carbon sinks because the quantity of atmospheric CO_2 incorporated by productive autotrophic communities is greater than the quantity released by heterotrophic communities occupying unproductive ecosystems.

13.1.4. Oceans and the carbon biological pump

In the oceans, the phytoplanktonic biomass that occupies the upper layers plays the role of a CO_2 pump. The fixation of inorganic carbon by autotrophs transforms the CO_2 dissolved in the water into organic elements, supporting a drop in the partial pressure in the surface layers. This drop in partial pressure at the air-sea interface leads to a flow of atmospheric CO_2 towards the ocean. Inversely, oxidation of organic elements by heterotrophs leads to the formation of CO_2 and to a flow from the oceans towards the atmosphere. The flows between the two reservoirs are regulated by the distribution of wind zones on the surface of the ocean, the temperature at the surface, and other factors.

However, the CO_2 fixed in the superficial layers can also be carried into the intermediate and deep layers following the movements of general circulation in the oceans (see Chapter 15) or by sedimentation of dead organisms. The export of organic matter from the euphotic zone towards the deep ocean, followed by a biological oxidation of organic carbon, leads to an inverse gradient of concentration of inorganic carbon: the concentration of inorganic carbon below 500 m in the ocean is greater than that observed at the air-sea interface. In the ocean, this flow is presently around 16 Gt C/yr, or around one-third of the total production (Falkowski et al., 1998). Once the organic carbon has sedimented under the thermocline, it is effectively sequestered from the atmosphere for centuries, or even millennia. This phenomenon is called the *biological pump* of carbon. It is the process by which part of the carbon of the superficial layer is exported towards the deep ocean in the form of sedimenting organic particles, or organic carbon transported by currents. This carbon is eventually remineralized in dissolved inorganic form by bacteria at the depths. The biological pump is considered to have been in equilibrium during the 20th century.

13.1.5. Carbon in the lithosphere: a long-term store of biological origin

Another way of extracting the CO_2 from the atmosphere or the ocean is to elaborate calcareous biological constructions. Living organisms producing calcium carbonate (CO_3Ca) by extracting Ca^{++} and HCO_3^{--} ions in the water or food are found in large numbers especially in warm climates: molluscs, echinoderms, corals, foraminifers, marine algae such as coccolithophoridae, etc. In the earth's crust or lithosphere, the carbonate sediments are formed essentially by the accumulation of debris of testa and shells or other structures of biogenic origin, containing considerable quantities of carbon that can remain stored for a very long time before being remobilized. Another part, of chemical origin, comes from the precipitation of calcium carbonate under the action of physical or chemical factors.

The storage of carbon in the sediments began very early in the history of the earth. More than 3000 million years ago, the first bacterial communities (cyanobacteria) reduced CO_2 to synthesize their organic constituents, but they also used atmospheric CO_2 to fabricate limestone. They thus elaborated enormous accumulations of carbonates called *stromatoliths*, the fossilized remains of which are found on all the continents. At the beginning of the Cambrian, more than 500 million years ago, new calcifying organisms appeared, which modified the capacities of calcium and carbonate extraction in the ocean. Then, about 200 million years ago, with the advent of Coccolithophoridae and other calcifying planktonic organisms, the biochemical control of carbonates by biomineralization actively contributed to the control of the quantity of carbon present in the atmosphere because the carbon thus fixed escaped the carbon cycle sometimes for a very long time.

Another important carbon reserve is also embedded in the earth's crust in the form of petroleum, carbon, bituminous schists, and gases, the result essentially of transformation of plants and bacteria. In general, the CO_2 used by plants to fabricate living matter is restored to the atmosphere by processes of decomposition, which involves the oxidation of organic matter and leads to the mineralization of carbon. However, if the organisms are embedded in sediments and substrates with oxygen activity, they carbonize and become, for example, fossil fuels (lignite, coal, petroleum, etc.). When plants invaded the continents during the Carboniferous Era, photosynthetic production increased considerably and large quantities of organic carbon synthesized by plants accumulated without being mineralized. In fact, organisms capable of degrading lignin and cutin probably appeared only later (Jupin, 1996).

In summary, at the beginning of the Secondary Era most of the carbon coming from CO_2 of the primordial atmosphere accumulated in two major reservoirs that were inaccessible to living organisms, or accessible with difficulty: one contained reduced carbon (fossil fuels) and the other contained oxidized carbon in the form of carbonate sediments (limestone of biogenic origin, dolomites resulting from $CaCO_3$ precipitation under the action of physical and chemical factors). It is estimated that carbon stored in the mineral form amounts to 34,000,000 Gt in the continental lithosphere and 14,500,000 Gt in the oceanic lithosphere. The mass of organic carbon stored in fossil fuels of the lithosphere is estimated at 16,000,000 Gt (Jupin, 1996). The reservoir made up of carbonate sediments and fossil fuels is almost definitively removed to the atmosphere and would have participated in only a limited way in the present flows if human beings had not discovered how to use it as a source of energy.

13.1.6. The missing carbon

The flows of CO_2 released into the atmosphere because of human activities is relatively well known. The increase in atmospheric CO_2 is the result of emission of 6.5 Gt C/yr resulting from the consumption of fossil fuels and industrial activity, and 1.6 Gt C from deforestation (IGBP, 1998). On average, 3.4 Gt accumulates in the atmosphere each year. If we estimate that the carbon cycle is in equilibrium, in the absence of inputs of anthropic origin, we are confronted with a simple mathematical question: where is the missing 4 Gt of carbon? What happens to the carbon released by human activities and how is it integrated in the natural carbon cycle? What is the role of the two major reservoirs, the oceans and the terrestrial biomass? Will they conserve their regulatory role for a long time and absorb a part of the supplementary flow? This question of the missing carbon calls for a verification of the hypothesis proposed so far, a new task for scientists.

Only piecemeal information is available to us. Some of the CO_2 of anthropic origin is absorbed by the terrestrial biomass. According to some studies, between 0.5 and 1.5 Gt was stored each year during the 1980s and this figure reached 2.6 Gt C/yr in 1993. The same studies showed that the tropics were a net source of carbon, which implied a higher rate of storage in the middle and upper latitudes (Melillo et al., 1996). The oceans play an essential role in the storage of carbon of anthropic origin. It is estimated that the oceans together absorb only 40% of the CO_2 released by the human race, or 2 ± 0.5 Gt. However, the capacity of the superficial layer is limited and, to maintain its regulatory role, the ocean must cede part of the superficial CO_2 to the deep ocean, which has a high storage capacity and thus a long reaction time. The result is that if the CO_2 is absorbed by the ocean, it remains there for a long time, whereas CO_2 assimilated by plants and soil is in principle cycled in a few decades.

13.2. THE NITROGEN CYCLE

Nitrogen is an essential nutrient for plants and animals as well as a pollutant. It is involved in the composition of amino acids, the building block of proteins and nucleic acids (DNA and RNA). In comparison to the carbon cycle, where only CO_2 plays a central role, *the nitrogen cycle is one of the most complex biogeochemical cycles*. It involves a large number of compounds:

— mineral-ionized forms: molecular nitrogen N_2; ammonia NH_3; nitrite ion NO_2^-, nitrate ion NO_3^-, and ammonium NH_4^+; nitrous oxide N_2O; nitric oxide NO;

— organic nitrogen in the form of many small molecules sometimes directly assimilated by plants or microorganisms (amino acids, urea, uric acid, etc.).

13.2.1. Stages of the nitrogen cycle

The principal reservoir of nitrogen is the atmosphere, which contains in the gaseous state around 10^6 times the nitrogen stored in biomass. The other major reservoirs are soil organic matter and the oceans. The nitrogen cycle comprises many stages (Fig. 13.6).

Biological fixation of atmospheric nitrogen (N_2) corresponds to its combination with hydrogen, carbon, and/or oxygen to form proteins and compounds that can be used by living organisms. The *principal reservoir is the atmosphere*, which contains close to 80% of the gaseous nitrogen (N_2) because the nitrogen fossilizes little in comparison with carbon, calcium, or phosphorus. Paradoxically, *this reservoir of nitrogen is not used directly by plants, which mostly*

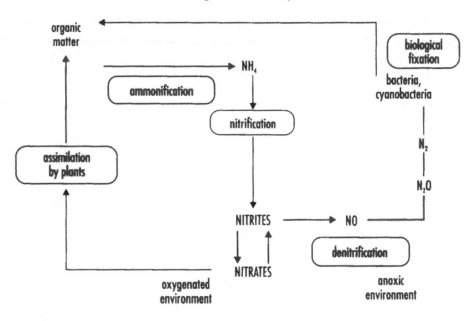

Fig. 13.6. Principal processes involved in the nitrogen cycle

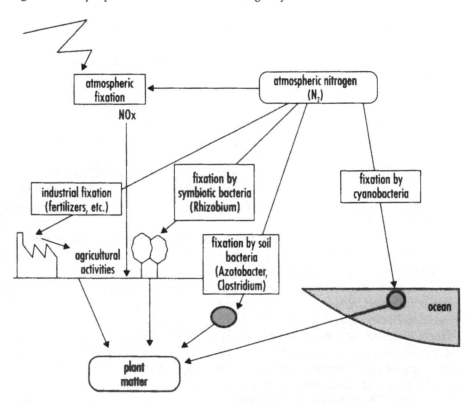

Fig. 13.7. Different modes of atmospheric nitrogen fixation before use by plants

assimilate nitrogen in the form of anions NO_3^- (nitrates) and cations NH_4^+. There are many routes that lead to the fixation of atmospheric nitrogen (Fig. 13.7).

Biological fixation is the principal source of nitrogen in terrestrial systems (around 50%) (Berner and Berner, 1996). It is estimated that biological fixation by free bacteria contributes around 25 kg N/ha/yr. Input from symbiotic bacteria is much greater and may reach values of several hundreds of kilos per hectare per year. In soils, atmospheric nitrogen is reduced by anaerobic heterotrophic microorganisms (*Clostridium*) or aerobic heterotrophic microorganisms (*Azotobacter*), as well as by bacteria that live in symbiosis with higher plants, such as *Rhizobium* symbiotic with members of the family Leguminosae, or *Frankia* symbiotic with *Casuarina* and alder. Lichens can also fix atmospheric nitrogen to satisfy their metabolic needs while liberating ammonia, which becomes available to allow other organisms to synthesize their biological compounds.

In aquatic systems, nitrogen is biologically fixed by free or symbiotic cyanobacteria such as *Anabaena*, which play a role similar to that of soil bacteria. The reduction of atmospheric nitrogen leads to the synthesis of ammonia molecules that are involved in the elaboration of amino acids and then proteins.

Phytoplankton, autotrophic bacteria, and fungi assimilate mineral nitrogen by preference in the form of ammonium ion, while macrophytes use nitrates and nitrites.

The use of large quantities of nitrogenous fertilizers to increase crop yields has distorted and modified the progress of the nitrogen cycle. The fertilizers are synthesized in the form of NO_3^- and NH_4^+ by reduction of atmospheric nitrogen (N_2). The production of fertilizers is estimated at 85 Mt (megatonnes) N/yr, and this figure will increase.

It is estimated that nitrogen fixation is presently around 280 Mt N/yr in total, of which 158 Mt has an anthropogenic origin (e.g., fertilizers, use of fossil fuels). Human activities are thus responsible nowadays for more than 50% of the nitrogen fixation, i.e., the conversion of nitrogen N_2 into chemical or biological compounds (Berner and Berner, 1996).

The quantity of nitrogen used by net primary production (1073 Mt N/yr) is greater than the annual nitrogen input estimated (280 Mt N/yr). This difference results from the fact that a large part of the nitrogen used by primary production is cycled in the biosphere.

Nitrification corresponds to the decomposition of organic matter ending in the production of nitrates (Fig. 13.6). When organisms die, the organic matter is decomposed by bacteria to yield ammonium (ammonification), of which one part escapes in the form of gas (NH_3). For many plants, however, the most useful form of nitrogen in the soil is represented by nitrates. In well–aerated soils, the nitrifying bacteria can oxidize NH_4^+ into NO_2^- (nitrite) and NO_3^- (nitrate) and extract from them the energy needed for their chemosynthesis. These bacteria decompose organic matter to produce ammonium ion NH_4^+, which is transformed into nitrite ion (NO_2^-) by bacteria of the genus *Nitrosomonas* (nitrosation) and then into nitrate ion (NO_3^-) under the effect of *Nitrobacter* (nitratation). This process of *nitrification* provides a large part of the nitrogen assimilated by plants.

In poorly aerated or anaerobic environments, *denitrification* is a reverse process that can degrade nitrates into nitrites and at the end of the reaction release gaseous nitrogenous compounds that escape to the atmosphere (Fig. 13.6). There is little reliable information on whether denitrification fully compensates for the fixation of gaseous nitrogen. This denitrification is due to the activity of an anaerobic bacterial microflora (denitrifying and heterotrophic bacteria containing carbon) that act in an opposite direction to nitrogen-fixing bacteria. These bacteria (*Pseudomonas, Nitrococcus*) decompose nitrates into nitrites and also into nitrous oxide (N_2O) and gaseous nitrogen (N_2). Denitrification is a mode of anaerobic respiration in which nitrates or their derivatives substitute for oxygen and contribute to oxidation of organic matter according to the following overall formula:

$$5\,(CH_2O) + 4\,NO_3^- + 4\,H_3O^+ \longrightarrow 2\,N_2 + 5\,CO_2 + 11\,H_2O$$

Oxygen inhibits the functioning and/or synthesis of most of the enzymes involved in denitrification. In soil, the functioning of denitrification depends on the humidity level, which determines the aeration and level of anoxia.

Denitrification in the sediments of water bodies is a phenomenon that has been known for a long time (Mariotti, 1994). In lakes with gravel beds, the elimination of nitric nitrogen is commonly observed. This is the consequence of assimilation by algal and microbial flora and denitrification. In eutrophic lakes, the capacity of nitrate elimination is 1000 kg/ha/yr, or 225 kg of nitrogen.

The riparian woods or grassy banks of rivers are intermediate zones between soils and aquatic systems that have drawn a great deal of attention because of their decontaminating capacities vis-à-vis mineral and organic pollutants, and in which an efficient denitrification has been observed. These zones in fact offer conditions favourable for denitrification: clogging of soils that reduces their provision of oxygen; high levels of organic carbon; and constant input of nitrates by flows from the ground water. In riparian zones of the Garonne, the potential for denitrification is 1.3 kg/ha/d for nitrogen (Fustec and Mariotti, 1991). Even higher values of 11.2 kg/ha/d for nitrogen have been measured in New Zealand in soils that are particularly rich in organic matter, subject to heavy clogging, and receiving waters that are highly charged with nitrates.

13.2.2. Atmospheric phase of the nitrogen cycle

Ammonia (NH_3), nitrogen monoxide (NO), and nitrous oxide (N_2O) are the three primary nitrogenous compounds released or formed in the atmosphere from atmospheric nitrogen (N_2) by different natural processes or processes linked to human activities.

The principal sources of ammonia are the degradation of animal urea, the biological activity of soils, and combustion (Fig. 13.8). The role of ammoniac in the atmosphere is linked to its basic character, which allows it to buffer the acidic character of the atmosphere due to the formation of sulphuric and nitric acids.

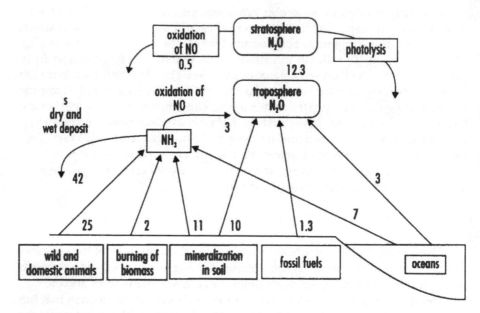

Fig. 13.8. Atmospheric cycle of ammonia (NH_3) and nitrous oxide (N_2O). The values correspond to millions of tonnes of nitrogen per year (modified from Chapel et al., 1996).

Nitrous oxide is produced on the earth's surface by a large number of sources, including the activity of bacteria associated with the mechanisms of denitrification or nitrification in soils. However, the oceans are also a significant source, as well as effluents of anthropic origin. The total production is around 15 Mt/y (Académie des sciences, 1998). It is a relatively inert gas in chemical terms. However, most of the N_2O undergoes photochemical oxidation in the stratosphere to yield nitrogen monoxide (NO), which combines with ozone (O_3) or other compounds to form nitrogen dioxide (NO_2). Nitrogen dioxide reacts in turn with the radical OH to form nitric acid (HNO_3). This reaction is not an important part of the process of acidification of rain but contributes significantly to the destruction of the stratospheric ozone (Berner and Berner, 1996).

Nitrogen monoxide (NO) has other origins such as microbiological processes in soils. It is also formed in the atmosphere by the reaction of atmospheric nitrogen (N_2) and oxygen at high temperature, i.e., by lightning. It subsequently reacts with ozone to yield NO_2, then nitric acid. Once nitric acid is formed, it is leached by rain, with which it dissociates in the form of NO_3^- and H^+. In this way, nitric acid may contribute to the acidification of rain water. However, at the same time, these nitrate inputs from rain may stimulate plant growth in nitrogen-poor zones. In fact, nitrates, and to a lesser extent ammonium (NH_4), are important elements in rain: the inputs in dissolved form or through dry deposits on land are estimated at 60 Mt N/yr and inputs in the oceans are estimated at 30 Mt N/yr (Berner and Berner, 1996). Only a third of nitrates present in rain water is of natural origin, the rest being attributable to human activities (particularly burning of fossil fuels).

13.3. PHOSPHORUS

Phosphorus enters the constitution of energy-rich molecules (ATP, ADP, AMP) or information molecules (nucleic acids). It is also involved in the composition of bone. The cycle, which is fairly simple, is largely dominated by geochemical rather than biological reactions. Phosphorus, in comparison to nitrogen and carbon, does not have stable gaseous forms and there is thus no atmospheric reservoir. On the other hand, this element is found in different mineral or organic forms that are soluble or insoluble. Soluble organic phosphorus is made up of molecules derived from living or dead organisms.

13.3.1. The phosphorus cycle

In terrestrial systems, the phosphorus cycle is relatively simple. *The major source is the alteration of apatite rocks or fossilized sediments.* Mineral phosphorus is found partly in the soluble state, essentially in the form of orthophosphates (PO_4^{-3}, HPO_4^{-2}, $H_2PO_4^-$) and partly in the insoluble state in the form of calcium phosphate or apatite. Phosphate ions are more easily soluble in water with pH between 6 and 7. In this form phosphorus is directly used by plants. For higher or lower values of pH, phosphate becomes insoluble and it is generally immobilized rapidly in soils or sediments, in the form of phosphates of iron, aluminium, or calcium, which are not accessible to plants.

Mineral phosphorus is a rare element in the biosphere and it therefore tends to become a limiting factor of productivity in many ecosystems. Nevertheless, a significant source of phosphorus today is environmental pollution of anthropic origin by effluents containing phosphates (e.g., detergents, fertilizers). This phosphorus comes from the exploitation of sedimentary soil, where it was blocked since the Secondary Era.

The phosphorus cycle, as observed at present (Fig. 13.9), is characterized thus by a nearly unidirectional flow of phosphorus from land to fresh or sea

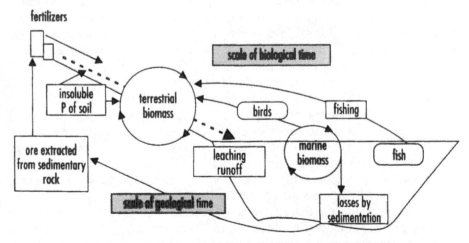

Fig. 13.9. Simplified phosphorus cycle as currently modified by human activity

water. The biological part of the cycle, even though it is essential to life, involves only a minimal fraction in relation to the quantities stored in the sediments of the lithosphere. In the oceans, a large quantity of mineralized phosphorus is lost by sedimentation of organic matter in the depths of the ocean. Phosphates are thus removed from cycling in the biosphere and end up being fossilized in the form of phosphate rocks. However, part of the phosphorus from oceanic and freshwater systems returns to terrestrial systems via the practice of fishing and the excrement of aquatic birds, which when abundant constitutes guano.

13.3.2. The phosphorus cycle in freshwater ecosystems

In freshwater ecosystems, phosphorus is generally considered the principal limiting factor for the production of plant biomass. The phosphorus cycle begins in the watershed, where water courses alter and erode the rocks and transport phosphorus in dissolved form (5 to 10%) but mostly in the form of particulate phosphate (90 to 95%). This transport occurs mostly during high water periods. In agricultural and industrial regions, these proportions are considerably modified by inputs of soluble phosphorus from fertilizers, waste water, and excrement.

Plants can only use the dissolved mineral form of phosphorus. The quantity of phosphorus available for plant production in aquatic environments thus depends on direct inputs in the mineral form and the rate of transformation in the soluble mineral form of soluble or particulate organic phosphorus. This mineral phosphorus is rapidly assimilated by plants, so that concentrations in unpolluted waters are generally very low. This is not to say that there is no mineral phosphorus, but it is rapidly reused. The *bioavailability* of phosphorus is a concept that expresses the greater or lesser aptitude of this element to leave the particulate phase for the soluble phase, in a form that can be assimilated by plants. It is the expression of the overall dynamics of the system and of all the physicochemical and biological processes involved in the phosphorus cycle.

In lake environments, phosphorus is present in water in different forms (Fig. 13.10): a dissolved *mineral* form (orthophosphate) that comes mostly from soluble mineral phosphorus imported by runoff from the watershed; a soluble *organic* form that comes from excrement or the decomposition of living organisms; and a *mineral or organic particulate* form. These different forms constitute what is called total phosphorus. Generally, the levels of dissolved phosphorus are low in natural waters, about 0.01 mg/l for orthophosphates (PO_4^-) and 0.025 mg/l for dissolved phosphorus including organic forms. In the pelagic zone of lake environments or in humid zones, most of the total phosphorus is present in the particulate form, i.e., stored in living or dead organisms (bacteria, plants, animals), or in clay particles and minerals.

Phosphorus is essentially cycled by excretion of soluble and organic phosphorus by living organisms and by bacterial decomposition of organic matter. Zooplankton and fish excrete large quantities of phosphorus that can be used by plants. For example, zooplankton can excrete 10% of the phosphorus stored in its biomass each day, in the form of phosphate and organic phosphorus

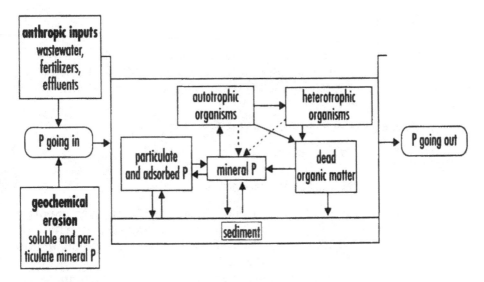

Fig. 13.10. Diagram of the major components of the phosphorus cycle in a lake environment

in equivalent quantity. This excretion varies according to the temperature, rate of consumption, composition of the plankton, and other factors. Fishes contain a large proportion of the total phosphorus of an aquatic system and their excretion contributes also to the enrichment of an environment in mineral phosphorus, in a proportion that is generally lower, however, than that of zooplankton, because the biomass turnover rates are much lower.

However, a certain quantity of phosphorus is lost to biological production. In fact, part of the organic matter sediments on the bottom, where, in well-oxygenated environments, the phosphorus combines with other elements (iron, aluminium, calcium) to form insoluble compounds that precipitate to the sediment level. The sediment functions as a trap that accumulates the particulate phosphorus. There is also adsorption of phosphate ions on clays, organic constituents, and especially metallic hydroxides and oxides (iron, manganese, aluminium). Nevertheless, this trapping is not always definitive: the immobilization of dissolved phosphorus in insoluble forms and the future of the particulate phosphorus trapped in the sediment depend essentially on oxygenation and pH conditions in these environments. In reducing conditions, i.e., in a period of oxygen deficit, for example, of the immobilized phosphorus is solubilized, notably by reduction and solubilization of oxidized forms of manganese and iron, and there is hydrolysis of iron and aluminium phosphates leading to the release of adsorbed phosphorus. These phenomena intervene in the anaerobic sediments of the hypolimnion. The hypolimnion is rich in dissolved mineral phosphorus. In the lakes, in a period of destratification, waters of the hypolimnion are mixed with those of the epilimnion and enrich them in mineral phosphorus. The phosphorus previously trapped in the sediment and remineralized in anaerobic condition thus becomes available for primary production.

13.4. SULPHUR

As with phosphorus, the principal natural sources of sulphur are volcanic deposits and sedimentary deposits such as gypsum ($CaSO_4$) and pyrite (FeS_2). However, like nitrogen, sulphur is present in several chemical forms in the natural environment:

— in the atmosphere in the form of H_2S, SO_2, H_2SO_4, etc.;

— in the biomass and in soils in the form of sulphur, etc..

13.4.1. The terrestrial sulphur cycle

In terrestrial and freshwater ecosystems, the leaching of rocks is an important source of inorganic sulphur represented by water-soluble sulphates. The sulphates are absorbed by plants, which reduce them and elaborate sulphurous amino acids (cystine, cysteine, methionin).

Organic wastes containing sulphurous proteins are decomposed by heterotrophic bacteria. In an *aerobic* environment, there is production of sulphates and in an *anaerobic* environment there is production of sulphides and, ultimately, hydrogen sulphide (Fig. 13.11). Besides, sulphur-reducing activity of bacteria of the genus *Desulfovibrio* leads in an anaerobic environment to the reduction of sulphates and the production of hydrogen sulphide. In shallow environments such as marshes H_2S is released, while in deep aquatic systems this compound reacts with iron from sediments to yield precipitates of black ferrous sulphur. This reaction, which liberates phosphorus, is reversible. However, it is in the sediments at the bottom of the sea that the largest reserves

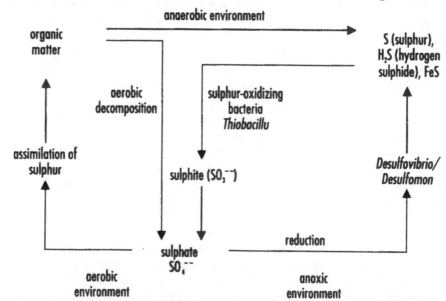

Fig. 13.11. Major processes involved in the sulphur cycle in *aerobic* and *anaerobic* environments

of sulphur are found. As they become compacted, these sediments form rocks such as gypsum or pyrite.

Inversely, there are sulphur-oxidizing phototrophic bacteria in aquatic environments that can reoxidize hydrogen sulphide or other sulphurous substances into sulphates. *Thiobacillus* is an autotrophic bacteria that fixes carbon dioxide by synthesizing all its biochemical substances from the energy produced by oxidation of hydrogen sulphide. In waters and on the surface of anoxic mud the sulphur-oxidizing bacteria thus oxidize the sulphur cation produced by mineralization of organic matter and sulphate reduction.

13.4.2. Atmospheric phase of the sulphur cycle

In natural conditions, the sulphur present in the atmosphere has three principal origins:

— The formation of sea spray when the wind blows introduces sulphates (Na_2SO_4) into the atmosphere in the form of aerosols.

— Volcanism produces mainly hydrogen sulphide (H_2S) and sulphur dioxide (SO_2). This is transformed in the atmosphere into sulphuric acid by oxidation and then into sulphates by reaction with various cations such as ammoniac.

— Biogenic fermentation occurs because of the intervention of specific bacteria, whose sulphate-reducing activity leads to the formation of H_2S depending on complex reactions. An important source of biogenic sulphur, especially in the oceans, is the production of dimethyl sulphur (DMS) (see Chapter 16).

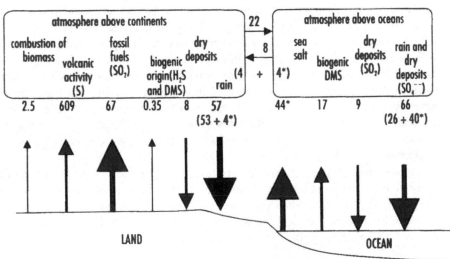

Fig. 13.12. Atmospheric cycle of sulphur. The values are in megatonnes per year. The figures marked with an asterisk involve marine salts (values according to Berner and Berner, 1996). The dry deposits in the terrestrial environment are made up of SO_2 and SO_4^{--}. Rain water contains SO_4^{--}. (As for the other biogeochemical cycles, there may be sometimes significant differences between the various sources of information concerning the quantitative aspects of the sulphur cycle.)

The sulphur cycle (Fig. 13.12) is profoundly modified by human activities, particularly the use of fossil fuels (coal, petroleum). The quantities released into the atmosphere are much higher than those resulting from natural biogeochemical processes, so that the sulphur cycle is one of the most disturbed by the action of man. Increasing SO_2 emissions are the source of acid rain and its ecological consequences, such as the depletion of forests and the acidification of lakes. However, the sulphur cycle has also been studied extensively because of the impact sulphurous aerosols may have on the climate. Apart from the albedo effect, sulphate particles influence the physics of cloud systems: the more sulphate particles there are, the greater the number of droplets in a cloud, and the more the cloud reflects the solar radiation into space (see Chapter 15). The increasing SO_2 emissions, that represent more than two-thirds of sulphur emissions in the Northern hemisphere, may thus cause a cooling of the climate that partly counterbalances the warming due to the greenhouse effect of other gases (Chapel et al., 1996).

13.5. OTHER ELEMENTS

Living organisms use many mineral elements other than those mentioned so far in this chapter. In this section we give two examples of biogeochemical cycles involving particularly important elements.

13.5.1. Calcium and buffering capacity

Calcium plays an important role in the constitution of bone tissue. The biogeochemical cycle is relatively simple. The principal reservoir is made up of carbonate sedimentary rocks. Leaching from dry land results in the presence of various calcium salts in the interstitial and surface waters. The dissolution of limestone is considerably increased by the presence of dissolved CO_2, which leads to the formation of highly soluble calcium bicarbonate that could, depending on environmental conditions, be transformed into insoluble calcium carbonate. The calcium cycle is thus associated with that of carbon and silicon because it is found essentially in the form of carbonates and silicates in the lithosphere.

A major role of calcium in continental waters is the *buffer effect*, which tends to stabilize the pH by neutralizing carbonic acid. In fact, unlike oxygen, carbon dioxide concentrations develop according to complex rules that result in the existence of several chemical forms. Carbon dioxide may hydrate and form carbonic acid according to a reversible reaction:

$$\text{dissolved } CO_2 + H_2O \longleftrightarrow H_2CO_3 \text{ (carbonic acid)}$$

Carbonic acid is then dissociated into bicarbonate ions (HCO_3^-) and carbonate ions (CO_3^-) as a function of the pH, i.e., the acidity of the water:

$$H_2CO_3 \longleftrightarrow H^+ + HCO_3^- \quad \text{and} \quad HCO_3^- \longleftrightarrow H^+ + CO_3^-$$

However, because of the laws of chemistry, the relative concentrations of free carbonic gas, carbonate, and bicarbonate are based on a state of equilibrium between the pH, the level of dissolved carbonic gas, and the derivative forms of carbonic acid. In particular, calcium, abundant in many aquatic environments, plays a key role in these processes of equilibrium of derivative forms of CO_2. In fact, the presence of carbonic acid considerably increases the dissolution power of water, which can therefore tackle calcium carbonate or calcite ($CaCO_3$), a compound poorly soluble in water. This calcite is transformed into calcium bicarbonate ($Ca(HCO_3)_2$), which is much more soluble in water according to a reversible reaction:

$$CaCO_3 + CO_2 + H_2O \longleftrightarrow Ca(HCO_3)_2 = Ca^{++} + 2\,HCO_3^-$$

The chemical reactions are reversible and tend to resist variations in the environment. When the release of CO_2 is predominant, there is dissolution of calcite and formation of carbonates and bicarbonates. The waters are thus called "aggressive". On the contrary, when photosynthesis is predominant, there is consumption of dissolved CO_2 and transformation of soluble calcium bicarbonate into calcium carbonate, which precipitates.

In the oceans, a large quantity of calcium is fixed and fossilized by the biogenic pathway is important. In shallow waters, there is formation of aragonite, while in deeper ones there is calcite of planktonic origin.

The rates of sedimentation in the oceans are presently estimated at 520 Mt Ca/yr (or 1300 Mt of $CaCO_3$) in shallow water and 440 Mt Ca/yr (or 1100 Mt of $CaCO_3$) in deep water. Most of the inputs into the sea come from rivers (550 Mt Ca/yr) and reactions between sea water and volcanism (191 Mt Ca/yr). The result is that sedimentation in the ocean currently exceeds the inputs, notably because of the significant deposit of $CaCO_3$ in shallow zones as a result of the rise in sea level since the last glaciation (Berner and Berner, 1996).

13.5.2. Silica

Some plants such as the Graminaceae or algae such as the Diatomae contain significant proportions of silica.

The silica cycle is relatively simple. Most of the inputs of soluble silica into an aquatic system come from the erosion of rocks, especially feldspar. In water, silica exists in different forms, but only silicic acid (H_2SiO_4), which is partly dissociated, can be used by algae. When diatoms die, their frustule, made up of opaline, partly dissolves but part of it sediments and is buried in the sediment.

Most of the inputs of silica in the ocean sediment under the effect of biogenic activities (Table 13.2). This biogenic sedimentation is estimated at 200 Mt Si/yr.

Table 13.2. Budgets of some chemical elements in the oceans (Berner and Berner, 1996)

Elements	Ca Tg Ca^{++}/yr	Bicarbonate Tg HCO_3^-/yr	Cl Tg Cl^{--}/yr	Na Tg Na^+/yr	S Tg S/yr	Si Tg Si/yr	K Tg K^+/yr	Mg Tg Mg^{++}/yr
Total inputs	778	2125	308	269	160	236	105	137
•input by rivers	550	1980	308	269	143	180	52	137
—natural			215	193	66			
—anthropic			93	76	77			
• deposit and rain					17	56	53	
• others	228	145						
Total loss	960	2920	65	79	47	190-203	105	137
• atmospheric transfers	0.5		40	21	8		1	3
• deposits	960	2920			39	190-203		15
• others			25	58	39		104	119
Long-term budget								
• inputs	760	2053	215	193	70			137
• outputs	729	2070	215	193	70			137
—deposits	729	2070	163	106	42			
—others				58				

13.6. INTERACTIONS BETWEEN BIOGEOCHEMICAL CYCLES

In the real world, the cycles of chemical elements are closely related (Fig. 13.13). In fact, elements such as C, N, P, and S, fundamental constituents of biological molecules, must necessarily be present at the same time for these molecules to be elaborated.

The organic part of the cycle of these elements (the living or dead biomass) is identical. Their release from organic mater is controlled by the same processes of decomposition (Scholes et al., 1999). The cycles then diverge, given that certain elements are stored in the atmosphere or in the lithosphere in sometimes very different forms.

13.6.1. Biogeochemical cycles, the water cycle and the oxygen cycle

The cycles of C, N, P, and S are linked with the water cycle to the extent that biological process of soils, such as decomposition, depend on the soil humidity. The processes of chemical transformations, and thus the products that result from decomposition, depend on the oxygenation controlled by the level of water

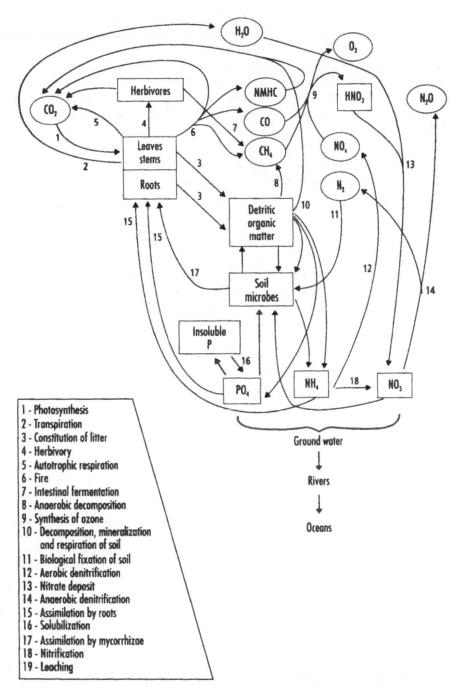

1 - Photosynthesis
2 - Transpiration
3 - Constitution of litter
4 - Herbivory
5 - Autotrophic respiration
6 - Fire
7 - Intestinal fermentation
8 - Anaerobic decomposition
9 - Synthesis of ozone
10 - Decomposition, mineralization
     and respiration of soil
11 - Biological fixation of soil
12 - Aerobic denitrification
13 - Nitrate deposit
14 - Anaerobic denitrification
15 - Assimilation by roots
16 - Solubilization
17 - Assimilation by mycorrhizae
18 - Nitrification
19 - Leaching

Fig. 13.13. Relationships between carbon, nitrogen, phosphorus, and water cycles in a terrestrial ecosystem. The sulphur cycle has not been incorporated but is similar to the nitrogen cycle in many aspects. The existence of common reservoirs, processes, and controls shows that a given cycle is constrained or governed by the disturbances in the other cycles (Scholes et al., 1999).

saturation of the soil. If aerobic processes are dominant, the principal gaseous emissions are carbon dioxide and NO, while methane is oxidized. In anaerobic systems, on the contrary, nitrates or sulphates are used as a terminal acceptor of electrons in place of oxygen. The anaerobic decomposition releases nitrogen in the form of N_2 and N_2O, carbon in the form of methane (CH_4), and sulphides such as H_2S. Anaerobic decomposition is slower than aerobic decomposition, so that the systems saturated with water (wetlands, peat bogs) are the principal reservoirs of carbon.

13.6.2. Threshold effects: limitation of the carbon cycle by nitrogen and phosphorus availability

The availability of one element may influence the cycle of others by the threshold effect. It has also been observed that the proportions of four principal elements in different categories of organic mater are relatively fixed (Table 13.3). The predictability of ratios between elements allows us to establish links between the cycles without necessarily knowing the biological species involved or the flow pathways. If we hypothesize that nitrogen availability is a limiting factor in terrestrial ecosystems, then we can predict the net primary production if we know the rates of mineralization of nitrogen and the composition of the vegetation.

Table 13.3. Relative proportion of chemical elements in various biological materials (Scholes et al., 1999)

Biological matter	C	N	P	S
Tropical deciduous forest	100	5.4	0.05	0.09
Tropical grassland	100	2	0.06	0.09
Woods	100	1.2	0.07	0.05
Temperate soil organic matter	100	9.4	0.4	0.3
Tropical soil organic matter	100	4.2	0.2	
Phytoplankton	100	13.6	0.85	

Nitrogen is often the limiting element for the productivity of natural terrestrial ecosystems, given the many routes by which it can escape the system. On the one hand, nitrates are highly soluble and easily leached in the soils. On the other hand, many bacteria use nitrogenous compounds as a source of energy, liberating several gaseous forms of nitrogen. Inversely, only some organisms are capable of extracting atmospheric nitrogen (N_2) and the energy cost of the process is high.

In tropical savannahs, the dry grasses constitute a highly flammable matter. With fires of natural origin or caused by man, about 40 to 90% of the nitrogen stored in the grasses is released into the atmosphere, which results in a still greater limitation of productivity of these environments by nitrogen.

In aquatic systems, it is phosphorus that is most often the limiting element, given its low solubility. Phosphorus forms insoluble complexes with iron, which further reduces its availability, particularly in the tropics. The phosphorus cycle is greatly disturbed by inputs of fertilizers, which represent a transfer to the biosphere from highly localized sources that are normally not accessible to plants. This phosphorus penetrates freshwater and coastal ecosystems through runoff and generates the eutrophication of water bodies, i.e., overproduction resulting from the addition of nutrients and leading generally to oxygen deficits in the water and a loss of biodiversity.

13.6.3. Importance of micronutrients: iron in the carbon cycle in the marine environment

While phytoplankton plays an important role in the regulation of atmospheric CO_2, the regulation of oceanic primary production is the consequence of interactive biogeochemical networks with climatic feedback effects that are not well known (Falkowski et al., 1998). Oceanographers have long acknowledged that phosphorus availability is the limiting factor of primary production in the oceans. They subsequently showed by observation and experiment that phytoplankton productivity is in fact limited by the availability of fixed inorganic nitrogen. In the ocean today, the vertical profiles of the two elements invariably demonstrate a deficit in fixed N in relation to P in the deep ocean. In the past few years, moreover, it has been discovered that nitrogen fixation is itself limited by another factor, probably iron (see Chapter 4).

Even though it is the fourth most abundant element in the earth's crust, the bioavailability of iron in the ocean is low. In fact, reduced or ferrous iron is highly soluble in sea water, but oxidized or ferric iron is virtually insoluble. Before the appearance of photosynthetic organisms, the ocean was anoxic and slightly reductive. Iron was present in high concentrations. With the development of photosynthesis producing oxygen, iron was oxidized and precipitated so that it is now present in only low concentrations. This massive precipitation of iron by autotrophic photosynthetic organisms is a good example of negative feedback in the evolution of biogeochemical cycles on earth.

At present, the major source of iron in the oceans is wind-borne dust. This flow of dust is considered to control photosynthesis over most of the Pacific Ocean. According to information obtained from ice cores, the iron input into oceans was probably of a greater order of magnitude in the glacial period than in the interglacial period. According to some hypotheses, this input of iron probably stimulated biological fixation of nitrogen in the ocean, simultaneously favouring carbon fixation until phosphorus became the limiting factor. There is palaeo-oceanographic proof that flows of biogenic detritus in the deep southern ocean (the biological pump) increased during the glacial period (Kumar et al., 1995).

Part IV
Global Ecology

Chapter 14

Global Ecology: Dynamics of the Biosphere

The biosphere is the only region of the earth's crust occupied by life. It is in the biosphere, a thin external layer on our planet, that life is concentrated; all the organisms are found there and are always separated from raw ambient matter by a clear and impermeable limit. No living organism has ever been generated by raw matter. In its death, its life, and its destruction, the organism constantly restores its atoms to the biosphere and takes them back again, but living matter imbued with life is always capable of its own genesis within life itself.

Vernadsky (1997: 74)

Strictly speaking, the biosphere is all the living organisms, animals and plants, that live on the earth's surface. Nevertheless, in a wider sense the biosphere is often defined as the superficial layer of the planet, which contains living things and in which life is always possible. This space includes the *lithosphere* (the earth's crust), the *hydrosphere* (all the ocean and the continental waters), and the *atmosphere* (gaseous envelop of the earth). Some ecologists also use the term *ecosphere* to designate jointly the lithosphere, hydrosphere, atmosphere, as well as living organisms (biosphere strictly speaking) and the sun (the photosphere), the energy source that controls the functioning of the system as a whole. However, this terminology is rarely used nowadays.

In any case, *the concept of biosphere is above all a functional concept: it emphasizes the interrelationships between organisms and their environment on the planetary scale. It is a kind of ultimate ecosystem in the present state of our knowledge.* What is new and important in comparison with older ideas is the recognition of the active participation of the living world in the creation and dynamics of its physicochemical environment (Charvolin, 1994).

14.1. THE ORIGINS OF THE BIOSPHERE CONCEPT

The term *biosphere*, which chronologically precedes the term *ecosystem*, was used for the first time in 1875 by the Austrian geologist Eduard Suess in a study on the geological structure of the Alps (Suess, 1883). However, it was the Russian geochemist Wladimir Vernadsky (1863-1945), considered one of the founders of global ecology, who developed the concept and devoted an entire book (Vernadsky, 1926). The term *biosphere* designates the complex system at the earth's surface made up of environments having specific physicochemical characteristics (oceans, lithosphere, atmosphere, etc.) and all the living things: the biosphere is the only region of the earth's crust occupied by life.

In Vernadsky's view, the biosphere is not a static concept but the result of reciprocal interactions between living things, the terrestrial environment, and energy flows from the sun: life is not a phenomenon external or accidental to the earth's surface, according to Vernadsky. It is closely linked to the structure of the earth's crust, is part of its mechanism, and serves in it functions of primary importance that are necessary to the very existence of that mechanism. Life is thus a constant and continuous powerful disturbance of the chemical inertia on the surface of our planet (Vernadsky, 1997). *In his time, Vernadsky thus expressed a transdisciplinary and at the same time a revolutionary concept of the interactions between life and the chemical processes of the planet.*

Vernadsky's theories stand at the crossroads of several scientific traditions:

— *Pedology,* the founder of which, the Russian Dokouchaev, was Vernadsky's teacher. Dokouchaev wrote that the soil was the cover of the surface rock formed by the joint action of all the climatic factors, with the necessary cooperation of plant and animal organisms, micro- and macroscopic (Boulaine, 1984). The notion of biosphere elaborated by Vernadsky owed a great deal to the chemical study of soils and, therefore, anchored ecology to the earth sciences, unlike the English approach in which ecology was rooted in geobotanical concepts.

— *Chemistry of living things* from France and Germany following the works of Lavoisier and, later, the agronomists Justus Liebig and Jean Baptiste Boussingault. Boussingault in 1841 spoke of the mysterious circle of organic life on the earth's surface (Deleage, 1997).

— *Physics of the early 20th century,* as well as ideas from Timiriazev (1903) and other Russian naturalists, for whom the biosphere was a huge thermodynamic machine that drew its energy from solar radiation and transformed it into new and varied forms of free energy that could work on the environment: the biosphere, according to Timiriazev, could essentially be considered a region of the earth's crust occupied by transformers that change cosmic radiation into active terrestrial energy, electric, chemical, mechanical, thermal, etc.

The concept of biosphere, which was long considered hardly practical, has now become a relevant field of study because of the evidence of the global impact of human activities, and especially the worldwide consequences of activities carried out on a local or regional level. *To analyse the biosphere as a giant ecosystem thus makes sense, especially since it is an ideal ecosystem in thermodynamic terms because it does not exchange matter with the external world: except for energy, the matter flows and cycles are closed.*

14.2. THE EARTH SYSTEM AND GLOBAL ECOLOGY

The idea that the living world and the physicochemical world are interactive systems is not new. As early as in 1802, Lamarck noted in his work on hydrogeology that a good terrestrial physics must include meteorology, geology, and biology. However, in reality, under the pressure of reductionism that emerged in the physical and mathematical sciences, the biological sciences developed well away from the geological and physical sciences.

At the same time as Vernadsky, the American physicist Alfred J. Lotka (1925) opened up a new field of research. He was interested in the modelling of systems and cybernetics and he embodied another origin of global ecology. Starting from the geochemical tradition, he pointed out the important role of the living world in biogeochemical cycles. He introduced quantitative assessments of mineral cycles between the three compartments that constitute the world: the atmosphere, lithosphere, and hydrosphere. He was also able to show the influence of industrial activities on the major cycles of nature. Lotka's approach was that of a physicist who develops a practical conception of the biosphere. He anticipated in a premonitory way the idea of a necessary planetary management that has been acknowledged in our time (Deleage, 1997).

Lotka and Vernadsky had a fairly small impact on ecology in the mid-20th century. Their ideas remained marginal in comparison to the currents of thought derived from the physical sciences that considered, to put it simply, that the physicochemical context determines the biological content. Thus, within the sphere of influence of Galilean science, meteorology was essentially based on the physical and mathematical sciences and almost completely ignored the role of the living world in the water cycle. However, the American ecologist G.E. Hutchinson grasped the fundamental importance of the ideas of Vernadsky and Lotka and set out to investigate them. Before the publication in 1942 of a pioneering article by Lindeman on the trophodynamic aspects of ecology, Hutchinson had published in 1941 a paper in which he proposed to describe the dynamics of living things in an ecosystem using energy transfers. Hutchinson especially influenced Lindeman, to whom he was a teacher and a friend.

It is only recently that the legitimacy of a *bio*geochemical approach to functional processes of the biosphere has been recognized. The global ecological, biogeochemical, and planetary conception of the biosphere formulated by Vernadsky was reactivated in the early 1970s by the rapid development of global ecology, i.e., the ecology of the earth as a whole. The Gaia concept is the rediscovery with new arguments of the great holistic tradition of scientific ecology—a thermodynamic, biochemical, geophysiological, and cybernetic vision of the biosphere. The intuition of Vernadsky, Lotka, or Lovelock is the evidence that living organisms and the earth form a dynamic system, controlled for and by living things. We may or may not share these ideas, but the approach is nonetheless stimulating.

Global ecology is thus a new interdisciplinary and holistic science contemporaneous with the era of space travel. It is based on the fact that the different elements of the biosphere (oceans, dry land, atmosphere, and lithosphere) interact at several scales of time and space, forming a huge interconnected network. These interactions generate or modulate the diversity of climates, seasons, the distribution of biodiversity on the earth's surface, and other phenomena. The network as a whole constitutes the earth system (Rasool, 1993).

In the 1980s, the idea that the biosphere could be seriously disturbed by a conflict or nuclear accident became a subject of international concern. Multidisciplinary studies initiated by ICSU-SCOPE envisaged the consequences of a nuclear winter as well as the impact that human activities could have on the entire planet (SCOPE, 1985). However, the biosphere is also threatened by other ecological changes such as the thinning of the ozone layer, increase in the CO_2 concentration in the atmosphere, and deforestation. In fact, the use of land, water, and natural resources has increased in very high proportions over the past few centuries and will intensify still further with the growth of the population and technological development. Large-scale chemical transformations are at work, as well as changes in the natural transfers of energy and matter throughout the world. It became clear during the 1980s that increase in the concentration of greenhouse gases in the atmosphere could cause significant climatic changes (a warming of the planet) in the decades to come (see Chapter 15).

These observations brought scientists to consider the earth as a single ecosystem and to investigate seriously the interrelationships existing between the different components of the biosphere. They observed, for example, that the climate, global cycles of carbon and water, and the structure and functioning of natural ecosystems are closely linked. Any significant change in one of these factors would have an impact on the others, as well as for life on earth. They also showed that the metabolic activity of living organisms, including microorganisms, has modified the terrestrial environment for thousands of millions of years. Studies of energy flows have revealed that the global thermal equilibrium as well as albedo are modulated by the presence of the living world on the earth's surface (see Chapter 16). The global changes, at the level of primary productivity, due to human activities affect the energy budget of the earth by modifying the composition of carbon dioxide and various other gases. There is thus a reciprocal interaction of organisms and their environment and not just a one-way relationship of cause and effect. To understand these complex reactions, scientists relied on modelling, which has improved greatly over the past two decades particularly because of increasingly efficient information technology.

In a way, it was the catastrophic scenarios about the future of the planet, which obliged scientists and politicians to launch various means of study that lead to a dynamic and global vision of the biosphere, that paved the way for this natural extension of ecosystem ecology into global ecology.

14.3. THE GAIA HYPOTHESIS AND GEOPHYSIOLOGY

How could we explain that terrestrial conditions have allowed life to survive for some 3,500 millions of years? It is unlikely that the climate and chemical properties of the earth were always adequate for living organisms. However, some scientists suggest that living things help overall to reduce the variations of the terrestrial environment and thus to favour the perpetuation of life itself. In the 1950s, the British chemist James E. Lovelock proposed the Gaia hypothesis (after the Greek goddess of the earth), which proposed that the biosphere is a self-regulated entity that controls the temperature and composition of the earth's surface and thus maintains conditions favourable to life on the planet (Lovelock, 1990). According to this author, living organisms and their inanimate environment behave as a single unit. In other words, Gaia is a cybernetic system that maintains the planetary homeostasis.

The Gaia hypothesis was rapidly adopted by the "New Age" ecological movement, which was involved in *deep ecology*, which attempted to revive the myth that man is not the master of the world but that Gaia, in her infinite wisdom, is the one who determines the future of our planet. This analogy compromised the ideas of Lovelock at first and led to severe criticism. Still, several scientists saw in it a truly scientific theory (Westbroek, 1998). All other things being equal, Lovelock's idea recalls that of Clements, who at the beginning of the 20th century compared living communities to a superorganism (see Chapter 3). The biosphere as seen by Lovelock, a result of multiple interactions between biotic and abiotic components, is no longer fundamentally different from Tansley's concept of ecosystem. Lovelock used the metaphor of the organism to carry the idea of a unity between biological and abiotic factors. He found in the energy flows a powerful concept. Lovelock moreover put forth the hypothesis that life forms tend to stabilize the temperature of the earth at levels compatible with the survival of life. In this context, he developed a virtual simulation called the "world of daisies" in which, to put it simply, the relative proportion of black daisies and white daisies helped to maintain a constant temperature on the planet (see Chapter 10). However, it is difficult to confirm or disprove this hypothesis with our current knowledge.

Because the word *Gaia*, cornered by the New Age movement, was considered contaminated for use in scientific literature, the term *geophysiology* was proposed as a substitute (Westbroek, 1997). This word was invented in the 18th century by the Scottish geologist James Hutton, who considered the earth as a superorganism. In 1875, at a meeting of the Royal Society of Edinborough, Hutton declared that he turned to physiology in studying the earth, comparing the cycling of nutrients in the soil and the movement of water in the oceans to the circulation of blood in the body.

The Geophysiology Society, created in 1996, studied the interactions between life forms and the rest of the earth. The rationale of Peter Westbroek, who for a long time headed the geobiochemistry group at the University of Leiden, was the following: scientific communities have forgotten that life has played an extremely important role in the dynamism of the earth. For a long time, geologists regarded the earth only as a physicochemical system: life could only adapt to this context, without sending feedback to the system. This simplified approach was useful for the development of geology, but the problem of climate change forces us now to regard the earth as a whole and to study the influence of life on the dynamics of the earth.

In this context, biochemistry is undoubtedly the science that acknowledges that many chemical reactions are influenced, even controlled, by the living world. This biogeochemistry allows us to unify many research activities, including studies on energy flows and biogeochemical transfers, which often tend to be carried out separately. More than a third of all chemical elements are recycled by living organisms, and some of those organisms can concentrate rare chemical elements. All organisms, for example, use and recycle oxygen, hydrogen, carbon, nitrogen, phosphorus, sulphur, calcium, potassium, magnesium, sodium, silica, iron, manganese, cobalt, copper, zinc, and other chemical elements involved in particular biological functions. These biological reactions are responsible for the formation of considerable deposits of mineral elements such as silica, calcium, sulphur, and carbon.

14.4. GLOBAL CHANGES

The changes observed in terrestrial or marine ecosystems are the cause and at the same time the result of modifications in the characteristics of the global environment (Lubchenko et al., 1991). The expression "global change" often calls to mind climatic changes. However, although the reality of climatic change cannot be denied, the expression includes a whole set of phenomena that can be classified in four major categories: changes in land use and vegetation cover; changes in the composition of the atmosphere; changes in climate; and alterations in the composition of natural communities and the loss of biodiversity (Walker and Steffen, 1999). This term also covers the impacts of economic development and the growth of the world population on the main compartments of the system Earth—i.e., the atmosphere, the soils, and aquatic systems—as well as the processes involved in matter and energy exchanges between those compartments.

Two points ought to be emphasized:

- The average values used most often mask the great variability of situations: the consequences of human activities or the natural climatic variability do not occur with the same intensity in various regions and periods.

- Global changes are the result of a large number of processes that act permanently on the surface of the planet, at different scales of time and space, and sometimes in different directions, so that the prediction of consequences and these global changes is difficult. Here we face one of the fundamental principles of ecosystem ecology: we cannot understand the dynamics of a system without taking into account the nature and importance of interactions between the environmental factors that act on this system. Figure 14.1 gives a very simple diagram of the most important interactions.

14.4.1. Changes in land cover and use and the use of water

Over centuries, land use has transformed natural systems into a variety of modified systems: cultivated fields, prairies, timber plantations, urban and

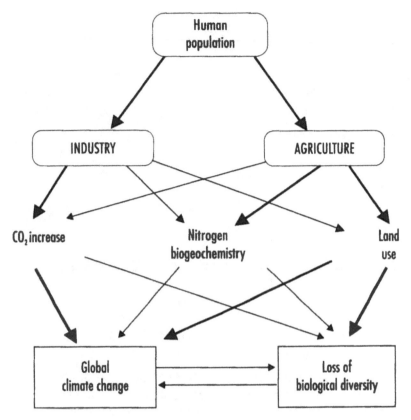

Fig. 14.1. Interactions between the principal factors responsible for global changes showing the influence of human activities (modified from Vitousek, 1994)

industrial environments. It is estimated that half of the cultivated land has been created since the beginning of the 20th century and that agriculture now occupies more than half of the potentially exploitable area of the earth (Fig. 14.2). Throughout the world, the use of fresh water has increased, for agriculture, domestic or industrial use, and hydroelectricity generation. Such use has many consequences for the chemical composition of water as well as the flood regimes of rivers.

Land cover and use

In scientific literature, the term *land cover* is distinguished from *land use*. Land cover refers to major biophysical categories such as forest, prairie, aquatic system, and colonized zones. These major categories of land cover are subject to climate fluctuations and more or less random events such as fire or flood. This land cover has been modified by human activity, principally for agricultural purposes. That activity is what is considered land use, and it relates to the way in which the biophysical characteristics of land are manipulated or exploited.

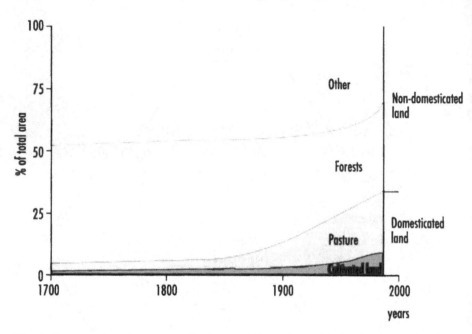

Fig. 14.2. Estimation of changes in land cover from 1700 to 1995 (*Global Change Newsletter*, March 1999)

Changes in land use and water use have an important impact on the functioning of ecosystems whether by elimination of certain species or introduction of exotic species, the modification of biogeochemical cycles, or the change in plant cover and its direct consequences for the local or regional climate. Changes in land use can also modify the condition of soils: e.g., loss of arable soils by urbanization, loss by erosion, overgrazing, and compaction due to agricultural practices.

14.4.2. Changes in the composition of the atmosphere

The alteration of the chemical composition of the atmosphere due to human activities is another important component of global changes. It consists specifically of the increase in carbon dioxide concentration resulting from the burning of fossil biomasses, as well as many other elements (phosphorous, sulphurous, nitrogenous, chlorinated, and other compounds) that circulate between the different compartments of the biosphere (atmosphere, oceans, vegetation, soils) and that come from industrial activities. Such biogeochemical cycles partly have a natural origin but emissions of anthropic origin are today responsible for the intensification of certain processes, the effects of which are beginning to be felt on the planetary scale (the additional greenhouse effect, for example).

Chemical compounds of industrial origin such as chlorofluorocarbons (CFC) are responsible for the destruction of the stratospheric ozone, which results in the increase of UV-B radiation on the surface of the earth. Industrial emissions are also the cause of acid rain (see Chapter 15).

14.2.3. Climatic changes

The biosphere participates in and responds to the regulation of the climatic system by physical, chemical, or biological feedback mechanisms. An important challenge for ecologists is to understand the processes at work in order to be able to predict the response of ecosystems to climates that may exist in the future.

Generally, biological systems participate in the control of thermal and hydrological budgets through albedo, evapotranspiration, the rugosity of the surface, soil humidity, etc. Ecological processes also control the release and assimilation of greenhouse gases (CO_2, CH_4, N_2O), which retain infrared radiations from the soil surface, thus contributing to a modification in the thermal balance of the earth. Although there is a consensus in the scientific community in recognizing a general warming of the climate, there is debate about its causes. Many scientists feel that human activities now have a measurable effect on the climate, either by the increased emission of greenhouse gases or by the emission of aerosols that will tend, on the contrary, to cool the surface of the planet. In 1996, the international community, in the framework of the GIEC recognized that a combination of elements suggests that there is a perceptible influence of humans on the global climate and that the climate will continue to change, but in a way that is different and presently unknown, depending on the region. These modifications are expected to affect the temperature, precipitation, rise in sea level, the occurrence of extreme events, air quality, penetration of UV light, and other phenomena.

However, palaeoclimatic studies have shown that the temperature of the earth has undergone regular cycles of glaciation and deglaciation, and that the natural climatic variability needs to be taken into account. The glacial archives indicate also that there have been significant variations of temperature over periods of a few decades. The causes of such rapid changes are still unknown. However, a significant present example of such short-term variability is the El Niño phenomenon, which starts in the western Pacific but affects the entire planet. The difficulty that scientists face is to assess the role of variations linked to natural changes of the climate and the role of the impact of human activities.

The ecological responses to climatic changes are complex. They can be felt primarily at the individual and population level: alteration of photosynthesis, modification of behaviour, genetic modifications. Changes in the structure and composition of communities may result from the direct action of abiotic factors on species (dispersal, physiology) as well as modifications in the interactions between species (competition, mutualism, predators, etc.). The question is made still more complex by the many introductions of exotic species that may profoundly modify the relations between species.

14.4.4. Changes in biological diversity

Alterations in the distribution of living organisms and the composition of plant and animal communities on the earth's surface are among the other elements of global changes. The disappearance of overexploited species and the modification of land cover are irreversible and unprecedented in that it is human societies that are now responsible for them. The extinction of species, however, is difficult to quantify precisely for most groups of organisms, with the exception of those that have been extensively studied (mammals, birds) but may not necessarily be representative of the processes that are occurring. We must also remember that the planet has experienced other periods marked by massive extinctions, the causes of which are often not well established. However, one of the most marked phenomena at present is undoubtedly the introduction of many exotic species on all the continents and the fact that fauna and flora have become common and "globalized".

The extinctions, like introductions of species and biological invasions, have the consequence of modifying the functioning of terrestrial and aquatic ecosystems, notably the biogeochemical cycles, but this domain is still not well explored.

14.5. THE BIOSPHERE AS AN OBJECT OF STUDY

Given the complexity of the phenomena involved in global changes, it is necessary to take a "system" approach to the planet as a whole. This is the object of the International Geosphere-Biosphere Programme. In the present state of our knowledge, there are some important characteristics of the biosphere that flesh out the idea that it could constitute a vast self-regulated system:

— The biosphere is a dynamic system whose constituents interact closely with one another: atmosphere, cryosphere, hydrosphere, living world, etc.

— It is a large thermodynamic machine that draws most of its energy from the sun and dissipates it in the form of heat.

— In comparison to the other planets of our solar system, the earth is characterized by the abundance of the H_2O molecule, which exists in three physical states: water vapour, liquid water, and ice. The changes in state are associated with significant thermodynamic changes so that the water cycle is also a favoured vehicle for transfer of thermal energy.

— The relative stability of the system is ensured by biogeochemical cycles of the elements and molecules (C, N, S, P, H_2O, etc.) that determine the development phases of life in the various compartments of the biosphere.

— The greenhouse effect resulting from the absorption of terrestrial infrared radiation by the atmosphere maintains a temperature favourable to life on the earth's surface. This greenhouse effect, which ensures the thermal protection of the planet, is due essentially to some gases present in traces in the atmosphere, while the three major constituents (nitrogen, oxygen, argon, making up 99.9% of the volume) are practically transparent to solar radiation in the visible and infrared parts of its spectrum.

— There is a protective atmosphere that serves as a filter with respect to certain radiations harmful to living things, such as ultraviolet radiation.

The International Geosphere-Biosphere Programme

Begun in the early 1980s by the American National Academy of Science, the Global Change study was set up on the international scale in 1986 within the framework of the ICSU (International Council of Scientific Unions) and later became the International Geosphere-Biosphere Programme (IGBP). The objective fo the programme was to describe and understand the interactions between the physical, chemical and biological processes that regulate the earth system, the changes that occur in that system, and the way in which they are influenced by human activities. The IGBP is based on the principle that human activities have become a significant force affecting the functioning of the biosphere. It poses six fundamental questions:

• How is the chemistry of the atmosphere regulated and what is the role of biological processes in the production and consumption of trace gases? (International Global Atmospheric Chemistry Programme—IGAC)

• What are the consequences of global changes for terrestrial ecosystems? (Global Change and Terrestrial Ecosystems—GCTE)

• What are the interactions between the vegetation and physical processes of the hydrological cycle? (Biospheric Aspects of the Hydrological Cycle—BACH).

• What are the consequences of changes in land cover on coastal resources? What will be the consequences of changes in sea level and the climate on terrestrial ecosystems? (Land-Ocean Interactions in the Coast Zone—LOICZ)

• How do biogeochemical processes in the ocean influence and respond to climatic changes? (Joint Global Ocean Flux Study—JGOFS)

• What were the principal changes in the climate and environment in the past and what were their causes? (Past Global Changes—PAGES)

The objectives assigned to the programme are the following:

— comprehension of interactive processes that regulate the geosphere-biosphere system;

— explanation of changes that occurred during the history of the earth;

— exploration of routes by which the processes are or may be influenced by human activities; and

— identification of the impact of man on the functioning of the system.

Numerical modelling is the best approach for analysing the complex interactions of the earth system. The ultimate objective of the IGBP is thus to combine models developed for specific processes (e.g., ocean-atmosphere or

PHYSICAL CLIMATIC SYSTEM

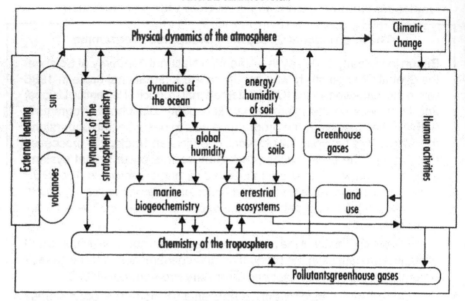

BIOGEOCHEMICAL CYCLES

Fig. 14.3. The global modelling approach followed by IGBP consists of linking biogeochemical systems, the physical climatic system, and anthropic impacts in order to develop our capacity to predict the behaviour of the earth system as a whole (IGBP, 1992)

vegetation-climate models) and then to integrate all these models in general models of circulation of the climatic system (Fig. 14.3).

Chapter 15

The Climatic System and Its Variability

Heated at the equator and frozen at the poles, our planet resembles an immense thermal machine: the winds and currents act on its surface to redistribute energy and establish on average an equilibrium that characterizes the climate.

Joussaume (1993)

The climate is the principal determinant of the dynamics of the biosphere in the wider sense. Although the sun is the essential source of energy, the "climatic system" of the earth is an extremely complex system, regulated by a set of physical, chemical, and biological processes that operate over a wide range of spatial and temporal scales, with multiple interactions between the biosphere in the strict sense, the lithosphere, the atmosphere, the cryosphere, and the hydrosphere.

The functioning of the planet earth can be analysed by the transfer of solar radiation. For a long time, the energy machinery of the earth system was thought to be controlled essentially by physicochemical processes. And for a long time ecological science rested on the dogma that the climatic framework defines the composition and functioning of ecosystems. It is only in the last two decades that scientists have begun to consider exchanges in terrestrial and oceanic ecosystems as a cause as well as a consequence of climatic changes. *In other words, the living world also controls certain physical and chemical parameters of the climatic system.* This intellectual mini-revolution is partly the result of an increasingly better understanding of the functioning of the climatic system. It also corresponds to a more dynamic interpretation of the ecosystem concept.

The contributions of ecological systems to climatic changes are complex and we are still far from understanding all the mechanisms. They are involved, for example, in the control of water budgets and budgets of surface energy, which are the main drivers of the global climate: albedo, evapotranspiration, soil humidity, and the rugosity of the surface depend on the characteristics of marine and terrestrial environments. The biosphere also contributes significantly to the chemical composition of the atmosphere, exerting an indirect influence on the climate (see Chapter 16). Besides, climatic changes have feedback effects on the functioning of ecosystems and on the biodiversity, thus modifying biogenic emissions. Generally, however, quantification of the biosphere's influence on the dynamics of the earth's climate still remains very crude. To minimize the actual uncertainties to construct a model, it would be necessary to better understand the feedback mechanisms involved, which can amplify or reduce the climatic response (those linked to water vapour, clouds, soil hydrology, etc.). Only simple (or simplified) models have so far been available, and they do not give a sufficiently precise representation of phenomena.

According to what we now know, the earth's climate depends on various parameters (Fig. 15.1):

— External factors, such as variations in the solar radiation that reach the surface of the earth and/or the relative position of the earth with respect to the sun. These two phenomena may be partly linked. The theory of Milankovitch suggests that there are cyclic variations in characteristics of the earth's orbit that could be responsible for climatic changes observed in the past.

— Factors linked to the morphology and activity of the earth, especially volcanic eruptions that may release aerosols and dust that reduce the penetration of solar radiation over short or moderately long periods. In addition, there are thermal exchanges between the ocean and the atmosphere on the one hand and terrestrial systems and the atmosphere on the other.

— Biogenic factors that modify the chemical composition of the atmosphere (greenhouse gases) or its structure (aerosols). The plant cover of soils also plays an important role with respect to terrestrial albedo. For some decades, human activities have modified relationships between the living world and the atmosphere and have mobilized chemical elements stored in the lithosphere (fossil fuels, phosphates, etc.).

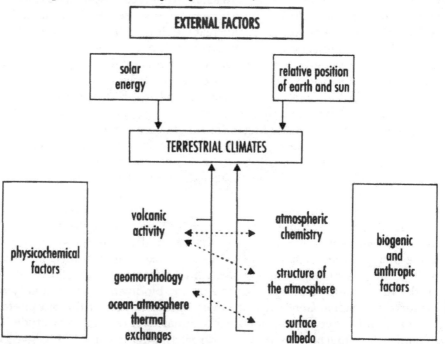

Fig. 15.1. Principal factors involved in the dynamics of the earth's climate

15.1. THE EARTH'S ENERGY BALANCE AND THE GREENHOUSE EFFECT

The planet earth is a good example of a thermodynamic system: it exchanges energy with the external environment but recycles matter within the system (see Chapter 5). The primary function of the system is the degradation and distribution of solar energy in different states—transformation of a significant part (70%) of the incident solar radiation into heat, distribution of this energy flow between the surface and the atmosphere, and radiation of the same energy flow towards space—so as to ensure a global equilibrium.

15.1.1. Radiation and the energy balance

The climatic system draws practically all its energy from the solar radiation that warms the earth's surface. Solar energy originates in nuclear fusion (fusion of hydrogen atoms into helium), and a small part of it is transformed into electromagnetic energy that radiates from the surface of the sun towards space. From the time it is captured by the earth to the time it returns to space in the form of infrared rays, that energy undergoes many transformations. The ocean may heat up or cool down by a transfer or accumulation of sensible heat. It is also cooled by evaporation, and water vapour drawn into the atmosphere will transfer its "latent" energy to the atmosphere when it condenses to form clouds.

In thermodynamic terms, the planet behaves like a system that transforms solar energy into radiative energy of a long wavelength in thermal infrared, which is also called *telluric* radiation. The radiation balance of the earth corresponds to the difference between the flows of solar energy (short-wave radiation between 0.2 and 4 microns) absorbed in the atmosphere and on the surface of the earth, and the flows of thermal radiation (long-wave radiation, between 4 and 100 microns) emitted by the earth towards space.

Sensible heat and latent heat

Sensible heat is the heat accumulated in a body that can be measured by a rise in temperature. The flow of sensible heat at the ocean-atmosphere interface is an exchange of heat associated with a direct transfer of temperature by conduction or convection. *Latent heat* is the energy that is stored when a substance changes state from a liquid to a gas, for example, changes in the phase of water: during evaporation, energy is required to complete the liquid-vapour transition (2.5×10^6 J/kg). This energy is ultimately restored to the atmosphere during the condensation of water vapour to form clouds.

In the course of a year, the earth receives on an average energy corresponding to 340 W/m² coming from the sun as short-wave radiation. However, the atmosphere and the earth's surface reflect more than 30% (100 W/m²) of this incoming radiation particularly because of clouds and atmospheric dust: this

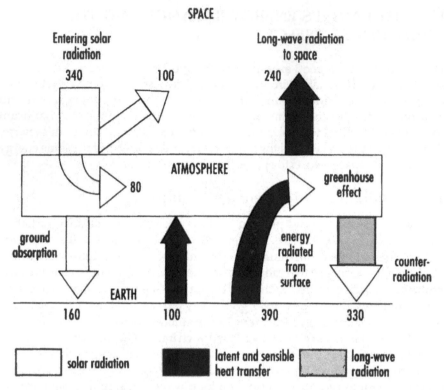

Fig. 15.2. Radiation balance of the earth. The values (W/m²) are averages for the entire surface of the globe.

is terrestrial albedo. The 240 W that is not scattered back into space serves to raise the temperature of the planet: around 80 W/m² is absorbed by the atmosphere, while the rest (160 W/m²) heats the oceans and the continents (Joussaume, 1993). The solar radiation absorbed by the earth is transformed into heat in such a way that the terrestrial system in turn emits long-wave radiation. Around 240 W/m²/yr leaves the earth's surface toward space, which keeps the planet in a thermal equilibrium: in theory, the planet does not heat up or cool down (Fig. 15.2).

Nevertheless, the radiation balance of the earth's surface is not nil, but reaches +100 W/m² on global average. This is compensated by non-radiative flows of heat from the surface to the atmosphere. These flows correspond to the flows of sensible heat and in large part to the flows of latent heat transferred to the atmosphere by evaporation. Because of this, the energy cycle in the climatic system is closely linked to the water cycle.

15.1.2. The greenhouse effect

The global energy budget actually disguises a more complex phenomenon. In fact, solar energy received by the earth suffices only to bring the average temperature of the surface to –20°C, according to theoretical calculations. The average temperature of the planet, however, is 15°C. This difference is due to the atmosphere, which plays an important role with respect to solar and terrestrial radiation. Certain gases called *greenhouse gases*, such as water vapour (the most abundant) and CO_2 as well as methane and sulphur and nitrogen compounds present in traces in the atmosphere, can absorb long-wave radiation from the terrestrial surface, so that only a small part of those (60 W/m²) go through the atmosphere and escape to space. The atmosphere heated by terrestrial long-wave radiation in turn radiates in all directions, including the earth's surface (counter-radiation), with a consequent rise in the ground temperature. The atmosphere thus serves as a glass roof that allows visible solar light to penetrate but is relatively opaque to infrared radiation from the earth's surface, which cannot entirely escape. This is where the term *greenhouse effect* comes from.

This "trapping" of the infrared energy is important because it results in an additional warming of the surface by 330 W/m², over the direct input by insolation. The average temperature of the surface thus reaches 15°C for a total flow of 490 (330 + 160) W/m². A large part of this energy is re-emitted in the form of long-wave radiation (390 W/m²), or used for evaporation and to heat the lowest part of the atmosphere (flows of latent heat of 100 W/m²). The 240 W/m² emitted by the earth into space corresponds thus to 80 W/m² of solar radiation absorbed by the atmosphere, plus the 60 W/m² of loss by long-wave radiation, plus the 100 W/m² lost as latent heat.

In reality, the energy balance of the planet depends on the presence and concentration of several gases, including rare gases in the atmosphere. Without a "natural" greenhouse effect, which owes nothing to human activities, the earth would be a vast desert of ice. This greenhouse effect created the conditions for water to have a liquid phase on the surface of the earth and thus made it "inhabitable" a long time ago. To evaluate the role of the atmosphere in the temperature regulation of the earth's surface, we can compare it to the situation prevailing on other planets in the solar system. Venus, which has an atmosphere a hundred times as dense as that of the earth and made up almost entirely of CO_2, would have a temperature of –20°C in the absence of greenhouse gases, while the actual temperature of the surface is about 460°C. Mars, which has an atmosphere of CO_2 that has one-hundredth the density of earth's, has only a limited greenhouse effect (5°C) and the average temperature of the surface is only –55°C (Rasool, 1993). The primitive atmosphere of the earth, made up of CO_2, methane, nitrogen, and hydrogen sulphide, would have led to an intense greenhouse effect at a time (about 3000 million years ago) when the solar radiation was only 70 to 80% of the actual energy. This greenhouse effect would have allowed the earth to have positive temperatures and the presence of water in liquid form (Chapel et al., 1996).

Of all the greenhouse gases, water vapour is the most important because it has the highest concentration in the air. In the absence of water vapour, the outward infrared flows would increase by about 100 W/m², while they only increase by 50 W/m² if we remove all the CO_2. Water vapour thus has a particularly essential role in the thermal equilibrium of the planet, especially through the formation of clouds, which modify the radiation balance.

However, this greenhouse effect constitutes a source of worry because for some decades it has been observed that the greenhouse gas level in the terrestrial atmosphere has increased rapidly because of human activities (see Chapter 16). Hence the insistent question posed by scientists: what will be the climatic consequences? In fact, slight variations of the average temperature of the earth may correspond to significant climatic changes: during the glaciations of the past 160,000 years, the average temperature of the earth was lower by only 5°C, but the continents of North American and Europe were covered with a thick layer of ice and the sea level was lower than the present level by over 100 m.

15.2. THE THERMAL MACHINERY

The input of energy is not uniform on the earth's surface: at low latitudes, the incoming solar radiation exceeds outgoing long-wave radiation, while the reverse is true in high latitudes. Besides, the degree of warming is partly subordinated to the nature of the terrestrial surface. The oceans and dry land heat at different rhythms, and soils covered with vegetation do not absorb or reflect heat in the same way as deserts and ice caps. This is the consequence of albedo (see further).

The earth thus behaves like a huge heat engine in which the local disequilibrium of energy balances is compensated by displacements of air and water masses that tend to homogenize the distribution of heat. The energy surplus of equatorial regions, for example, is transported by general atmospheric circulation towards the poles, and the circulation pattern of the oceans constitutes a sort of giant conveyor belt that moves the heat energy accumulated in tropical areas poleward (Foucault, 1993).

15.2.1. Atmospheric circulation

The air moves in different ways. On the one hand, there is convection or upward movement of heated air: warmer air tends to rise because it is less dense. This is called thermal convection. On the other hand, there are horizontal movements called winds generated by differences in atmospheric pressure from place to place. We will not go into the details of dynamics of atmosphere and the factors responsible for displacements of air, which are complex phenomena. We are only highlighting some aspects of atmospheric circulation that are relevant to the thermal regulation of the earth ecosystem.

First of all, a major atmospheric circulation tends to establish itself between the equator and the poles to ensure the transport of energy toward the poles. These are, it must be remembered, "average" phenomena on the scale of a year, for example, and they characterize the "general circulation" of the atmosphere. The meridian atmospheric circulation (from the equator towards the poles) is structured in convective cells: the Hadley cells between the equator and about 30°N or S, the Ferrel cells between 30 and 60°N or S, and the polar cells (Fig. 15.3). In very simple terms, in the tropical zone the warm and humid air, which is lighter, rises above the equator and separates into two branches, each of

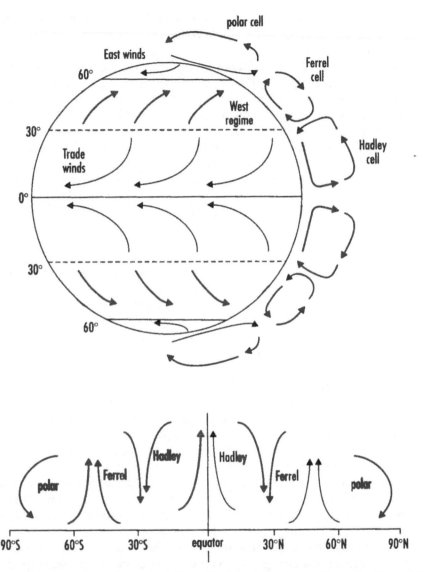

Fig. 15.3. Meridian atmospheric circulation with the principal convection cells. The middle latitudes have zones of great variability in which weather forecast is difficult.

which moves towards a pole at the level of the troposphere. The air masses that rise above the equator cool down at high altitude and lose their humidity, which is the cause of tropical rain. The air does not rise uniformly along the equator. There are major zones of ascendance above South East Asia, Africa, and South America. The dry air then descends to the low layers of the atmosphere around the subtropical regions, towards 30° latitude N and S. The subtropical regions that receive the dry air masses that have circulated at high altitude correspond to the belt of large deserts observed in the two hemispheres. These air masses then return towards the equator, becoming charged with water vapour. These are the trade winds that converge in each of the two hemispheres towards the intertropical convergence zone, which is close to the equator but moves away as a function of the seasons.

In the middle latitudes, between 30 and 60°, the meteorological conditions are different from the tropical regions. There is a significant meridian temperature gradient, with meteorological fronts in which the warm tropical air meets the cold air from high latitudes (Chapel et al., 1996). The air masses move from high pressure zones to low pressure zones. A series of depressions and anticyclones create highly variable conditions. The middle latitudes are dominated by west winds at low and high altitude. At high altitude the intensity of winds increases, creating an unstable and variable jet current. Heat is no longer transferred towards the poles by convective cells as in the intertropical zone, but rather by horizontal exchanges.

15.2.2. The ocean as regulator of climate

As with the winds, which help to redistribute heat and especially to transport surplus energy from tropical regions towards the polar latitudes, the ocean is an essential element of the climatic system. It constitutes the principal reservoir of heat on the earth's surface because it is a powerful thermal stabilizer and a fluid that can transport heat over long distances (Sadourny, 1994). This transport is effected by surface currents as well as by thermohaline circulation (Minster, 1994).

The ocean currents are generated largely by winds and follow the principal directions of the general atmospheric circulation. Since the ocean covers nearly three-quarters of the globe, currents are responsible for half the thermal transports from the equator to the poles, or a proportion equivalent to that of the atmosphere. However, in transporting heat, they in return help modify the wind regime. The dynamics of the atmosphere and the ocean are thus closely coupled. *One of the regulatory functions of the ocean is to slow the manifestation of disturbances in the climate. For example, following variations in the intensity of solar radiation, or following an increase in the greenhouse effect in the atmosphere, the ocean absorbs a large part of the excess heat at the surface and stores it in the depths.*

Exchanges at the ocean-atmosphere interface modify water temperature and salinity and cause changes in volume mass that induce pressure gradients generating or enforcing currents. This is what is called thermohaline circulation (Minster, 1994) or the "conveyor belt" model (Sadourny, 1994), which animates all the oceans of the world. The pump of this movement is located in the Norway Sea, near Greenland and the Labrador Sea, at the edge of the ice, where the surface water, cooling down on contact with the air, becomes denser. Part of this water, forming ice, expels the salt that it contains to sea water, making it heavier, so that it sinks to depths of 2000 to 4000 m. It thus starts a long route towards the south in the depths of the Atlantic, turns the Cape of Good Hope, and continues to descend until around 60°S latitude, rejoining the Antarctic circumpolar current turning from west to east (Fig. 15.4). The characteristics of the water evolve gradually as it mixes along its trajectory. This transport of deep

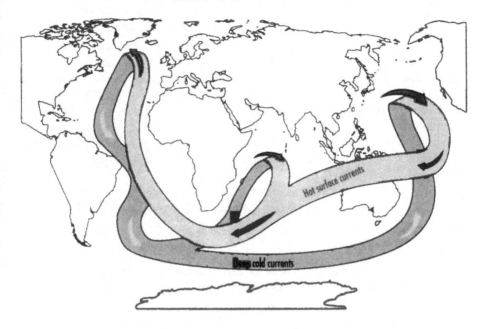

Fig. 15.4. Global oceanic circulation (thermohaline circulation) associated with the submersion of water in the North Atlantic under the effect of evaporation and cooling. This water descends to the Antarctic then rises slowly towards the surface in the Indian Ocean, and in the Pacific Ocean, where it warms up. These warm waters move towards the Atlantic, where they rise towards the North to begin the cycle again. A round trip on this "conveyor belt" takes about 1000 years (Chapel et al., 1996).

water follows its route towards the east at the bottom of the Indian Ocean and the Pacific Ocean. The water then rises slowly to the surface in the middle of these two oceans and warms up. This grand circulation returns by the intermediary of warm currents near the surface that pass between the Indonesian islands, cross the Indian Ocean, go around Africa, and then go up the Atlantic Ocean. In the Atlantic, the surface current, called the Gulf Stream, moves along the North American coast, then turns to the middle of the North Atlantic Ocean to divide and again sink to the depths near the Norway Sea. There is thus a huge oceanic "conveyor belt", one of the commonly known effects of which is to warm the North Atlantic and thus western Europe, which has particularly mild winters. The duration of this circulation is estimated to be a thousand years (Chapel et al., 1996).

Some indexes suggest that this thermohaline circulation is not stable. It could have weakened, even disappeared, during certain climatic episodes in the past, for example during the massive melting of polar ice during the last deglaciation.

Among the movements of masses of oceanic water, upwellings deserve particular interest in climatic and biological terms. Coastal upwellings are rises of deep cold water rich in nutrients along the continental plateaux that allow the development of a considerable biological production. On the Atlantic coast of Africa, for example, the trade winds, which blow more or less parallel to the coast, are responsible for constant or seasonal upwellings that develop along the coasts of the Gulf of Guinea up to Mauritania (Roy, 1991). It is a physical process resulting from the action of the wind, which moves the surface layers of the ocean along the coasts and moves them towards the open sea, leading to a vertical flow along the continental plateau to compensate for the disequilibrium at the coast. The result is the formation of a steep temperature gradient from the coast to the open sea. However, the deep water that rises to the surface is also richer in nutrients than the surface water. This input of minerals allows the development and maintenance of high biological productivity in the zone concerned.

15.2.3. The role of vegetation in the water cycle and climate equilibrium

How can the biosphere partly control the climate? Very simply because vegetation serves as an intermediary for exchanges of water and energy between the earth's surface and the atmosphere. For example, it regulates the water and carbon cycles, soil fertility, and levels of nitrous oxide, sulphur, and phosphorus in the lower atmosphere. As we know that the atmosphere is partly a biogenic creation whose composition is controlled by a constant flow of gas released by the biosphere (see Chapter 16), the changes in the extent and structure of ecosystems thus affect the climate itself.

Scientists have stated sometimes contradictory opinions on the role of vegetation and forest, particularly in the control of the climate. According to Walter and Breckle (1985), the climate is the principal factor that influences all the others: the soil, the vegetation, and, to a lesser extent, the fauna. These factors will in turn have an effect only at the microclimate level. This hypothesis prevailed for a long time but now seems to be shaken. Hayden (1998) suggests that the climate controls the biosphere but that the biosphere also actively controls the climate at all the scales. To understand the dynamics of the earth system, and to model correctly the trends of our environment, we must take into account these factors of control, which are not always considered in predictive models.

a) Albedo

The composition and structure of the vegetation influences one of the important parameters of the radiation balance, the *albedo*. Albedo is a measure of the reflectivity of a surface, in this case of the fraction of solar radiation reflected by the earth's surface towards space. The vegetation absorbs the visible fraction of solar radiation (short-wave radiation from 0.4 to 0.7µm) but reflects the infrareds (long-wave radiation from 0.7 to 1 µm). The proportion of solar energy reflected by the terrestrial surface varies from 80% for snow-covered land to 35% for the

Sahara, 12% for the temperate forest, and 5% for dense forests (Shuttleworth et al., 1991). Regions with dense vegetation absorb a large part of the incident solar energy and lose relatively little heat by long-wave radiation. The reverse is true for desert regions.

The consequences of albedo may accompany the changes induced by climatic variations and have feedback effects. The warming of high latitudes may lead in the northern hemisphere to the development of forest in zones that are at present occupied by tundra. It is possible that these changes will increase the warming of the northern hemisphere by close to 50% over a period of 50 to 150 years, because of a weaker albedo of forests during the snowy season.

b) Control of temperature ranges

Minimum and maximum temperatures are partly controlled by evapotranspiration and the emission of greenhouse gases. The water vapour emitted each day forms clouds in the troposphere that regulate the temperature, first by increasing the albedo of the atmosphere and second by contributing to the greenhouse effect. There is more water vapour in the atmosphere and the loss of nocturnal radiant energy is limited. This is why nocturnal cooling by loss of radiative energy varies from place to place as a function of the atmospheric content of water vapour. In a desert without vegetation, the solar energy received during the day heats the earth's surface. During the night, in the absence of clouds, the soil surface releases heat to the atmosphere and rapidly cools. The result is a wide variation in the temperature during the day and night. The presence of abundant vegetation modifies these thermal contrasts because during the day a large part of the solar energy serves to evaporate the water contained in the plants so that the atmosphere above a plant cover is more humid and cloudy, favouring a nocturnal greenhouse effect. So, during the night there is less cooling.

c) Consequences of modifications of the plant cover

Changes in the structure of terrestrial ecosystems can in turn affect the climate itself. The idea that deforestation can have an influence on the climate and especially on the rainfall is not new. It is said that even Christopher Columbus posed this hypothesis after having observed the desertification of the Azores and the Canary Islands following their deforestation during the period of Portuguese colonization. It is likely that many similar observations were made but never recorded or made available to us.

In any case, human activities have profoundly reshaped the earth's surface. Since the third millennium BCE, clearing and cultivation and the use of wood as fuel and for construction led to deforestation of the Mediterranean region that later intensified and spread throughout the world. In the Middle Ages, a large part of Europe was deforested, and then it was the turn of North America during the 19th century. Deforestation is now in progress in the tropical forests, the Amazon region, Indonesia, and Africa. For a long time scientists did not realize that the combination of these local disturbances could have more global consequences and in turn affect regional climates. To tell the truth, the scientists did not have the knowledge or the technical means to tackle these questions. They had to wait till the means of observation (satellites, for example) and information processing technology (computers, models) were sufficiently evolved to address these questions in a quantitative manner. Deforestation of the planet leads, in principle, to a global cooling linked to the increase of albedo. It is difficult, in the absence of sufficient data, to assess whether this was the case in the past few centuries. However, massive deforestation will probably have more significant climatic consequences in the tropics because the radiative energy at work there is the greatest.

The principal effect of a change in the plant cover is the modification of the water balance. In 1977, the American meteorologist Jules Charney proposed the hypothesis that the desertification observed in the Sahel could be the consequence of excessively intensive grazing. This overgrazing led to a partial disappearance of the vegetation, which increased the albedo. The resulting energy deficit favoured the descent of masses of dry air from the upper layers of the atmosphere to just above the region concerned, a phenomenon that in turn inhibited precipitation. There was a subsequent trend of chronic drought that amplified the desertification and disappearance of local vegetation. This explanation remains plausible even though other causes, such as variations in the temperature of the Atlantic Ocean, have since been shown to have a strong influence on Sahelian aridity (Sadourny, 1992).

Deforestation also results in the release of carbon stored in the aerial part of the biomass and in the soil. Farmland occupies around a fifth of the world's land area. A significant part of that area was earlier covered with forest and thus contained large quantities of carbon in the soil and the trees. The conversion of such land released that carbon. The IPCC (1994) estimates that the net emission due to deforestation in tropical environments was 1.6 ± 1.0 Gt C/yr between 1980 and 1989.

15.3. SPATIAL AND TEMPORAL VARIABILITY OF CLIMATE

Climate changes on the surface of the earth are obvious to everyone. However, that it could vary over time in a single place, sometimes greatly, is a phenomenon relatively less well known. In a single place, for example, temperatures and rainfall vary from one year to another with more or less harsh winters and more or less dry summers. The natural variability of the climate is a reality that has been largely demonstrated over different time scales, from a decade to a million years (Duplessy and Morel, 1990).

One of the causes of climatic variability lies in the variations of solar radiation received on the earth's surface. It may result from changes in the orbital parameters of the earth relative to the sun and affect scales of time of 10,000 to 100,000 years. Another cause lies in the changes in solar activity that occur at different time scales. But the radiation balance may also be modified by atmospheric aerosols, a phenomenon that remains difficult to understand, given the great diversity of their characteristics and their great spatial and temporal variability.

In our time, however, it is not always easy to differentiate between long-term trends of natural origin and those that result from the impact of human activities. For some centuries, in fact, anthropogenic activities have considerably modified the composition of the atmosphere and seem to have a significant impact on the thermal balance of the planet. Scientists have thus been faced with a new task: to identify the respective roles of these two factors. This is one of the reasons why the scientific community has organized itself to carry out large-scale international research.

International research programmes

Since the 1980s, the question of climatic change has led to the establishment of several international programmes:

— the World Climate Research Programme;

— the International Geosphere-Biosphere Programme (IGBP) in 1986;

— the International Human Dimensions Programme on Global Environmental Change in 1995.

Apart from these, in 1988 a group of intergovernmental experts on climate change was created under the aegis of the World Meteorological Organization and the United Nations Environment Programme, the mission of which was as follows:

— to evaluate the scientific data on climate change;

— to evaluate the ecological and socioeconomic consequences of those changes; and

— to formulate strategies of adaptation to those changes.

15.3.1. Rising temperature

It has been observed that average world temperatures have increased by about 0.6°C to 0.8°C since the end of the 19th century (Fig. 15.5), the values varying slightly depending on the periods concerned and the methods used. Three phases of this warming can be distinguished:

— Between 1910 and 1940, the average temperature increased by 0.4°C.

— Between 1940 and 1975, there was a plateau, or even a slight cooling.

— Since 1975, the average temperature has increased by 0.4°C with slight fluctuations.

Given that the level of atmospheric CO_2 increased regularly from the beginning of the 20th century, it is difficult to explain the 1940-1975 episode. These variations could be linked partly to El Niño (in the Pacific) and the North Atlantic Oscillation, the effects of which on regional climates, or even global

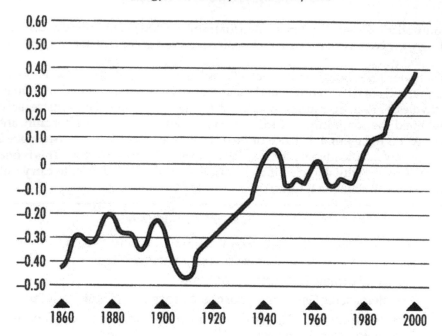

Fig. 15.5. Variations of average temperatures around the globe (air and sea, in °C) since 1860

climates, are now established. It is also possible that the variations have a "natural" cause.

The global warming trend is not the result of a uniform increase in temperature but an average between regional anomalies of cooling and warming, with a predominance of the latter. On the historical scale, we also have some information on climatic variations.

Between 1450 and 1880, the glaciers of the Alps and North America stood well beyond their present limits. It is estimated that the drop in average annual temperature was around 1°C in France during this "Little Ice Age", but it could have reached 3 to 4°C in northern Europe. In any case, this cold period ended the prosperity of Greenland, which the Vikings colonized and developed during a relatively mild climatic period between the years 900 and 1200 (Joussaume, 1993).

The Little Ice Age

The reconstruction of annual summer and winter temperatures in middle England from documents and meteorological records indicates the cooling down of the climate during the Little Ice Age between the medieval optimum, which occurred around 1300, and the recent warming that occurred near the mid-19th century (Lamb, 1977). In addition to the fairly long-term temperature records in England (which date back to 1659), there is a significant record made by English sailors, including logbooks that record the events of voyages (storms, surface temperatures, etc.). These observations have been confirmed for western Europe. Many historical sources have been used to reconstruct the history of major periods of cooling and warming in the past few centuries (Pfister, 1999). Generally, the hot and dry summers between 1530 and 1565 gradually became cold and humid from 1570 onward. The analysis of registers that note the starting date of grape harvests (Le Roy Ladurie, 1967) has shown that, during the coldest decades of the Little Ice Age (1620-1629 and 1690-1699), the temperature from April to September was lower by one degree than the average temperature of the period 1901-1960.

During the Little Ice Age, average annual temperatures dropped by 1 to 1.5°C in the eastern part of the North Atlantic. The frequency of northern and northwestern winds increased throughout the British Isles. The Gulf Stream followed a more southern track than today, while the polar water progressed further south, leading to a cooling of the Norway Sea and the North Sea. In Europe, the winters were characterized by a greater penetration of polar flows carrying cold and dry Arctic air. The pictures of Brueghel offer an indication of more severe conditions (Magny, 1995).

According to several recent studies, there was a retreat of the immense Arctic ice cap, which extends over 14 million km². Observations recorded by submarines indicate that the thickness of the pack was reduced by 40% in about 30 years. The average thickness changed from 3.1 m to 1.8 m, while the area diminished on average by 37,000 km² a year. The causes of this phenomenon are still not known. Was it due to global warming? Or to a climatic cycle that we still do not understand? Or a disturbance of thermohaline circulation with intrusion of a warm and saline current from the Gulf Stream? All questions reveal the extent of our ignorance in this field.

15.3.2. The Sahelian drought

The abundance of precipitation is one of the most difficult climatic variables to measure because of its great spatial and temporal variability. Despite the existence of many sources of error in the measurements that may affect their interpretation, large-scale studies have shown over the past few decades a trend of higher precipitation at middle latitudes and lower precipitation in the tropical regions of the northern hemisphere, while there is a general increase of precipitation in the southern hemisphere (Diaz et al., 1989; Bradley et al., 1987). There is, however, a great regional variability with, for example, a continuous increase of precipitation above the Soviet Union and a decrease over the Sahel, while Europe does not seem to have undergone significant changes.

The aridity of the Sahel became apparent in the 1970s and lasted till the early 1990s. However, observations collected since the beginning of the 20th century about the flows of some large Sahelo-Sudanese rivers such as the Senegal, the Niger, the Chari, and the Nile indicate that similar episodes have occurred during the 20th century, particularly in the 1910s and 1940s. For the second half of the 20th century, the most humid period in the Sahelian region was observed between 1951 and 1970, with a maximum in 1962-1963 (Mahe, 1993;

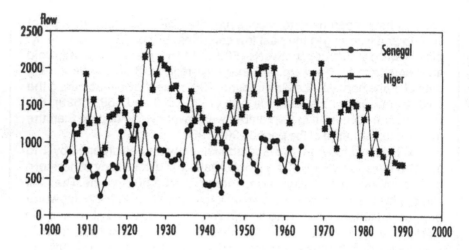

Fig. 15.6. Flows of the Niger and Senegal rivers since the beginning of the 20th century (m^3/s)

Mahe et al., 1990). The annual peaks of precipitation show a tendency to a particularly acute drop especially in 1968 in the Sahelo-Sudan region. In the 1980s, the extension of arid regions in central Africa became more marked and the major rivers of tropical and equatorial Africa were in turn seriously affected by very low flows, with a minimum in 1983. Although a recent improvement has been observed, the annual precipitations remain low (Olivry et al., 1993).

The influence of drought on the flow of the African rivers is expressed not only in a reduction of high water flow (Fig. 15.6) but also in a reduction of low water flow resulting from a drop in the water table, which is no longer fed as abundantly as in the past. In addition to the climatic drought, which is expressed in low annual flows, there is a "phreatic drought" resulting from impoverishment of the water table, which is manifested by a significant modification of the drying-out regime. In some cases, permanent water courses have become intermittent. This phenomenon is actually only an extension of a process already observed in the north of the Niger and Chad basins, in which the hydrographic networks are now fossilized (Lévêque, 1997). The result, generally, is a degradation of the Sahelian hydrographic network, which has consequences for the aquatic communities as well as for the landscape in general.

15.3.3. El Niño

The El Niño phenomenon (El Coriente d'El Niño, or the current of the Infant Jesus), which is a warm current that occurs with some regularity around Christmas time along the coasts of Peru, is now a well-known climatic fluctuation that triggers interactions between the atmosphere and the surface of the tropical ocean (Voituriez and Jacques, 1999). El Niño involves the equatorial Pacific but takes on a global dimension by its climatic and ecological repercussions. The mechanism at work in what is also called the ENSO events

(El Niño Southern Oscillation) is now quite well known. Normally, the surface waters of the equatorial Pacific, moved by trade winds from the southeast, drift slowly from the Peruvian coasts westward toward Indonesia. During this transit they warm up to 30°C and accumulate in the Pacific west, causing a depression of the thermocline. This displacement, which leads to a 30 cm rise in the water level close to the Indonesian islands, is compensated by upwelling of deep cold water along the coasts of Peru. These cold waters are rich in nutrients, stimulating the development of plankton and fish. The temperature difference of water masses between Peru and Indonesia also maintain the trade wind regime. An atmospheric convection loop is created that generates significant amounts of rain above the western Pacific. The dry air moves higher towards the east, cools, and descends again over the eastern regions of the equatorial Pacific, where the weather is generally hot and dry (Fig. 15.7).

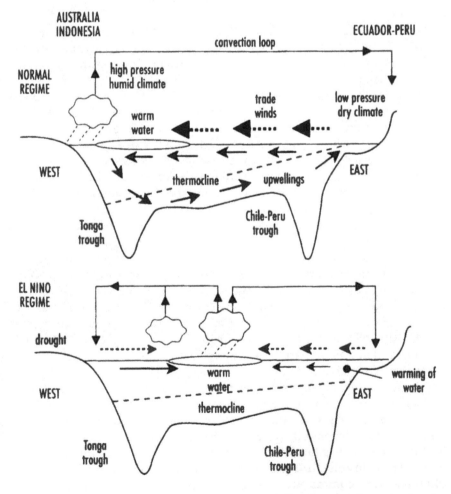

Fig. 15.7. Wind regime, movements of oceanic water and climatic events in a normal regime and El Niño regime in the equatorial Pacific

If for various reasons the trade winds weaken, the masses of warm air accumulated in the west return to the east along the equator, leading to a rise in the level of the Pacific Ocean at the centre of the basin. The atmospheric convection is thus very active above the central and eastern Pacific and there is high precipitation on the islands in the middle of the Pacific as well as on the South American coasts. Inversely, drought is severe in Indonesia and northern Australia. The weakening of the trade winds in the eastern part disrupts the upwelling of cold water on the western coast of South America, which intensifies the warming of surface water and the trade winds are partly replaced by westerly winds, which constitutes a positive feedback. Along the Peruvian coasts, the warm waters replace the usually colder waters, and this warming sometimes lasts more than a year. The warm waters, which are poor in nutrients, do not stimulate the development of food chains and the ecological functioning of the marine ecosystem is profoundly disturbed. Given the diminution of primary production, the fish disappear from the Peruvian coasts and shelter further south. The economic consequences for the fishing sector are serious. An ENSO event is generally followed by an inverse situation, with more intense trade winds and an abnormally low ocean temperature, called La Niña.

El Niño had a considerable amplitude in 1982-1983. It was partly responsible for the rainfall deficit observed in West Africa in 1982-1983. The 1997-1998 event reached a comparable amplitude. El Niño manifested itself very early in 1997 (in March-April). From July onwards, the warm anomalies of surface temperature reached 4 to 5°C in comparison with the seasonal averages along the South American coasts. A persistent drought from June-July, the most severe in 50 years, occurred in Papua New Guinea and the Indonesian archipelago. From November onwards, tropical storms and cyclones affected the regions that were normally spared, such as the islands of the central Pacific (Cook, French Polynesia). They caused significant damage on several atolls. The cyclonic activity was severe along the coasts of Central America from October 1997 to March 1998. From November 1997 there was torrential rain on the coasts of Peru and the Equator. Argentina, Chile, and Paraguay also experienced heavy rain, while there was drought on the Atlantic zone of Central and South America. Fires in some parts of northeastern Brazil occurred with dramatic amplitude, while the cyclonic activity of the Caribbean region was reduced to its minimum in 1997. In Africa, Kenya and Somalia recorded very high rainfall in October-November 1997.

The consequences of the El Niño phenomenon in 1997-1998 were studied in the equatorial Pacific Ocean (Chavez et al., 1999) by means of buoys. In late 1997, and early 1998, the ocean surface was greatly impoverished in nutrients and chlorophyll concentrations were among the lowest recorded in this region. This situation resulted from the absence of upwellings and low nitrate levels limited primary production. In early 1998, the phytoplankton community was restored and by mid-1998, because of the rise of nutrient-rich water due to upwellings, the system was as productive as the coastal environment.

El Niño is not a unique phenomenon. Like its homologue in the Pacific, the North Atlantic Oscillation is characterized by a huge rocking movement of a warmer mass of water (in the east-west direction) and of the atmosphere (in the north-south direction) over the North Atlantic. This oscillation, which is felt mostly in winter, between December and March, accentuates the differences of pressure between the anticyclone of the Azores and the depression of Iceland. When the differential is positive, the wet west winds dominate, and the winters are mild and humid in Europe but colder on the northeastern coast of America. A negative differential has the opposite effect.

15.3.4. The astronomic theory of climates

The Yugoslavian scientist Milankovitch proposed in the 1920s a theory that the glacial-interglacial cycles are governed by the position of the earth with respect to the sun. In fact, the attraction of other planets slightly alters the movements of the earth around the sun, which modifies the energy received at different latitudes during the course of the seasons. Therefore, there would be oscillations of different periods depending on the processes at work (Magny, 1995):

— *Variations of the eccentricity*. The shape of the earth's orbit varies between an ellipse and a near-circular track. In consequence, the distance from the earth to the sun varies, following cycles of about 100,000 years and 400,000 years. The existence of characteristic periods for the earth's orbit around the sun has been recorded in the sedimentary marine ridges, thus confirming the Milankovitch's astronomic theory about ice ages (Berger, 1996).

— *Variations of the obliquity*. The tilt of the earth's axis with respect to the plane of ellipsis varies at an angle between 22° and 25°. This variation follows a cycle of 41,000 years. When the tilt is steepest, the summers are hotter and the winters are colder.

— *Precession of equinoxes*. The earth's axis of rotation describes a circular movement around an axis perpendicular to the plane of ellipsis. At a given time of the year, the earth is not always at the same point in its orbit and its distance from the sun varies. The distance from the earth to the sun is thus modulated by variations of the eccentricity and the precession of equinoxes according to cycles of 23,000 and 19,000 years.

This theory, which was refuted in the 1950s, has been rehabilitated because it was corroborated by climatic records in marine sediments. If it is the case, we are presently in an interglacial period and there will be a new glaciation in some 60,000 years. However, climatic changes are not only a response to changes of insolation. The biosphere, atmosphere, and oceans also interact in such a way that in the meantime we may experience rapid episodes of cooling and heating, whether or not they are due to human activities. This complexity makes it rather difficult to make climatic predictions (Joussaume and Guiot, 1999).

15.4. CLIMATIC SYSTEM MODELLING

If we wish to answer questions about the possible consequences of the greenhouse effect due to human activities, we must first understand the functioning of the climatic system and the many interactions existing between all its components. Modelling, which best uses our understanding of physical laws regulating atmospheric and oceanic phenomena, allows this integration at different scales of time and space. It is a favourite tool of research on the climate change and at the same time may contribute to an unbiased assessment of the risks linked to an increase of the greenhouse effect.

According to the terminology defined by experts in the intergovernmental group on climate evolution, "climatic change" refers to changes of the climate attributed directly or indirectly to human activities altering the composition of the world atmosphere, which may add to the natural variability of the climate observed over comparable periods. Climatic changes of anthropic origin are an additional constraint for many ecosystems that are already affected by land use, pollution, or increased exploitation of resources.

Modelling of the climate necessarily limits the system to a set of relevant environments and processes that are manageable as well as exhaustive for the problem considered (Sadourny, 1996). The first attempts to model the climate, in the early 1960s, used global models at a time when weather forecast models, based over the very short term, were limited to the northern hemisphere. Still, it was a virtual revolution in climatology, which could thus mature from the descriptive to the quantitative. In simple terms, it consisted of considering the atmosphere as a stack of boxes (links) for each of which we could mathematically describe the evolution to a condition by fixing the conditions at the outset (winds, temperature, etc.) and knowing the forces (flows of solar heat) and the conditions at the limits (state of the earth's surface). At the end of the 20th century, climatic models became powerful tools of numerical experimentation thanks to impressive advances in calculation and data collection techniques.

Climatic changes: What do we know?

• **Greenhouse gases**

—Carbon dioxide, methane, nitrous nitrogen, and water vapour are greenhouse gases that retain the radiation emitted by the earth's surface. Because of this process, the earth's temperature is 30°C higher than it would be in their absence.

—The CO_2 levels in the atmosphere have increased by 25% during the past two centuries, from around 280 ppm to 356 ppm.

—The methane level of the atmosphere has doubled in 100 years.

—The increase in CO_2, methane, and nitrous nitrogen levels is the result of human activities such as the burning of fossil fuels, transportation, and agriculture.

• **Air temperature**

—The average global temperature increased by 0.6°C over the past 130 years, but it has not increased as much as we might expect from the increase in CO_2 level. It is believed that small particles present in the atmosphere and coming from volcanic eruptions or industrial activities reflect the solar radiation and help to cool the atmosphere.

—The Arctic peninsula has warmed up by around 2.5°C between 1945 and 1990 (Vaughan and Drake, 1996).

—The isotherm 0°C has risen by around 110 m in high altitude at tropical latitudes (15°N to 15°S) in the 1970s and 1980s (Diaz and Graham, 1996).

• **Precipitation**

—The duration of the snowy season and the quantity of snow in the Swiss Alps has clearly diminished since the mid-1980s (Beniston, 1997).

—There has been a slight increase in precipitation (1%) on the continents during the past century, but with regional differences. The precipitation has increased in high latitudes in the northern hemisphere but has declined since the 1960s in the subtropical and tropical regions from Africa to Indonesia (IPCC, 1995).

• **Oceans**

—The sea level has risen by 10 to 25 cm over the past 100 years.

—The oceans are particularly important in the control of climatic change because they serve as major reservoirs of heat, which they then redistribute through the circulation of oceanic masses.

• **Frozen sea ice**

—Since 1990 there has been a reduction of the frozen sea area in the Arctic, with an acceleration between 1987 and 1994. The extent of frozen sea was 9% less between 1990 and 1995 than between 1979 and 1989 (Maslanik et al., 1996).

Models of general atmospheric circulation that have been developed simulate the atmospheric flows (wind, temperature, humidity, clouds) over the entire globe on the basis of equations of fluid mechanics on one hand and radiative transfer on the other. These models, composed of several hundreds of interdependent equations, are based on rigorous principles but they explicitly represent only large-scale mechanisms. There are also many mechanisms (convection, nebulosity, turbulence, etc.) that act at the small scale but play a fundamental role in the control of large-scale processes. These mechanisms can be studied with detailed local models, but they can be integrated into general circulation models only in the form of very simple approximations or "parametrizations", the objective of which is to represent better the physics of the phenomena within the meshes of the model.

Models of general atmospheric circulation are not very different from weather forecast models. However, they are based on long time periods, so that the initial state of the atmosphere is practically forgotten. The important thing is to better simulate the atmospheric circulation. Besides, the physical processes that must be taken into account are much more numerous and complex than for weather forecast. Despite the extreme complexity of climatic models, they represent the essential atmospheric processes only very crudely: the frozen sea is represented in a simplistic fashion, and clouds and aerosols are still not well represented.

When we take into consideration sufficiently long periods of time, the interactions between atmosphere and ocean become crucial. Oceans are involved in the transfer of heat on the planet at a slower rhythm than the atmosphere, but simulation of the ocean complicates the model. The most simple way of simulating the interannual variability is to use atmospheric models supported by observed oceanic temperatures (Sadourny, 1996). Thanks to better information technology, the natural trend today is to abandon global atmospheric models in favour of combined ocean-atmosphere models that explicitly simulate the atmospheric and oceanic circulations, evolution of the ice cap, and the interactions of these different environments. Such models now realistically reproduce alternations of warm and cold events of the El Niño type, even though they still do not predict the occurrence and duration of such events.

Many uncertainties remain. In particular, the identification and quantification of the influence of ecological processes on the climatic equilibrium is a field still not well explored that probably holds new surprises. The medium-term response of the climatic system may also depend on processes we still do not understand. Variations in the deep oceanic circulation, for example, may cause climatic modifications much greater than those simulated by most present-day models. We may also ask about the role of the ocean and the vegetation in the regulation of CO_2 inputs. These uncertainties owe a great deal to the complexity of climatic mechanisms, as well as the possible existence of feedback mechanisms and the difficulty of measuring precisely over the entire planet and over long periods of time very small variations of temperature that do not result exclusively from natural variability.

Global changes: many uncertainties

• Carbon balances: Even though real progress has been made, we still do not know the quantitative balance of carbon exchanges between the atmosphere and the terrestrial reservoirs (soil, vegetation, oceans, etc.).

• Temperature: It varies from year to year and over long periods as a function of the natural variability of the climate. To the extent that serious data have been available for only a century, it is not easy to actually evaluate the long-term trends when the variations are weak, nor to attribute to them natural or anthropic causes.0141

• Solar radiation: It varies because of physical changes in the sun, with cycles of 11 years, for example. The consequences of such variations over long periods with respect to global warming are still unknown.

• Clouds: They can reflect solar radiation and thus prevent warming, or capture terrestrial radiation and thus add to global warming. These positive or negative effects depend on their height and properties, which vary over time and in space. Clouds are an important source of uncertainty.

• Plant growth: It may increase if the CO_2 level increases, thus absorbing more atmospheric carbon and increasing the storage of this element. Still, we must consider other interactions, with nitrates, for example.

• Ice cap: The ice caps will diminish with global warming, but this melting may be partly counterbalanced by greater snowfall in the polar regions, which could, in transitory conditions, contribute somewhat to cooling by their albedo.

Researchers using models have experimented with simulation of "interactive terrestrial biospheres". Their experiments aim essentially to take into account the effects exerted on vegetation by a modification of the soil water content. These effects have repercussions on the rugosity of the surface, the quantity of heat absorbed by the soil, and the speed with which the soil moisture evaporates. The diversity of the biosphere adds to the complexity of the modelling process, and its interactions with the climate remain blemished with great uncertainty.

15.5. THE CLIMATE TO COME

One potential application of modelling involves attempts to predict the climate. The climate of the earth has always changed and will undoubtedly continue to change. We cannot know what the future climate will be, but we are giving greater attention to its evolution. In 1996 the best international experts confirmed, and scientists today do not dispute it, that the levels of CO_2 and methane in the atmosphere are rising because of human activities. According to the laws of thermodynamics, this situation must induce or favour a rise in the average global temperature. The first simulations carried out in the late 1970s with primitive numerical models that gave way to rather alarming predictions of a warming of several degrees in the century to come. However, with improved models and the consideration of other phenomena, scientists are now talking of a less severe warming, with, however, more contrasting climatic consequences depending on the regions.

In temperate latitudes, the predictions are still diverse, even contradictory. Some regional simulations show deviations of polar fronts towards the south, which could increase the thermal contrasts in the Atlantic and produce more intense depressions, i.e., more storms. On the other hand, the ice of Greenland, as it melts, could modify the circulation of warm salt water on the surface. If the Gulf Stream, which makes the climate of western Europe mild, disappears or is deflected, Europe may become colder. Almost all models indicate a tendency to greater variability of atmospheric phenomena.

Climate change: What is likely to happen in the future?

• If greenhouse gas emissions follow the present levels, the climatic models forecast a doubling of the CO_2 level between the pre-industrial period and the end of the 21st century. If we add to this the increase in water vapour, the average temperature will rise by 1 to 4°C, the most probable value being 2.5°C.

• Generally, most models predict that the rise in greenhouse gas concentrations will diminish rainfall in southwestern Europe. Inversely, there will be an increase in the winter precipitation at high latitudes, which will result in an increase in the quantity of water vapour transported towards the poles from the low latitudes. Since the beginning of the 20th century, it has been observed that the quantity of rainfall has increased in the high latitudes of the northern hemisphere, while it has diminished in the tropical and subtropical zones, particularly in the Sahel and in Indonesia. In the northern part of the North American continent and in Eurasia, where temperatures are usually below zero, the snowfall has increased in the past few decades.

• The sea levels could rise on an average by 50 cm during the 21st century, which will lead to the inundation of low-lying coasts and certain habitats, such as coastal marshes.

• We can expect more extreme climatic events (heat waves, floods, droughts, tornadoes), but it is difficult to predict where or when they will occur with the present models.

• The vegetation zones will be considerably modified, with a blurring of limits between forests, savannahs, and shrub zones.

• The deserts will become hotter and desertification will intensify.

• Half of the world's glaciers may melt and the polar ice may shrink perceptibly in area.

• There will be changes in temperature, river flows, and lake levels, with consequences for the biodiversity, water resources, and water quality. The pressures on accessible water resources will intensify.

• Human and animal diseases will spread to new regions.

One of the persistent questions of prediction involves the reliability of the process. Today, we have sufficient mastery over the numerical prediction of weather, for which knowledge of the initial state of the atmosphere plays a determinant role. Besides, the quality of short-term predictions can be tested and improved daily by comparison with the evolutions observed. The advantage of meteorological forecast models is that we can keep constant, given the period of time considered, the value of slowly changing factors, such as the concentration of greenhouse gases. This is not the case with modelling of the climate, which is not concerned with the instantaneous behaviour of the system but with the way in which its average behaviour may evolve as a function of slow changes in certain factors. At the time scales in question, predictions become very sensitive to the way in which the critical processes of radiation balance are modelled, such as the atmospheric water cycle and the radiative properties of clouds. At present, the greatest uncertainties concern accounting for the complex role of clouds, which reflect as well as absorb solar radiation and reflect infrared thermal radiation towards space. If we can better understand the distribution of clouds and the water or ice content of the cloud cover we can undoubtedly improve our attempts at simulation.

To verify whether a climatic model correctly simulates the present climate, we have only to compare the outputs (seasonal charts of temperature, winds, rain, snow) to the reality observed. However, the basic problem remains: How do we validate today our predictions about the climate of a few decades from now? How do we validate a model simulating the sensitivity of the climate to the greenhouse effect? We can expect that the "experiment" will be carried out, i.e., that climatic warming will occur, and then compare the predictions to reality. But the problem does not end there. We need to have relatively reliable predictions in order to anticipate the phenomenon in socio-economic terms. Some researchers use data on past climates as a basis for numerical models.

For all the reasons mentioned above, we would be deluding ourselves if we were to pretend to make precise predictions about the future evolution of our global environment. Still, models are our only tool of analysis and expertise, and our only way to estimate the "climatic risk". They are valued greatly for their use in better anticipating the regional consequences of global changes because it is at the local or regional level that we are concerned with the eventual changes of the climate. They are likely to be improved in the years to come by the taking into account of new components. The present development of research shows, in fact, at what point our planetary environment must be understood as a unique system, where interactions between the atmosphere, oceans, ice caps, and the oceanic or continental biosphere, are constant. We may believe, as some do, that this diversity of interactions and processes leads to a capacity for self-regulation or homeostasis of the climatic system. The moderate level of global warming so far, given the rising content of greenhouse gases in the atmosphere over the last few decades, pleads in favour of this capacity of self-regulation. However, this apparent moderation of the climate change may also be one consequence of the slowness of the response of the oceans. It could also result from parasitic processes such as the emission of dust by industrial activities, which could mask the initial warming. In short, there are many sources of error and they may upset all our elegant intellectual constructions of models.

Chapter 16

Biosphere-Atmosphere Interactions and their Consequences for Global Equilibrium

Life does not start or stop the water cycle. By transpiration and other mechanisms, it modulates, diversifies, intensifies, and exploits the water cycle. Life changes the terrestrial environment so as to make it more favourable to life.

P. Westbroek (1998)

For the scientific community, which was long cloistered in its academic disciplines, the climate was in the past a topic for physicists, or perhaps chemists. Meteorology was structured essentially as a physical science even though the influence of the plant cover on the local and/or regional climate has been known for a very long time. However, during these past few years we have had to recognize that there is a close interaction between climate and life on earth: *the climate influences life on earth, and life in turn influences the climate, and more generally the global environment.*

This recognition was achieved not without some difficulty because the physical and biological sciences do not share the same paradigms. It is indeed a new challenge for the ecological sciences, which were primarily interested till now in the relationships of living things and their terrestrial or marine environment, but paid limited attention to the possible interactions of the planetary environment with the biosphere. A comparison of the earth's atmosphere with that of Mars and Venus confirms that *life has played a decisive role with respect to the composition of gases in the earth's atmosphere*: the high oxygen level and the low CO_2 level reflect the photosynthetic activity of bacteria, algae, and plants for hundreds of millions of years. Methane also exists in higher concentration than it would be in the absence of life.

Knowledge of the interactions between the biosphere and the atmosphere is a key element of our comprehension of the dynamics of the climate. It is a domain of research that requires mobilization of new competences to treat increasingly complex questions about the chemistry and physics of the atmosphere. It is also a highly sensitive domain since different human activities are known to be capable of modifying the composition and structure of the atmosphere, with possible consequences on the medium-term evolution of the climate.

16.1. STRUCTURE AND COMPOSITION OF THE ATMOSPHERE

The atmosphere is made up of a combination of gases and particles that surrounds our planet. It extends for some hundreds of kilometres above the earth's surface but it is made up of different concentric layers that can be characterized by temperature and chemical composition:

— The *troposphere*, between the earth's surface and on average 10-11 km altitude, contains 85% of the total mass of the terrestrial atmosphere. It is proportionately the thickness of the skin of an apple (Joussaume, 1993). This troposphere is characterized by a negative temperature gradient (around –50°C at 10 km altitude) and the presence of water in its three phases: gaseous, liquid, and solid. The orographic influences are largely felt here. This is the layer of the atmosphere in contact with the terrestrial surface that is principally concerned by the surface processes, such as evaporation, as well as by meteorological phenomena.

— Between the troposphere and the stratosphere, there is an isotherm layer of around 10 km thickness, the *tropopause*, which is the site of jet streams. In the *stratosphere*, which extends from 20 to 50 km altitude, the temperature rises progressively until it reaches around 0°C. The ozone present in the tropopause and the troposphere absorbs UV radiation, which explains the warming observed.

— The lowest temperatures (–90°C) occur at the summit of the *mesosphere*, which extends between 50 and 80 km altitude.

The dry air is made up of 78% nitrogen, 21% oxygen, 0.95% argon, and various gases in trace levels (Table 16.1). Nitrogen, an inert gas in geochemical

Table 16.1. Average composition of the atmosphere (percentage of volume) up to an altitude of 20 km

Gas	Formula	% volume	Time of residence	Cycle
Nitrogen	N_2	78.08	106 years	biological and microbiological
Oxygen	O_2	20.95	10 years	biological and microbiological
Water vapour	H_2O	0 to 4	10 days	physicochemical and biological
Argon	Ar	0.93		no cycle
Carbon dioxide	CO_2	0.036	15 years	biological and anthropic
Neon	Ne	0.0018		no cycle
Helium	He	0.0005	10 years	physicochemical
Methane	CH_4	0.00017	7 years	biogenic and chemical
Hydrogen	H_2	0.00005	10 years	biogenic and chemical
Nitrous oxide	N_2O	0.00003	10 years	biogenic and chemical
Ozone	O_3	0.000004		chemical

Gases with a variable level are indicated in italics.

terms, accumulates in the atmosphere over the course of time. The same is true for argon, another inert gas produced by the radioactive disintegration of potassium 40 in the earth's crust and released into the atmosphere by volcanoes (Ingersoll, 1996). Oxygen is a migrating gas that passes from the atmosphere into the biosphere and the ocean, and vice versa. Another important constituent of the atmosphere is water vapour. While the composition of dry air is constant throughout the world, the quantity of water vapour varies greatly and may reach 4% in volume.

16.2. ORIGIN AND EVOLUTION OF THE ATMOSPHERE

The terrestrial atmosphere has not always had the composition that we know at present. Its changes over time are a good example of the long-term interactions between the biosphere and the atmosphere, to the extent that biogenic activity is partly the cause of changes in the composition of the atmosphere since the earth was formed.

16.2.1. Influence of the biosphere on atmospheric composition

More than 4000 million years ago, the atmosphere did not contain free oxygen. The primitive atmosphere must have been made up of inert gases emitted by volcanoes and thermal springs, such as hydrogen, helium, neon, or argon. Once the planet was sufficiently cooled, the principal gases that constituted the atmosphere were, according to what we now know, carbon dioxide (CO_2), carbon monoxide (CO), nitrogen (N_2), water vapour, and, no doubt, hydrogen sulphide (H_2S). Only the abundance of water distinguished the primitive atmosphere of the earth from that of Venus or Mars. It is nevertheless likely that this primitive atmosphere contained enough ammonia and methane to generate organic matter (Allegre and Schneider, 1996). Oxygen could be formed transitorily by photodecomposition of water vapour and CO_2, but it would rapidly recombine.

The first living organisms (prokaryotic bacteria) were thus anaerobic and drew their energy from the fermentation that produced glucose. *How did this initial biosphere without oxygen become the present biosphere?*

Photosynthesis was to appear about 3000 million years ago, probably in cyanobacteria with phototrophic forms able to fix CO_2 as well as atmospheric nitrogen to synthesize their own substances. These cyanobacteria elaborated large deposits of calcium carbonates, the stromatoliths that were the first bioconstructions. We still find them today in waters of warm regions.

Photosynthesis liberates oxygen but, given that many oxidizable compounds were present in the ancient oceans (iron, for example), oxygen produced by the first living creatures was rapidly used to oxidize these compounds rather than reaching the atmosphere. Simultaneously, the level of sulphates of oceans increased progressively, allowing the development of sulphate-reducing bacteria that proliferated in the absence of oxygen. With time, photosynthesis developed and the production of free oxygen became higher than the rate of oxidation of terrestrial matter.

The transition from an atmosphere without oxygen to an atmosphere rich in oxygen occurred around 2500 million years ago, as indicated by oxidized sediments on the continents (Turekian, 1996). However, the accumulation of oxygen in the atmosphere was slow.

Close to 700 million years BCE, the oxygen concentration reached its present value. On the one hand, this oxygen would be the origin of the formation of the ozone layer, which protected life forms from ultraviolet solar radiation, and on the other hand it allowed the appearance of new types of metabolism because living organisms had to adapt to the presence of oxygen, which is a highly reactive substance. Eukaryotes would draw part of this poison-oxygen, elaborating an energy metabolism based on oxidation, a considerable source of chemical energy. With respiration, the energy provided in a form useful to cells is much greater than that from fermentation. This is one explanation for the invasion of land and sea by plants (Foucault, 1993) (Fig. 16.1).

These theories attribute great importance to the living world: Oxygen is a product of the biosphere. The abundance of oxygen and the low level of carbon dioxide in the present atmosphere are thus the result of photosynthesis by plants.

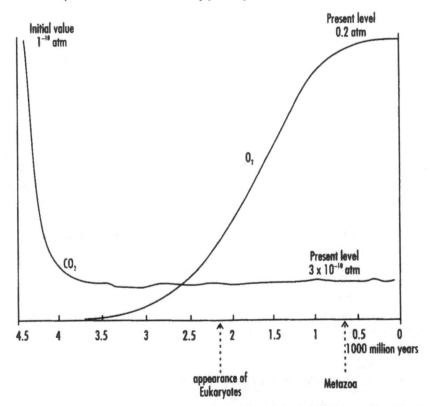

Fig. 16.1. Comparative evolution of CO_2 and oxygen in the atmosphere during the history of the earth (Turekian, 1996)

16.2.2. Stratospheric ozone and protection against ultraviolet radiation

The rise in oxygen levels in the atmosphere created conditions for what appears to be another essential element for the development of life on earth: the stratospheric ozone. In fact, the ozone molecule (O_3) is a derivative of oxygen that forms normally in the stratosphere. Under the effect of solar radiation, some of the oxygen molecules break into two highly reactive atoms (photolysis) and the atomic oxygen thus formed recombines with molecular oxygen to yield one molecule of ozone (triatomic oxygen). The presence of ozone in the atmosphere is thus not linked to a direct emission in the form of ozone molecules but to a set of complex physicochemical processes given the number of reactions at work. Most of the ozone is found in the stratosphere between 15 and 40 km altitude. Ozone is generated mainly at low latitudes, but the movements and circulation of the atmosphere cause an accumulation of ozone at the two poles, where the values reach a maximum during the spring.

Despite its low abundance in the atmosphere, the stratospheric ozone plays an essential role for the biosphere. It absorbs part of the solar radiation corresponding to ultraviolet rays in the range of short wavelengths (between 0.20 and 0.33 microns). *The anti-UV filter effect of ozone could have allowed life to colonize the terrestrial environments, eliminating the short wavelengths that could destroy cells and inhibit photosynthesis.* Ozone continues to serve this function at present, whence the concern about its destruction by gases resulting from human activities.

16.3 BIOGENIC AND ANTHROPOGENIC TRACE ELEMENTS

The chemical composition of the terrestrial atmosphere is largely controlled by processes of exchanges between the atmosphere and the biosphere. Biological processes and more recently human activities are principal sources of several trace elements found in the atmosphere (Table 16.2), such as sulphurated and nitrogenous compounds: nitrous oxide and ammonia, methane, and other hydrocarbons. These trace elements have an important, but not always well-known, role in the control of the overall climate of the earth. Some contribute actively to the greenhouse effect (see Chapter 15), while others act on the thickness of the stratospheric ozone, which protects the earth and its living organisms from UV rays. We must emphasize that our understanding of the dynamics of these gases and the interactions between the processes at work is still very limited.

16.3.1. Methane and other hydrocarbons

The presence of methane in the atmosphere has been known since 1940 but the first measurements date only from the late 1960s, when sufficiently sensitive

Table 16.2. Effects of gases in low concentrations in some atmospheric phenomena

Effect	CO	CO_2	CH_4	NO and NO_2	N_2O	SO_2	CFC
Greenhouse effect		+	+		+	−	+
Destruction of stratospheric ozone	+/−	+/−	+/−	+/−			+
Acid deposits				+		+	
Total emissions (million t)	2000	5500	550	30 to 50	25	150 to 200	1
Time of residence	some months	100 yrs	10 yrs	some days	170 yrs	some days	60 to 100 yrs
Conc. (pp 1000 million) 100 years ago		290,000	900	0.001	285	0.03	0
Avg. present conc. (pp 1000 million)	100 to 200 in northern hemisphere	350,000	1700	0.001 (pure air) to 50 (polluted sites)	310	0.03 (pure air) to 50 (polluted sites)	~ 3 chlorine atoms

+ indicates an aggravation of the effect and − indicates an amelioration.
+/− indicates that the effect depends on local conditions.
Quantities emitted and average concentrations of greenhouse gases in the atmosphere (Crutzen and Graedel, 1996).

techniques of analysis became available. For studies of long-term trends, relevant information can be obtained from analysis of air bubbles trapped in polar ice.

Methane (CH_4) has various origins: e.g., bacterial decomposition of organic matter in anaerobic conditions (flooded rice fields, marshes, oceans), discharges, rumination of herbivores (a cow releases on average 200 g methane a day), termites, vegetation fires, released gases from mining activity, or escapes during transport of natural gas. The release of methane in marshes, which sometimes burns spontaneously, is the source of legends of will-o'-the wisp.

The sources of methane are balanced by two principal sinks: a chemical sink corresponding to photochemical oxidation of methane in the troposphere and the stratosphere, which leads to the production of another greenhouse gas, ozone, and a biological sink linked to decomposition by microorganisms in soils.

The assessment of the quantities of methane emitted each year is still imprecise, about 525 Mt (IPCC, 1996) (megatonne = 10^6 tonnes), with estimates varying between 210 and 710 Mt. Around 70% of that is linked to human activities. Its concentration in the atmosphere is presently increasing by 0.7% to 1% a year (around 16 ppmv), which suggests that the present concentration could double in about a century. In reality, the methane concentration of the atmosphere has multiplied 2.5 times in a little more than a century (Fig. 16.2), and 5 times if we refer to the last glacial period, about 15,000 years ago (Lambert, 1992). It began to rise rapidly around 1850, from 70 ppmv to 172 ppmv in 1990. The potential greenhouse effect of methane is 20 to 30 times as great as that of

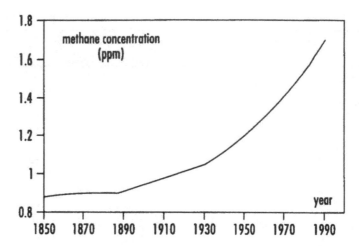

Fig. 16.2. Average concentration of methane in the atmosphere (ppm) estimated from the composition of air bubbles contained in polar ice

CO_2 but the life span of a molecule in the atmosphere is relatively short and varies from 7 to 10 years. Its contribution to the rise in world temperature is about 20%.

Apart from methane, hydrocarbons produced by plants, such as terpenes, isoprenes, and aromatics, also contribute to the regulation of the global temperature. The principal emissions of isoprenes are due to deciduous hardwood forests: oak, maple, etc. These emissions are diurnal and seasonal, with a low level in winter. Isoprene reacts with the OH radical during the day to form ozone. Monoterpenes are emitted continually, with a nocturnal rise in the forests. The terpenes and hemi-terpenes are hydrocarbons extracted from plants and used for many years in perfumes, insecticides, and solvents. It is estimated that the biogenic production of hydrocarbons is 10 times as high as that resulting from the use of fossil fuels. Around half of these volatile hydrocarbons agglomerate in the particulate form, while half remains in the form of vapour and helps absorb terrestrial radiation, like other greenhouse gases, but they have been studied very little till now (Hayden, 1998).

16.3.2. The nitrogen components

Ammonia (NH_3), nitrogen monoxide (NO), and nitrous oxide (N_2O) are the three primary nitrogenous compounds emitted into the atmosphere. Ammonia is released into the atmosphere by animal excreta (25 Mt/yr or about 25×10^9 kg/yr), the biological activity of soils (11 Mt/yr), and combustion (2 Mt/yr).

The ammonia released into the atmosphere reacts rapidly with water to form ammonium (NH_4^+) according to the reaction $NH_3 + H_2O \longleftrightarrow NH_4^+ + OH^-$, which contributes to the acidification of rain water.

Moreover, the ammonia reacts with aqueous aerosols of H_2SO_4 to form sulphate aerosols ($(NH_4)_2SO_4$. The conversion of gaseous ammonia into sulphate aerosols subtracts the ammonia, at least temporarily, from the biogenic nitrogen cycle. These aerosols can be transported over long distances (several thousands of km) before being deposited by rain or in the form of a dry deposit. The consequence is the dispersal of nitrogen in the form of a sulphate aerosol quite far from the sources of gaseous ammonia (Berner and Berner, 1996).

Nitrous oxide (N_2O) is a gas involved in the greenhouse effect that exists in infinitesimal traces in the atmosphere. Soils are the origin of 65% of N_2O emissions on the global scale (IPCC, 1996). Nitrous oxide is naturally produced by two major microbial processes that are part of the nitrogen cycle: *denitrification* and *nitrification* (Firestone and Davidson, 1989). In temperate climates, denitrification is the principal mechanism of N_2O production. Intensive agriculture and the use of nitrogenous fertilizers are largely involved (Bouwman, 1996). It has been calculated that the production of N - N_2O from farm land is around 1 kg/ha/yr, but it also results from the burning of vegetation.

Nitrification in the oceanic surface waters is shown to be a quantitatively significant source of a flow of N_2O towards the atmosphere (Dore et al., 1998). The atmospheric concentration of N_2O has grown from 285 ppbv around 1700 to 310 ppbv in 1990, or an average increase of 0.25% a year. It is a gas with a capacity for global warming that is 200 to 300 times that of CO_2. It contributes about 6% to the greenhouse effect. In the stratosphere, the abundance of nitrous oxide plays a role in the equilibrium of the ozone layer by participating in its destruction. The sources are said to be around 15 Mt/yr (IPCC, 1994), of which 3 Mt comes from the oceans. Its time of presence in the atmosphere is around 150 years.

Nitrogen monoxide results from bacterial activity in the soil (12 Mt/yr) and combustion of natural origin (8 Mt/yr) and anthropic origin (24 Mt/yr). The abundance of nitrogen monoxide in the stratosphere influences the equilibrium of the ozone layer, while by the intermediary of its oxidation in the troposphere it is the source of products such as nitric acid and ozone.

Tropospheric ozone and the pollution of the atmosphere

The concentration of tropospheric ozone (not to be confused with stratospheric ozone) produced at low altitude (0 to 10 km altitude) by the oxidation of carbon monoxide (CO) in the presence of nitrogen oxides and light has risen over the past few decades. This pollutant gas comes from agricultural and industrial activities, especially from motor vehicle exhaust, brush fires, and forest fires. Tropospheric ozone is involved in air pollution: its concentration has tripled since the end of the 19th century and its increase has been estimated at 10% per decade between 1970 and 1990 for the middle latitudes of the northern hemisphere. The principal reason is the increase in anthropic emissions of nitrogen oxides during the 19th century (three times the natural emissions today). It is responsible for the withering of some species of plants, the effects being seen in the appearance of small necrotic patches on the upper surface of leaves (Academie des sciences, 1993). Plants sensitive to the effect of ozone are those in which the stomata close slowly when they are exposed to high concentrations of ozone.

16.3.3. Chlorofluorocarbons, organic halogen compounds and the hole in the ozone layer

The organic halogen compounds, including chlorofluorocarbons or CFCs, are a minor subset of chlorine-containing compounds released at the earth's surface. The oceans and volcanoes emit each year around 210 Mt of chlorine equivalent into the atmosphere in the form of hydrochloric acid, sodium chloride, or molecular chlorine, compounds soluble in water.

a) Origin of organic halogen compounds

The only organic halogenic compounds of natural origin, which have a sufficiently long life span in the troposphere to contribute significantly to the additional greenhouse effect, are produced in the oceans by reaction between chlorine, iodine, or bromine ions contained in sea water and the reduced forms of sulphur produced by algae in the surface layers of the ocean. These products are chloride, iodide, and methyl bromide. This source produces around 5 Mt/yr of methyl chloride (CH_3Cl), the level of which in the atmosphere has not varied since the 1980s. The coastal zones of tropical islands are an important source of methyl chloride (Yokouchi et al., 2000).

These natural emissions have been augmented for the past 50 years by industrial production of organic halogen compounds, the CFCs, which are entirely artificial chemical compounds in which the molecules are highly stable and thus, in principle, harmless to living things. They have been widely used in industry: as propellant gas in aerosol pumps in extinguishers, refrigerants (in refrigerators and air conditioners), solvents, and in the manufacture of plastic foam. They are commonly known as Freon. However, these gases have an enormous power of absorption of infrared radiation, which makes their properties as greenhouse gases 14,000 to 17,000 times as efficient as CO_2 at an equal mass. Their implication in global warming is about 17 to 20%. They represent today a total of 16 Mt and may remain in the atmosphere for 75 to 110 years, depending on the molecules.

Close to 85% of chlorine-containing compounds measured in the stratosphere are of anthropic origin. The chlorine content of the stratosphere during the 21st century will continue to be dominated by CFC emissions from the years 1960-1990. The return to levels that existed before the appearance of ozone destruction in the polar regions will not occur before the years 2040-2050 because of the very long life of CFCs (Académie, 1998). Still, we must be vigilant and analyse carefully what happens over time.

b) The hole in the ozone

The behaviour of chemically active constituents that determine the equilibrium of ozone is regulated by a cycle that will affect "source" molecules (methane, water vapour, nitrous oxide, CFC) emitted at the surface of the earth and chemically inactive in the troposphere. When these molecules reach the stratosphere by diffusion, they are destroyed by intense solar radiation and result in the production of radical chemicals (HOx, NOx, ClOx) that react rapidly with ozone. They are eliminated from the stratosphere by "sink" molecules that are soluble in water and made up essentially of acid forms (hydrogen peroxide H_2O_2, nitric acid HNO_3, hydrochloric acid HCl). The increase of "source" constituents leads to the increase in radical molecules, which results in a reduction of the local concentration of ozone.

The "hole in the ozone" that appeared above Antarctica seems to be the first spectacular manifestation of the effect of human activities on the global physicochemical equilibriums of the atmosphere. It was soon suspected that chemical reactions could explain the phenomenon, and CFCs were incriminated. The stability of these products, which were industrially successful, had unpredictable consequences: they were not chemically broken down in the atmosphere and travelled to the upper atmosphere, where the ultraviolet rays could break up their molecules, liberating chlorine atoms. These atoms would attack the ozone molecules to yield oxygen and chlorine monoxide (ClO) (Cl + $O_3 \longrightarrow ClO + O_2$). When chlorine monoxide reacts with an atom of oxygen (produced by the photodissociation of another ozone molecule), it liberates chlorine, which can catalyse a new cycle (ClO + O $\longrightarrow$ Cl + O_2). The net result is a diminution of the ozone level.

The diminution of ozone concentrations in the lower stratosphere has several consequences for the energy equilibrium of the atmosphere. It leads to a cooling of the lower stratosphere of about 0.25 to 0.4°C per decade (Académie, 1998). It then induces a negative radiative heating on global average at the troposphere-stratosphere limit of about –0.1 W/m² over the past two decades that counterbalances, for around 20% in relative values, the positive radiative heating attributed to greenhouse gases in the troposphere.

c) Increase in UV radiation

Since the ozone layer protects the living world against UV rays, a reduction in ozone concentration poses a considerable danger to it. For some years, significant drops in the ozone concentration have been observed above the Antarctic. The diminution reached 30% in 1985 and the phenomenon has since been repeated and amplified. Today, the massive destruction of the ozone between 14 and 20 km altitude above Antarctica in the beginning of the southern spring leads to reduction of more than 60% in the thickness of the ozone layer during September and October. *In situ* measures indicate that the destruction of the ozone begins at the end of the southern winter, when the sunlight returns, and that it can reach a value of 50% in one month. Increases in the quantity of UV radiation reaching the earth's surface are clearly associated with the spring reduction of the Antarctic ozone (van der Leun et al., 1995). In the Arctic, since 1994-1995, destruction of the ozone has been observed between January and March that could reach close to 30% over an area comparable to that in the southern hemisphere.

Thinning of the ozone layer is accompanied by an increase in the UV-B radiation that reaches the earth's surface. The UV-B rays account for less than 1% of the solar energy reaching the earth's surface, but they could play an important role in the evolution of life. Their increase presents risks for humans and animals (burns and cancers of the skin, cataracts, etc.) and for forests and crops (reduction of photosynthesis and loss of productivity). In aquatic systems, exposure to UV-B has been shown to diminish phytoplankton growth, cause damage to DNA, and inhibit photosynthesis in phytoplankton, and thus the marine primary production (Denman et al., 1996). Given that species differ in their tolerance to UV, a change in the structure and composition of planktonic communities favouring the most tolerant species is probable.

UV-B rays also inhibit marine microbial activity (bacteria do not have pigmentation that protects them against radiation) and have varied effects on dissolved organic carbon by the photolysis of macromolecules. These are factors that could significantly modify the structure of ecosystems. In the regions of the Antarctic in which the ozone hole is large (diminution of nearly 50% in the spring), a drop in plankton productivity has been observed (6 to 12%) as well as changes in the species composition of algae: phytoflagellates dominate at present, while in the 1970s the centric diatoms were dominant (Demers et al., 1996). Nevertheless, studies on the effects of UV radiation on aquatic ecosystems are still rare and sometimes difficult to interpret with respect to observations in the natural environment.

In the absence of systematic and reliable measurements, we do not currently have much proof of an increase of UV-B rays over the past few years. However, measurements taken by spectrometer from the Nimbus-7 satellite of NASA show that the reduction in ozone concentration has caused an increase in the annual exposure to UV radiation of 6.8% in 10 years in the zone of 55°N latitude and of 9.9% in 10 years in the zone of 55°S latitude (Arques, 1998).

The available data also indicate that in New Zealand, in the summer of 1998-1999, the peak UV radiation was around 12% higher than during the early 1990s. This increase would be a consequence of the reduction in atmospheric ozone (McKenzie et al., 1999). The greatest increases were recorded for UV radiation that was capable of causing damage to plants and DNA.

The ozone hole and the Montreal protocol

If the hole in the ozone layer observed till now during the southern spring grows larger, the entire world could be subjected to a bombardment of ultraviolet rays. The uneasiness raised by this perspective has led the international community to take certain measures. The Vienna convention signed in 1985 proposed actions to protect the ozone layer. The Montreal protocol, adopted in 1987 by 81 states, stipulated that consumption and production of CFCs must be reduced by 20% in 1993 in relation to 1986 levels, and by 50% in 1998. The signatory countries decided to accelerate the processes and proposed that CFCs should no longer be manufactured or used from the year 2000. Manufacturers of CFC have found substitutes, but it is difficult to affirm at present that they are harmless.

The issue of CFCs is sometimes presented as a good example of collective awareness of the dangers incurred at the international level that can be resolved by international cooperation. But the situation is relatively simple: few countries are involved in the manufacture of CFC, which simplified the discussions, and substitute products were available that could quickly be used. Some even went so far as to say that as the patents on CFC were about to expire, it had become urgent to find other patentable products, and that the hole in the ozone layer was an opportunity for a major industrial economic operation.

16.3.4. Carbon dioxide and additional greenhouse effect

Human activities are presently involved to a great extent in the carbon cycle, especially with respect to the balance of exchanges between the atmosphere and the biosphere. The CO_2 level of the air has been measured continuously only since 1958, but earlier levels can be discovered from the air trapped in air bubbles at the time of ice formation.

At the poles, the ice caps have been preserved over periods that could be as long as some hundreds of thousands of years because of the near perpetual cold in these regions. Analysis of this "fossil" air indicates that CO_2 concentration increased about 27% from the pre-industrial age. The increase is attributed to the acceleration of CO_2 emissions from human activities, which increased significantly around 1800-1850 (Fig. 16.4).

Human beings today inject into the atmosphere, in the form of CO_2, around 7000 to 8000 million tonnes of carbon a year, or on average 4 kg/person/day (average 4 kg for France, 1 kg for developing countries, and 15 kg for an inhabitant of the United States: Sadourny, 1992). The principal artificial source of CO_2 is the burning of wood, coal, and petroleum. We extract and burn several thousands of millions of tonnes a year of fossil fuels that have been accumulated over millions of years in the depths of the earth. It has also been calculated that in the tropical regions the annual consumption of wood for fuel is around 465 kg/person/year (Andrea, 1991).

Moreover, in many countries, it is common to burn crop residues (straw, sugar cane wastes), or to light brush fires, or even to destroy forests by fire. It is estimated that 25 to 50% of the Sudanese savannahs and 60 to 80% of the Guinean savannahs are burned each year (Menaut et al., 1991). The contribution of biomass fires to planetary levels of CO_2 reaches 40% (Andrea, 1991).

The result is that the CO_2 level in the atmosphere has increased from 315 ppmv in 1955 to 355 ppmv in 1990. Over longer periods (Fig. 16.3), the data collected for the two last glacial-interglacial cycles (around 250,000 years) indicate that CO_2 concentrations have oscillated between 200 and 300 ppmv. The maximal values correspond to warm interglacial periods and the minimal values to the most severe glacial conditions.

These observations raise questions about the contribution of greenhouse gases to climate changes over the course of climatic cycles. Without being the initial cause, the greenhouse gases could have amplified the major climatic variations.

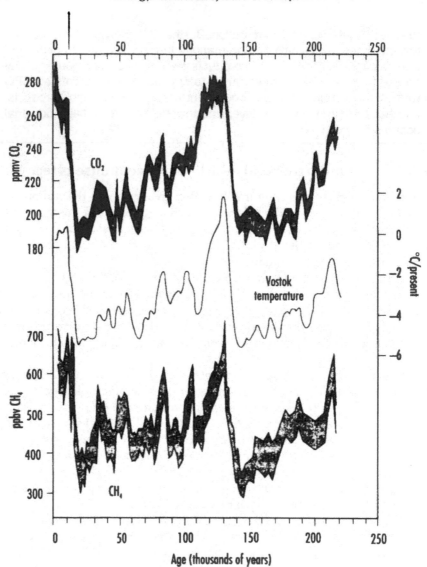

Fig. 16.3. The Vostok ice core extracted from the Antarctic ice reveals a close correspondence between the CO_2 concentration, methane concentration, and temperature during the last four glacial-interglacial cycles. The temperature curve was reconstructed from the concentration of deuterium in the ice. The CO_2 concentration was measured from air bubbles in the ice.

Storage of greenhouse gases

Instead of allowing the CO_2 produced by human activities to be released into the atmosphere, we could store it in natural reservoirs to limit the risks of warming of the atmosphere. Various means of storage have been proposed (Herzog et al., 2000):

—Planting trees, which absorb CO_2 during their growth and conserve it in their tissues for their entire lifetime. Reforestation and agroforestry have been proposed by the Kyoto treaty for the reduction of greenhouse gases. However, to compensate for the present emissions, we would have to plant each year an area equal to that of India.

—Injecting the CO_2 between 1000 and 2000 m depth in the oceans. The cold water of the ocean depths takes several centuries to rise to the surface. The ocean already absorbs a large proportion of anthropic emissions, and the process could be accelerated. There is, however, a risk of acidification of the sea water.

—Injecting the CO_2 into geological formations such as non-exploitable coal seams, sinks of petroleum or natural gas that are exhausted, and geological formations that contain saline water.

It is now known that changes in the concentrations of various greenhouse gases since the beginning of the 20th century have led to an increase in climatic warming estimated at 2.5 watts/m² (IPCC, 1996). Around 62% of this warming is due to an increase in the CO_2 concentration, 20% to increase in methane, 4% to increase in nitrous oxide, and 14% to increase in CFCs. The higher ozone concentration in the troposphere due to a rise in the emissions of methane, carbon monoxide, nitrogen oxides, and other compounds has led to a supplementary increase of 10 to 15% in the climatic warming since the beginning of the 20th century (WMO, 1998).

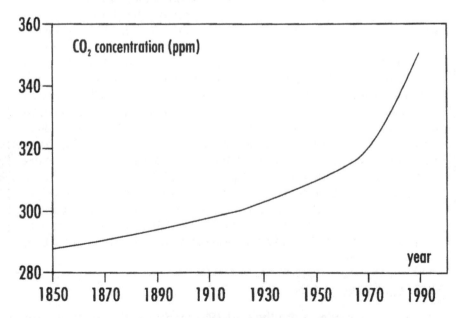

Fig. 16.4. Evolution of CO_2 concentration in the atmosphere. The CO_2 level was measured from 1958 onwards, but earlier values were obtained by analysis of air bubbles in ice in various places in the world.

The additional greenhouse effect

The genesis of the greenhouse effect theory goes back to the beginning of the 19th century. The physicist Joseph Fourier (1786-1830) published in 1824 his *Remarques générales sur les températures du Globe terrestre et des espaces planétaires*, in which he noted that the establishment and progress of human societies and the action of natural forces could considerably change, over vast regions, the state of the soil surface, the distribution of water, and the major air movements. Such effects might be able, over the course of several centuries, to modify the average heat of the planet. The idea of the greenhouse effect and the association of the physical phenomena with the economic development of humanity was thus launched (Grinewald, 1992).

In 1861, the Irish physicist John Tyndall (1820-1893) attributed the greenhouse effect essentially to water vapour, indicating that any variation in the water vapour in the atmosphere must produce a change in the climate. He added that this was also true for CO_2. The theories of Tyndall, that moderate changes in the composition of the atmosphere could have considerable climatic effects, were useful in explaining why the earth did not always have the same climate (Grinewald, 1992). However, Tyndall could not have imagined the impact of industrial development on the atmosphere.

In the early 20th century, the Swedish scientist Svante Arrhenius (1859-1927) took up the climatic theory of CO_2 outlined by Tyndall and definitively linked the trend of the greenhouse effect with the use of fossil fuels, on which the wealth of the industrialized nations was based. By a simple calculation, he showed that a doubling of the atmospheric load of CO_2 would translate into a global warming of the planet by about 6°C. The concept is perfectly founded, even though the quantitative result is arguable. Arrhenius was the first to propose quantitative estimations of the influence of the use of fossil fuels. He was not worried, however, about the increase of the additional greenhouse effect due to human activities; he saw in it, on the contrary, a way to counterbalance the next ice age. The ideas about the impact of deforestation on the dynamic equilibrium of CO_2 in the atmosphere were taken up in 1924 by Vernadsky in his work *La Geochimie*. However, the question of relationships between CO_2 of industrial origin and the climate did not worry people till 1972, when the MIT report on the "limits of growth" was published for the Rome Club. With the international geophysics year (1957-1958) began the systematic measurement of CO_2 in the atmosphere.

16.3.5. Acid rain

Acid pollution, recorded since the mid-19th century, is the result of a set of mechanisms that cause a transfer of acid or acidifying substances to the air-soil-vegetation interface. It is generated by pollutants (mainly sulphur, nitrogen,

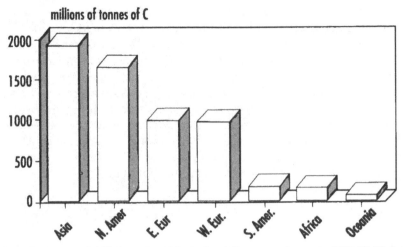

Fig. 16.5. Emissions of CO_2 from fossil fuels for different continents in 1992 (Siddiqi, 1996) millions of tonnes of C

and ammonia) that come essentially from emissions of anthropic origin such as the combustion of fossil fuels.

In the absence of pollution, rain water is slightly acidic, with a pH close to 5.6 on average. This acidity is the result of the dissolution of CO_2 from the air as well as the presence of sulphuric and nitric acid. Several chemical reactions result in the formation of sulphuric acid in the atmosphere:

— In the atmosphere, sulphur dioxide (SO_2), for example from volcanic emissions, yields sulphurous acid (H_2SO_3) on contact with water vapour. The unstable sulphurous acid is transformed into sulphuric acid (H_2SO_4). At present, a large quantity of sulphur dioxide comes from human activities.

— The SO_2 in the presence of radical OH is transformed into sulphur trioxide SO_3, which also yields sulphuric acid on contact with water.

— The combination of sulphur dioxide with nitrogen dioxide in the presence of water yields $SO_2 + NO_2 + H_2O \longrightarrow H_2SO_4 + NO$.

— Above the oceans, photochemical oxidation of dimethyl sulphur transforms it into sulphur dioxide and then into sulphates.

Although sulphuric acid is involved in more than two-thirds of the acidification of precipitation, nitric acid HNO_3 formed by the oxidation of nitrogen oxides from industrial combustion also plays a role. In addition to this there is chlorine, which comes form the combustion of PVCs. The sulphuric and nitric acids of the troposphere not only fall with the rain but are also deposited "dry" in solid particles.

It was in the 1950s that scientists began to be concerned by the problem of acid rain when observing the acidification of lakes in Canada. With the increase in emissions of sulphur compounds resulting from human activities (particularly wastes of sulphur dioxide from the use of fossil fuels and metallurgy of metallic sulphides), they observed a great increase in the acidity of rain water and acid deposits in various industrialized regions of the world, such as Europe, North America, and China, where the average pH of rain water has gone from 6 to 4 between 1955 and 1985. pH values as low as 2.4 have been recorded in the United States. This is a record, but in such conditions the rain water is more acidic than vinegar. This generalization of the phenomenon resulted from the circulation of air masses carrying the toxic substances far from the sites of emission. The consequences for aquatic life were considerable, with a rarefaction of all the fauna and flora. In the soils, acid rain mobilized toxic elements such as aluminium, which acts on plant roots. Acid rain also was implicated in the withering of forests, along with other gaseous pollutants.

16.4. AEROSOLS AND THE ALBEDO EFFECT

Greenhouse gas emission is not the only human activity that can modify the climate. There is also the emission of dust and aerosols (small solid or liquid particles in suspension in the atmosphere) from industrial activity (discharge of sulphurated molecules), automobile traffic, forest fires, or soil erosion. Among the sulphurated compounds present in traces in the free atmosphere, sulphur dioxide (SO_2) and dimethyl sulphur (DMS) play an essential role in the dynamics of the radiative balance of the earth because they undergo complex physicochemical transformations whose final products are submicronic aerosols.

One of the effects of aerosols with small particles is to cool the earth system by directly reflecting the solar radiation entering the atmosphere (albedo effect) or by increasing the number of nuclei of condensation in the clouds so that they reflect solar radiation. These direct and indirect effects could help cool the surface of the earth, in opposition to the greenhouse effect. In helping to reduce the temperature at the earth's surface, aerosols also limit photosynthesis, respiration, and methanogenesis. These sulphur compounds also contribute to the acidification of rain water and, when they are injected into the stratosphere, catalyse the destruction of the ozone above the poles.

Aerosols are in principle precipitated by rain water after a few weeks and thus do not accumulate in the atmosphere. Decontamination policies made by government, measures taken by industries, have considerably reduced SO_2 emissions in Europe and North America. However, the increase of aerosols since the beginning of the industrial age, which is difficult to evaluate precisely, has perhaps helped to reduce the radiative balance of the earth, partly masking the initial manifestation of the greenhouse effect (Le Treut, 1996).

The climate can also be disturbed unpredictably by volcanic emissions. Some scientists have attributed the fall of the average temperature by 0.5°C in 1991, as well as the slowing of the rate of increase in CO_2 and CH_4 in the atmosphere, to the eruption of Mount Pinatubo, which released into the atmosphere large quantities of dust and sulphate aerosols (Walker et al., 1999). This incident illustrates the sensitivity of biospheric processes to the temperature, and the fact that a localized event can have significant consequences on the global carbon cycle.

There are also natural aerosols such as continental dust, salt, or marine sulphates. The only significant natural biogenic source is DMS produced by oceanic plankton at the air-sea interface, which is the result of complex biological processes that are still incompletely explained (Academie, 1994; Bates et al., 1992). DMS is produced by the degradation of dimethyl sulphonium propionate (DMSP), a molecule acting as osmoregulator and cryoprotector for algae, especially dinoflagellates (Deman et al., 1996). The quantity of DMS available in sea water for ocean-atmosphere exchanges is controlled by biology. Seasonal peaks of DMS concentration have been recorded in the Pacific Ocean, the Indian Ocean, and the North Sea, but the mechanisms of sea-air exchanges of DMS are still poorly understood. It is thus not possible yet to determine the impact of potential climatic changes on its concentration in the sea water.

Still, once in the atmosphere, DMS is oxidized and gives rise to various products, including SO_2, methane sulphonic acid (MA), and sulphuric acid. The reactions that lead to products of oxidation of DMS are the origin of aerosol production (Andrea et al., 1995). A correlation has been found between DMS production and an increase in the nuclei of condensation in clouds. If it is confirmed that the changes in the flows of DMS and its derivatives have consequences for the climate, that will constitute another biosphere-climate feedback (Académie des sciences, 1994). An increase in DMS, for example, will be accompanied by an increase in the albedo of clouds, leading to a diminution of the solar energy absorbed by the planet, and a cooling of the soil. Here we have an illustration of the Gaia hypothesis (see Chapter 14), which states that living organisms maintain, by their activity, the conditions that are more favourable to them on the surface of the earth. This increase in DMS could also have a feedback effect of attenuating considerably the levels of UV that reach the ocean surfaces, helping to increase the nebulosity. Still, the production of DMS is influenced by the state of the sea and the processes of tides, and thus by the general circulation of the atmosphere.

At present, flows from the ocean towards the atmosphere are estimated at 25 Mt S/yr, as opposed to 10 Mt S/yr for the volcanoes and 75 Mt S/yr for the fossil fuels, which are the other sources of atmospheric sulphur producing essentially SO_2 (Deman et al., 1996). Human activities thus represent around 70% of the global emissions of sulphur towards the atmosphere and the majority of these emissions come from the northern hemisphere.

16.5. WATER CYCLE AND ENERGY EXCHANGES BETWEEN THE EARTH'S SURFACE AND THE ATMOSPHERE

The water cycle, driven by solar energy, plays a major role in the dynamics of the biosphere as well as in the dynamics of the climate. In fact, in its different forms (vapour, liquid, snow, ice), water is involved in exchanges of heat and humidity between the atmosphere and the earth's surface. Evaporation is responsible for half of the cooling of the surface, while atmospheric water vapour contributes to the greenhouse effect. The clouds control the climate by modifying the radiative balance, and more than a third of the thermal energy that governs the atmospheric circulation comes from the latent heat from condensation of clouds in the presence of particles (see Chapter 15).

It is known that the earth was formed more than 4500 million years ago by the collision and aggregation of rocky objects of various sizes that contained volatile elements, among them 1 to 3% water adsorbed on the surface or contained in hydrated minerals. After this accretion, which led to the formation of the present earth, the interior was differentiated into a metallic nucleus and a rocky mantle. The origin of the external envelopes (atmosphere, oceans) is disputed.

Certainly, the differentiation of the earth had the effect of transferring towards the exterior of the planet, by the intermediary of volcanic activity, the water and other volatiles (CO_2, nitrogen, rare gases). The atmosphere and hydrosphere would thus have been of internal origin and resulted from the release of gases from the mantle, which continues to this day by volcanoes.

According to other authors, the origin of the atmosphere and hydrosphere is linked to the arrival of meteor or comet matter rich in volatile elements during the last stages of accretion of the earth. In any case, the existence of sedimentary rocks among the most ancient rocks known attests to the presence of water in the liquid form on the earth's surface for at least 3900 million years.

16.5.1. Water cycle

Hydrology, the science of the water cycle, is barely three hundred years old. For a long time, in the absence of quantitative budgets, the origin of springs and rivers was the object of speculation. According to the Greeks, the water of springs came from the oceans after distillation due to internal fire. Plato taught that all the waters of springs come from the immense reservoir of Tartar that runs through the bowels of the earth, and that the waters of rivers and seas return to that reservoir. Aristotle, while denying that precipitation sufficed to maintain springs and rivers, believed that sea water was transformed into air, and that the air then became water again under the influence of cold in underground caverns, where the springs and rivers originate (Fig. 16.6).

It was only in 1674 that Pierre Perrault, brother of the storyteller, while measuring the level of rainfall in the upper basin of the Seine, realized that it represented six times the flow of the river. He demonstrated that precipitation alone fed the ground water, fountains, and water courses. He established the first hydrological balance and thus confirmed in his treatise, *De l'Origine des Fontaines*, the hypothesis proposed a century earlier by Bernard Palissy.

Water goes through a cycle on the earth's surface (Fig. 16.7). In broad terms, the cycle is relatively simple. Water is transformed into vapour as it evaporates at the surface of the oceans and continents. This vapour rises into the atmosphere, where it can condense in the form of clouds. The clouds, under the influence of different internal movements linked to the microphysics of clouds, precipitate

Fig. 16.6. Subterranean waters and canals connecting the seas to each other (P. Kircher, 15th century, *Le Mundus Subterraneus*). Photo by D. Decobecq.

in the form of rain or snow. Part of the precipitation runs off the soil surface, forming streams and rivers. Another part infiltrates the soil and feeds the ground water and underground reservoirs, which, in turn, end up feeding the streams and rivers. Finally, part of the water evaporates again into the atmosphere. In its greatest complexity, the water cycle thus runs a circuit from ocean to atmosphere to continent to soil to river to ocean, or from ocean to atmosphere to continent to soil to plants to atmosphere.

The water cycle, i.e., all the exchanges of water in its liquid, solid, or gaseous form, between the oceans, continents, atmosphere, and plants, mobilizes only 500 Tm³ (or 500×10^{12} m³) a year, or just 0.034% of the mass of the hydrosphere, which is equivalent to an average precipitation for the entire globe of about 1 m/yr. The cycling time of water vapour in the atmosphere is thus short, about 10 days. The continents receive more precipitation than the oceans and the

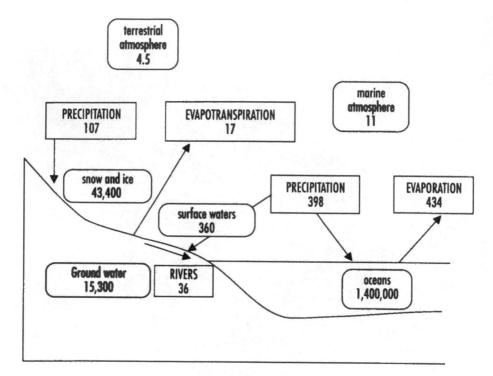

Fig. 16.7. Water cycle on the global scale. The volumes of water (grey boxes) and the flows (white boxes) are expressed respectively in $10^{12}/m^3$ and $10^{12}/m^3/yr$.

surface flow through rivers that run into the sea is 40 Tm³. The ocean provides the atmosphere with most of its water (425 Tm³/yr as opposed to 71 Tm³/yr for the soil-plant system). Despite the relatively low contribution of water evaporated from the continents, that water plays a major role in precipitation on the regional scale.

France annually receives 440 Gm³ (or 440 × 10⁹ m³) of water in an average year, or 750 mm rain, but only 330 Gm³ in a dry year. On average, 180 Gm³ of water flows, infiltrates, or runs off each year in nearly 270,000 km of water courses, or in the ground water. The difference between the rainfall inputs and the runoff waters, or nearly 60% of rainfall, returns to the atmosphere in the form of vapour.

16.5.2. Evaporation

Evaporation is the process that allows free water to leave the surface of the earth to join the atmosphere. In the case of plant surfaces, the roots draw water from the soil and the evaporation is called transpiration. In natural conditions, where there is nearly always plant cover, it is called evapotranspiration. A good part of the evaporation in a marine or terrestrial environment takes place in the intertropical zone, which is the hottest. The accumulation of energy in the lower layers of the atmosphere (sensible and latent heat) triggers, by instability, convection clouds that develop at very high altitude. These clouds generate high rainfall, higher if their supply through the humidity of lower layers is intense (monsoons in West Africa, for example).

The evaporation-precipitation (E-P) budget varies considerably from one place to another on the planet. Close to Lake Chad in the Sahelian region, the annual average precipitation is about 0.4 m, while evapotranspiration is 2.2 m/yr. But on the continents, generally, precipitation greatly exceeds evaporation. On annual average in dry land, P is 0.72, while E is 0.41 m. In Europe, the difference is high (E = 0.36 m and P = 0.60 m), while in Australia it is lower (E = 0.41 and P = 0.47) (Jacques, 1996).

Evaporation that occurs on the surface of the oceans and the soil requires energy to transform the water from the liquid state into water vapour. The energy required to evaporate 1 mm water over 1 m^2 corresponds to the latent heat of vaporization, which is 2.443×10^6 J/kg at 25°C. Inversely when the water vapour condenses in the form of water drops and snow crystals, energy is released in equal quantity to that consumed by evaporation. The water cycle thus helps cool the earth's surface (soil and oceans) through evaporation and helps heat the atmosphere in the course of condensation, thus attenuating the temperature differences between the layers of the atmosphere. We must also emphasize that water vapour is the most important greenhouse gas on the earth. Clouds in fact simultaneously cool the planet by reflecting incident solar energy (albedo effect) and warm the planet by retaining the infrared energy released from the earth's surface. If the water vapour were removed, the flows of outgoing infrared energy would increase by some 100 W/m^2 (see Chapter 15). Still, the amount of water vapour present constantly in the atmosphere is quite small: in the condensed state, it corresponds to a layer 2.5 to 3 cm thick over the entire surface of the planet, while the water cycle involves on an average the evaporation and precipitation of a layer that corresponds to around 80 cm of water in the course of a year (Joussaume, 1993).

16.5.3. Control of evapotranspiration: the soil-vegetation-atmosphere system

If the distribution of precipitation, temperature, and insolation largely determines the biological productivity and composition of the plant biomass, the plant cover in turn intervenes in the hydrological cycle. Plant transpiration is a process of water circulation across the soil-plant-atmosphere continuum, which is a key element of hydric balances. Around half the precipitation over terrestrial ecosystems comes from cycling of water by evapotranspiration, i.e., the combination of physical evaporation and biological transpiration.

Where the vegetation is abundant, a large part of the solar energy serves to evaporate the water contained in plants. The foliar transpiration of plants is a mechanism linked to photosynthesis. There is a demand for water that is retransmitted along the stem and leads to suction and absorption of water and mineral salts from the soil by the roots. The water follows its trajectory along the stem or trunk up to the stomatal cavities of leaves. The opening of stomata allows the water vapour to escape towards the atmosphere while allowing the atmospheric CO_2 to penetrate. This evaporation of water lowers the temperature of the plant. The tree cover thus allows the atmosphere to communicate with water reservoirs of the deep soil, while a bare soil dries rapidly on the surface, which interrupts the evaporation. The atmosphere above a plant cover will thus be more humid, favouring the nocturnal greenhouse effect and, consequently, the precipitation. All this contributes moreover by feedback to maintain conditions favourable to plant development.

As emphasized by P. Westbroek (1998), life does not start or stop the water cycle. By transpiration and other mechanisms, it modulates it, diversifies it, intensifies it, and exploits it. Its intervention changes the entire terrestrial environment, making it more apt to receive life. In the western part of the Amazonian basin, for example, it is estimated that 88% of the precipitation originates in evapotranspiration (Letteau et al., 1979). The air that enters the Amazonian basin from the North Atlantic contains around 18 g/m³ of water vapour, while that which reaches the foot of the Andes contains 25 g/m³, despite the substantial precipitation. Given that water resides on an average for 10 days in the atmosphere, this signifies that the water that falls in a given day is nearly the same as that which fell a few days before. As a comparative example, in the Sahel, evapotranspiration contributes only to 10% of the precipitation in June, but up to 48% in August (Brubaker et al., 1993).

Chapter 17

Responses of Ecosystems to Climatic Changes: Knowing the Past to Understand the Future

Recall, my children, that nothing is constant but change.

The Buddha

Given the diversity of processes at work, it is difficult to predict future manifestations of the climate, especially since human activities now play a significant role in the dynamics of the atmosphere. The untimely manifestations of El Niño, or the sometimes exaggerated predictions of the greenhouse effect, have helped these past few years to focus serious media attention on the question of climate changes in the medium and long term.

Whether the evolution of the climate is the result of natural trends or is accelerated by human activities, it remains true that the existence and composition of many ecosystems depends closely on local climatic characteristics (e.g., temperature, humidity, wind). Any modification of those characteristics could lead to more or less extensive changes in the structure and functioning of ecosystems. The argument is thus simple in essence: if we can predict the consequences of the greenhouse effect, we can anticipate the regional impacts of climatic change and set up ad hoc mechanisms in socio-economic plans. The operational implementation is more complex.

17.1. WHAT WE CAN LEARN FROM THE PAST: PALAEOCLIMATOLOGY, PALAEOECOLOGY AND PALAEOENVIRONMENTS

To elaborate predictive models about the effects of a predicted climatic warming, we can hypothesize that situations similar to those encountered today have probably existed in the past. If we know how and in what conditions the world has changed, we will certainly be able to better predict future evolutions of the climate and their ecological consequences. As a result, ecosystem ecology has been driven to consider spatial and temporal scales that go greatly beyond its usual studies. For this, the work of palaeoecologists, who some decades ago began to reconstruct palaeoenvironments using retrospective techniques, is particularly useful. Palaeoecology proposes actually a series of demonstrations on the reactions of regional or local ecological systems to various forcing factors, demonstrations that are useful in testing the reliability of predictions from the general circulation models (Davis, 1994). In fact, there has been a significant reinforcement of research in palaeontology and palaeoecology in order to better understand the responses of ecosystems to the last climatic cycles that governed the history of the planet.

Simultaneously, modern ecosystem ecology is directed towards long-term studies, acknowledging the need to collect long series of data to integrate the variability as well as the delayed effects that result from different reaction times of the constituents of the system. Moreover, some unusual events (El Niño, eruption of Pinatubo, droughts in the Sahel or in South America) have convinced ecologists, if they still needed to be convinced, that the impact of past or recent disturbances could have long-term and/or diverse repercussions on ecosystems. *In other words, knowledge of the history of ecosystems is a prerequisite to the understanding of their present structure and functioning.*

17.1.1. Palaeoclimatology

Palaeoclimatology seeks to reconstruct past climates of our planet. It is a science as well as an art, according to Duplessy and Morel (1990), because it requires an intimate understanding of nature, of the kind that would put even the most dedicated ecologists in the shade.

Over the past few decades, palaeoclimatology has allowed us to establish the bases of a quantitative description of past climates by means of fossil archives and indicators contained in marine, continental, and glacial sedimentary deposits or in fossils (Magny, 1995). These archives are of various kinds: e.g., air bubbles trapped in ice for tens of thousands of years, dust deposits, fossilized pollen, coral, plant and animal debris, products of erosion. The reconstruction of past climates usually includes a reconstruction of the flora and fauna, and the ecosystems in which they lived. Today, we are able to reconstruct climatic fluctuations of a very long period by relying on the various competences of geologists, pedologists, geochemists, climatologists, palynologists, and ecologists.

17.1.2. Palaeontology

Palaeontologists seek the origins of the living world. For a long time, fossils were studied for themselves, as an indicator of evolution or as biological markers that made it possible to date the geological formation. Palaeontologists, just like zoologists and botanists, described and classified thousands of species, many of which disappeared long ago. Palaeontologists, as Cuvier has put it, "resuscitate" living things of the past. Some of them investigate the conditions and living environments of those extinct organisms. According to Lamarck, we study fossils not only to retrace the genealogies of living things, but also to discover the nature of climatic changes in various regions of the world and reconstruct a history and geography of the past. The process of reconstruction consists of comparing the fossil species with the present species and concluding, from the possible resemblances, the identity or similarity of environmental conditions (Laurent, 1992). Observing that fossil debris of ferns, elephants, and crocodiles were found in Europe, Lamarck felt it fair to deduce that these organisms could only have existed in tropical climatic conditions similar to those in which they live presently. In this case, he was not wrong.

However, Lamarck went still further in linking the characters of species and fossil varieties with their geographic location, basing that on the modifications that living species undergo with the state of the places in which they live (Lamarck, 1802). He added about these living things that the diversity of circumstances brings about a diversity of habits, a different mode of living, and, consequently, modifications or developments in their organs and in the shape of their parts (Lamarck, 1801). Whether they were biologists or palaeontologists, transformists or fixists, many contemporaries of Lamarck were preoccupied with the attempt to discover relationships between the environment and living things (Laurent, 1992).

During the second half of the 20th century, palaeontology largely contributed to the knowledge of phylogenic relationships between organisms, in relation with the development of the cladistic approach, which attempted to reconstitute the phylogenetic relations of living things from the analysis of similarities between organisms. But one of the present directions of this discipline is the study of extinction, especially mass extinction linked to catastrophic events that have affected many groups of organisms in a short time. From the palaeontological data, we discover the major environmental crises that have repeatedly affected our planet (Buffetaut, 2000). This is the case, for example, for the major extinctions in the Cretaceous-Tertiary transition (65 million years ago) during which dinosaurs and other animals disappeared. These terrestrial environmental crises, no matter what their causes, had a considerable influence on the history of life on earth. From the reconstitution of past environments, with the construction of climatic models for increasingly ancient periods, we can thus understand the processes that were at work in the evolution of living things.

Advances in palaeontology have benefited from chance discoveries. Some groups appear suddenly in the geological strata, with relatively evolved forms, which suggests that the older representatives will be discovered. For example, the discovery of two small fish-like fossils in the Cambrian layer in China (Janvier, 1999) contributes new elements as to the origin of vertebrates and shows how phylogenetic hypotheses established for the primitive forms are fragile and can be modified by new discoveries. Contrary to what was thought till now, the common ancestor of all the vertebrates must lack a mineralized skeleton. However, the analysis of morphological characters of these fossils suggests also that there are still missing links in the chains connecting them to the hypothetical common ancestor, which takes us back to the Precambrian origin of vertebrates.

17.1.3. Palaeoecology and palaeoenvironments

Palaeoecology, conceived as the study of relationships between organisms and environments in the past (Pons, 1995), comes after palaeontology. Having described fossil remains, the palaeontologist seeks to reconstruct whole organisms, group them on a phylogenic basis, and investigate the climatic and physiographic conditions in which they lived, which determined their geographic distribution.

Given a set of fossils collected at a single site, the palaeontologist must respond to a primary question: are they ancient communities that were fossilized in that place, or are they accumulations of organisms transported from various other biotopes? *Palaeobiocoenosis* is defined as a set of populations having lived at a given time in a single environment. Unlike the present biocoenosis, it lacks a part of its components, notably the microorganisms and organisms with soft bodies that are rarely fossilized. In the *taphocoenosis*, on the contrary, there may be organisms of different age or coming from different environments found close together. The art of the ecopalaeontologist lies in determining the share of these two categories of associations (Gall, 1995).

The identification of a palaeobiocoenosis opens the route to the search for the palaeoenvironment in which it evolved. The term *palaeoecology* has been used since 1916, when the American botanist Clement first proposed it, but it was after World War II that palaeoecology was set up as a scientific discipline. The first congress of palaeoecology was held in Lyon in 1983. When the prefix *palaeo* is attached to *ecology*, a notion of time is introduced that could range from a few thousands to a few thousand millions of years. Theoretically, like ecology, this field addresses the relationships of living things to each other and to their environment. For the more or less recent sediments, palaeoecology may use information about present living forms to reconstruct the past. It thus provides access to an understanding of ecosystems and climatic conditions of the past and allows the field of present ecology to place itself in a temporal retrospective. By this process, palaeoecology dissociates itself from palaeontology to integrate itself totally with ecology (Pons, 1995). The temporal domain of this "ecological" palaeoecology can be defined as follows: from the present, it goes to the past until the living species and the major floristic and faunistic groups of the present are no longer found and the present climatic system of the planet is no longer evident. In practice, the past million years may be studied (Pons, 1995).

For more distant periods, palaeoecology can no longer rely on knowledge of the present. The palaeoecologist is faced with organisms that are most often extinct and environments that have disappeared. Palaeoecological data must therefore be interpreted by the analogical method, based on the actualist concept of Lyell that the same causes have always produced the same effects. The modes of life of extinct organisms (palaeobiology) are deduced from the interpretation of characters, structures, and organs having an adaptive significance in ecological terms. Thus, by the morpho-functional analysis of fossil remains, we can reconstitute the biological traits and behaviour of fossil species. This approach suffers, however, from a recurrent flaw: morpho-functional analysis is based on our knowledge of the organization of present organisms, so that the limits of the palaeoecological process are more imprecise when we are studying organisms that no longer are represented or that have no related forms in nature at present.

Reconstructing the modes of life and conditions of existence of fossil organisms and interpreting ancient environments and their dynamics involves a fundamentally multidisciplinary process. The approach of palaeoecology necessarily is based on many disciplines: geologists, geochemists, sedimentologists, botanists, zoologists, and others contribute, along with ecologists working on the present.

The physical and chemical analysis of sediments can provide accurate indications on the edaphic characters of the watershed of any lake or peat bog. The superposition of sedimentary beds also provides a relative chronology of species and palaeoenvironments. But this is not sufficient, because in the absence of absolute dating palaeoecology cannot develop a comparative approach between sites, nor evaluate the speed of the phenomena that need to be analysed (e.g., speed of spread of species). The possibility of dating fossils by physical methods has been a formidable thrust for palaeoecology. Among dating methods, carbon dating remains the most common. This method, invented in 1947, revolutionized the study of palaeoenvironments. Any living matter contains a small quantity of radioactive ^{14}C that comes from CO_2 of the atmosphere. When the organism dies, the exchanges with the atmosphere cease and the quantity of ^{14}C that it contains diminishes by half every 5570 years. By measuring the radioactivity of a sample of wood, we can evaluate the time that has passed between our measurement and the death of the tree that the sample came from. The dates are given in years BP (before present), i.e., by convention, in radiocarbon years before 1950. However, carbon dating is not reliable beyond around 40,000 years, which does not entirely meet the needs of a palaeoecologist. The uranium/thorium dating method allows dating up to 350,000 years. Other methods such as the potassium/argon method are being developed.

Transfer functions

The need for quantification and modelling has encouraged palaeontologists to develop the method of *transfer functions*. The method is based on calibrations established nowadays to model the relationships between environmental parameters and the species composition of communities. These functions are then applied to palaeontological data to discover, from floristic and faunistic assemblages, some characteristics of the ancient environment, such as temperature, salinity of water, or pH. They have been used in various groups of marine and continental domains: pollen, diatoms, foraminifers, ostracods, molluscs, etc.

Palaeoecology is also based on widely used techniques.

- *Palynology.* All plant species produce often considerable quantities of spores or pollen. Because of their highly resistant covers, these pollen are well preserved in sediments of lakes and peat bogs, favoured research sites for palynologists. Disseminated in very large quantities in the environment, pollen is preserved for hundreds of thousands of years. From its morphological characteristics, the parent plant can be identified and thus the surrounding plant cover at the time the pollen was deposited in the sediments can be reconstructed (Reille, 1990). The pollen spectrum of a temperate forest in western France is typically composed of 70% pollen of deciduous species (oak, birch, hazel, alder), 20% conifer pollen (pine), and 10% pollen of herbaceous species. We thus have a fairly reliable image of the vegetation. Thousands of pollen diagrams have been established in Europe, some covering several hundreds of thousands of years.

*   *Dendroclimatology* studies the thickness and density of rings in tree trunks in order to reconstruct, on an annual scale, the variations of parameters that determine tree growth (temperature, humidity, precipitation). In this way, we can reconstruct past variations of the climate, year by year. It is a highly precise method, but it is limited to the life span of a tree.

17.2. ANCIENT PALAEOENVIRONMENTS

Throughout the history of the earth, there has been interference between geological events and biological evolution (Gall, 1995). Life probably appeared in water about 4000 million years ago, when molecular structures were capable of self-reproduction and evolution. The first forms of life were phototrophic cyanobacteria capable of fixing CO_2 as well as atmospheric nitrogen. According to active theories, 2000 million years ago the primitive atmosphere, primarily reducing, had about 1% oxygen. *The biological activity linked to the appearance of photosynthesis would help modify the composition of the atmosphere, which would be enriched in oxygen and become progressively oxidizing.*

Simultaneously, biological activity led to the edification of the first bioconstructions, the *stromatolites*, which are virtual microcosms in which a wide diversity of microorganisms mingle. Stromatolites spread considerably between 2000 and 700 million years ago. They formed virtual microbial reefs probably because of the development of filamentous phototrophic cyanobacteria.

At first confined to water, life forms colonized dry land at the end of the Precambrian, around 700 million years ago. At this time, the first multicellular organisms also appeared. However, it was mainly at the margin of the Primary era, with the development of biomineralization, that there was a *decisive stage that would affect the transfers of chemical elements in the lithosphere.* From that time, various organisms captured from the external environment the Ca, Si, or P ions they needed to build their skeletons, shells, or carapaces. When the organisms died, these elements accumulated and formed sedimentary rocks. This retention of chemical elements by living things and their prolonged storage in sediments considerably modified their rates of transit within the biogeochemical cycles.

The first terrestrial plants date from around 415 million years ago. During the Primary era, plants were established on all the continents and helped to create the biosphere. The development of vegetation accelerated the processes of alteration of rocks and contributed to the formation of soils. The giant ferns and horsetails developed between 385 and 360 million years ago, and the first trees appeared around 350 million years ago (Jaeger, 1996). However, the Angiosperms (flowering plants) diversified only from 110 million years ago. With the spread of vegetation and the first terrestrial organisms, an organic soil (humus) very slowly developed. The formation of soils and their associated microbial activity initiated the chemical alteration of rocks while retaining at least partly the products of that erosion.

Burrowing organisms, which appeared 445 million years ago, accelerated the turnover of different soil horizons (Shear, 1992). Arthropods must have appeared during the Silurian and the Devonian, or around 400 million years ago, and adapted to the difficult physiological conditions of dry land. There are unfortunately a limited number of fossils. As for the vertebrates, they were first differentiated in an aquatic environment and it was the tetrapods that first colonized the terrestrial environment around 375 million years ago.

The toxicity of plants resulting from the accumulation of metabolites in the cells seems to have long prevented the appearance of herbivorous animals. When vascular plants appeared, animals could not profit from their produced organic matter. The communities of terrestrial animals were thus dominated for tens of millions of years by detritivores and predators before organisms emerged that could feed on the living tissues of higher plants (Shear, 1992).

Reconstruction of an ancient palaeoenvironment (Fig. 17.1)

During the lower Triassic, the Vosges were situated in the intertropical zone of Pangaea. From the characteristics of the sedimentation, we can reconstruct warm climates with contrasted seasons, including a marked dry season that caused significant mortality of the aquatic fauna. The fossils figure among the most ancient flora and fauna of the Secondary era. We find remarkable fossils of organisms with a soft body (jellyfish, annelids, etc.) as well as many insects, crustaceans, and molluscs. There are also several species of fish and tetrapod vertebrates. The flora is made up of large horsetails forming virtual wreaths, with ferns and conifers such as *Voltzia*. The biocoenoses in the flood plain waters were dominated by arthropods (crustaceans and limula).

The ephemeral nature of brackish-water bodies is confirmed by the accumulation of certain organisms (crustaceans, insect larvae) in dense tracks, similar to concentrations observed in residual pools during the drying up of aquatic environments. The aquatic fauna of clay levels, given the excellent state of preservation and the presence of organisms in a position similar to their living state, corresponds to a palaeobiocoenosis. The coexistence of adult and juvenile forms and exuvia prove that these animals developed on the site of their fossilization. The excellent state of preservation could result from the rapid development of a microbial film after the massive mortality of organisms resulting from rapid evaporation under a dry and hot climate. This veil would have protected the cadavers from decomposition, before the eruption of a detritic flow that buried the microbial tissue as well as the organisms.

Fig. 17.1. Reconstruction of an ancient palaeoenvironment. The biocoenosis of brackish water bodies of the Voltzia Sandstone delta (north of the Vosges). 1, jellyfish; 2, annelids; 3, lingula; 4, lamellibranches; 5, limula; 6, crustaceans (*Triops*); 7, estheria; 8, crustaceans (*Euthycarcinus*); 9, crustaceans (*Schimperella*); 10, crustaceans (*Antrimpos*); 11, crustaceans (*Clytiopsis*); 12, crustaceans (*Halicyne*); 13, fish (*Dipteronotus*); 14 and 15, horsetails; 16, conifers; 17, conifers (*Voltzia*); 18, spider; 19, scorpion; 20, myriapod; 21, insects (cockroach); 22, reptiles; 22, insect eggs (Gall, 1995).

At first, arthropods used microbial, protist, and fungal mutualists in order to detoxify plant matter and add calorific and nutrient value (Shear, 1992). Then they developed specialized mouth parts to pierce the ovules and consume the spores. It was not until around 290 million years ago (at the end of the Carboniferous era) that the first herbivorous vertebrates were differentiated.

Closer to our time, around 100 million years ago (during the Cretaceous period), a warm climatic period prevailed, as indicated by the remains of tropical fauna and flora (coral reefs, crocodiles, etc.) found far north of their present habitat. The average temperature must have been 6°C higher than at present and the sea level was 300 to 400 m higher than the present level, flooding nearly 20% of the land that is today dry (Joussaume, 1993).

Because of ongoing research, we have an increasingly precise knowledge of the climatic changes during the last million years as well as of the palaeoenvironments that are associated with them. In particular, the study of pollen over a period of 5 million years shows that the plant associations of Europe have greatly evolved (de Beaulieu and Suc, 1985). From 5 million to 3 million years ago, western Europe was covered with dense forests of trees that presently are found in warmer regions. On the shores of the Mediterranean Sea, there were plant associations with Taxodiaceae (trees related to the bald cypress) around coastal marshes, while in the less humid inlands there were groups of Juglandaceae and Hamameliadaceae, which indicate forests of evergreens with large leaves that are presently found in southern China in a wet tropical climate. In northern Europe, the many marshes were bordered with bald cypress formations as well as groups with *Sequoia* and deciduous species (beech, hornbeam, alder, oak) covering most of the land, similar to the mixed forests now found in China, Japan, and the eastern United States. During this period, the climate of western Europe was warm and humid, with a marked seasonal rhythm, as indicated by the Mediterranean elements present in this flora.

The Quaternary ice age in which we entered around 2 to 3 million years ago was characterized by many variations of the climate with glacial periods (characterized by the extension of ice caps) and interglacial periods (with a climate more or less similar to what we know today). Over the past million years, there have been many glacial cycles whose dominant periodicity is statistically 100,000 years.

Around 2.3 million years ago, there was a primary glacial period as well as a modification of the oceanic circulation that resulted in a floristic impoverishment in the north. This glacial phase caused in the Netherlands the extension of herbaceous groups (Graminaceae, Cyperaceae, Ericaceae) that suggest the tundras. Similarly, in southern Europe, there was an extension of herbaceous formations that recall the Mediterranean steppes and that are presently found in central Turkey or in North Africa. After this glacial period, there were periods of warming during which deciduous forests grew in northern Europe, alternating with periods of cooling marked by the return of tundra. The deciduous forests, vestiges of which survive presently south of the Caucasus and on the shores of the Caspian Sea, became gradually impoverished in exotic species during this period.

During the Quaternary, the beginning of which is by convention dated from 1.8 million years ago, about 20 cycles were identified, composed of a glacial period followed by warm periods. It was estimated even that the glacial regime was dominant for the past million years and that the more clement periods similar to that in which we live represent only 10% of that period. We thus live in exceptionally warm climatic conditions for the Quaternary (Joussaume, 1993).

17.3. ECOSYSTEM CHANGES DURING THE LAST GLACIAL CYCLES

Over the past million years, major glaciations occurred at an average rhythm of every 100,000 years. During these periods, northern Europe and America were covered with large ice caps that extended from the polar regions and mountainous reliefs. This huge storage of water resulted in a fall in the sea level.

Polar ice caps, archive of past climates

A team of French, Russian, and American glaciologists extracted an ice core 3623 m long at the Russian station of Vostok in Antarctica (Petit et al., 1999). This ice core provides continuous climatic records over a period of 420,000 years, i.e., the equivalent of four full cycles of glacial and interglacial periods.

The end of each of the major glaciations (310,000, 240,000, 135,000, and 15,000 years ago) was followed by a transition to a milder climate. The warming of the northern hemisphere and the massive fusion of ice caps occurred after a delay of 4000 to 9000 years with respect to the South pole. The analysis of the Vostok ice core showed a correlation between CO_2 and methane concentrations present in the atmosphere and the temperatures that prevailed in Antarctica. The warming periods (5,000 to 10,000 years) that followed the glaciations were more or less simultaneous with the increase in greenhouse gases, but the time gaps between the two functions were poorly defined given the precision of the information. Between cold and warm periods (an amplitude of variation of around 12°C was observed), the concentrations of CO_2 in the global atmosphere oscillated between 180 and 280 ppmv and those of CH_4 between 350 and 700 ppbv. The highest concentrations (300 ppmv of CO_2 and 750 ppbv of CH_4) were nevertheless much lower than those measured presently (respectively 360 ppmv and 1700 ppbv).

These glacial archives also indicate that the interglacial period that we live in is the longest of the four preceding ones: a relatively stable climate has prevailed for 10,000 years, while a reference to preceding cycles shows that the temperature should have already begun to drop (Petit et al., 1999).

Most of the research in palaeoclimatology involves the last glacial cycle, which ran from around 125,000 years BP to the Holocene interglacial period, which includes the last glacial maximum 20,000 years ago (Berger, 1996). In simple terms, the last complete climatic cycle was made up of a "warm" period (approximately between 145,000 and 75,000 years ago) and a "cold" phase (from 75,000 to 15,000 years ago) corresponding to the Wurmian phase of prehistorians. The temperatures were 10°C colder in temperate latitudes during the last glacial maximum, which was a period of high aridity.

The volume of ice increased in stages, but at the peak of the last glaciation, around 18,000 years ago, the Alpine massif as well as the Pyrenees were covered by ice. The ice caps covered most of northern Europe and North America and descended to the latitudes of Manchester and Berlin. The volume of accumulated ice reached more than double the present volume, so that the sea level was around 120 m below the present level (Fig. 17.2). The coast lines were very different from what they are today. One could have walked from Great Britain to France, and the exundated Bering strait (it is said) allowed exchanges between Asia and North America.

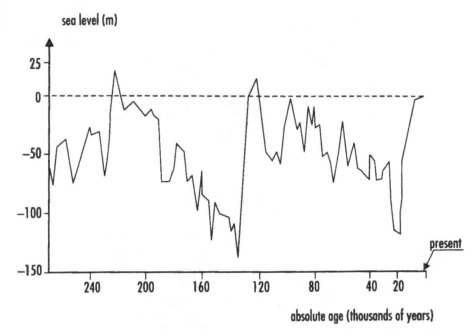

Fig. 17.2. Variations in sea level over the past two climatic cycles. The evolution of the sea level was reconstructed from the isotope composition of planktonic foraminifers of marine sediments.

One of the major contributions of palaeoclimatology over the past ten years is the discovery of sudden changes of great amplitude occurring at the scale of the human life span. These events are recorded in the marine and continental glacial archives.

Drilling in the ice of Greenland has shown significant warming throughout the glacial period that could reach a 10°C temperature rise over some decades, followed by gradual cooling (Jouzel and Lorius, 1999). These saw-tooth oscillations of variable duration (from 5400 to 2000 years) were repeated about 20 times in the interval from 75,000 to 15,000 years ago.

These instabilities involve the complex interactions between thermohaline circulation, the meridian heat transfers, the deep ocean convection at high latitude, the snow cover on the continents, the extent of large ice caps of the northern hemisphere, and other phenomena. In the tropics, such instabilities result in temperature variations and hydrological changes that could have great amplitude.

17.3.1. Temperate ecosystems

The last pleniglacial period was a sort of zero state for the vegetation of northern Europe, covered by the vast Scandinavian ice cap and thick layers of ice on the mountains and, at low latitudes, by a tundra vegetation near the Atlantic and a steppe vegetation in the southeast (Pons, 1993). The present ecosystems did not then exist. South of the ice cap, the ancient pollen shows the existence of tundra

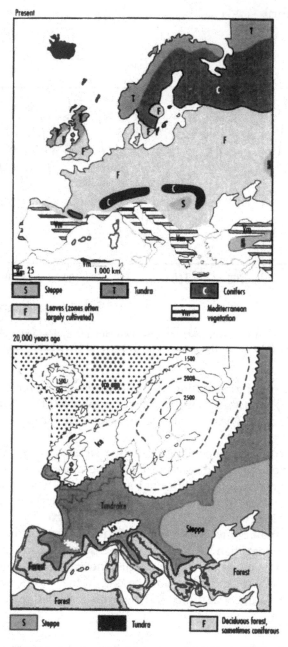

Fig. 17.3. Europe 20,000 years ago (below) and more recently before it was cleared by humans. Around 20,000 years ago, an inlandsis extended over the British Isles and Scandinavia. A glacier covered Iceland. Between the two, an ice cap was established. A vegetation characteristic of cold climates occupied most of Europe. The steppe (herbaceous plants) dominated the east and the tundra (herbaceous and shrubby species) dominated the north and west. Forests covered only the very southern part. The drop in sea level enlarged the continents and connected France to England. Above, a recent situation showing that the tundra is far to the north and the steppe far to the east. Most of Europe is covered by forest.

landscapes of the sub-Arctic type, landscapes similar to those of present Lapland, in which reindeer herds proliferated. Southern Europe was covered by cold grassy steppes scattered here and there with islands of forest (Fig. 17.3). In the southernmost part of Europe (southern Spain, Greece), the temperature of the coldest month (January) was lower by 15°C than its present average. In the region corresponding to present-day France, it was lower by 30°C. To this great cold was added a severe reduction in the water available for plants, which was 70% in southern France (Joussaume and Guiot, 1999).

During the present interglacial period, the vegetation reconstructions show a constant evolution of the plant cover. Generally, there was throughout Europe a similar dynamics of the plant succession, but with differences of synchronization.

The warming began around 16,000 years ago, with at first a phase of invasion by grassy formations of the steppe type. The tree recolonization was a function of the climate and the biological characteristics of the species. Relatively poor biomes were established in the temperate and cold regions, while richer Mediterranean forest systems developed in the south. Then, around 13,000 years ago, there was a fairly sudden warming followed by a cooling around 12,000 years ago (called the Recent Dryas). From 10,000 years ago, in the early Holocene, all the ice disappeared from North America and Europe and the invasion of Europe by temperate forest was rapid. Some 6000 to 800 years ago, the climate was warmer than the present climate (the climatic optimum) with a summer temperature 2 to 3°C higher than at present in western Europe. During that period, the deciduous forest (oak, elm, hazel) that developed after the deglaciation extended further north at the expense of the boreal forest, and further east at the expense of the coniferous forest, because of a climate 1 to 3°C warmer in winter. The present sea level was stabilized, with a few oscillations, only 7000 years ago. During the warming period, the rise of the sea level sometimes reached 2 m in a century, principally because of the fusion of large ice masses of the northern hemisphere.

In the middle latitudes of northern Europe, we can diagram the dynamics of forest species over a climatic cycle in the following way (Fig. 17.4). In the first part of the interglacial period, juniper (*Juniperus*) and then birch (*Betula*) and pine (*Pinus*) jointly invaded the land abandoned by the forest vegetation during the glaciation. Then, the temperate deciduous forest began to propagate itself across all of Europe around 10,000 years BP. The first trees to propagate themselves were hazel (*Corylus*), oak (*Quercus*) and elm (*Ulmus*). Later, there was ash (*Fraxinus*) and lime tree (*Tilia*) within a mixed forest dominated by hazel and oak. Alder (*Alnus*) invaded humid depressions. The timing of the appearance of these species in Europe depended mostly on the location of their refuge during the glacial period, their propagation rate and their ability to compete with other species within the ecosystems.

The last interglacial phase is a suitable period in which to study the processes of re-invasion of peri-glacial species by animal and plant species. It is known that during the last glacial period the tree species found refuge in southern Europe, where they could survive in the favourable climate of some

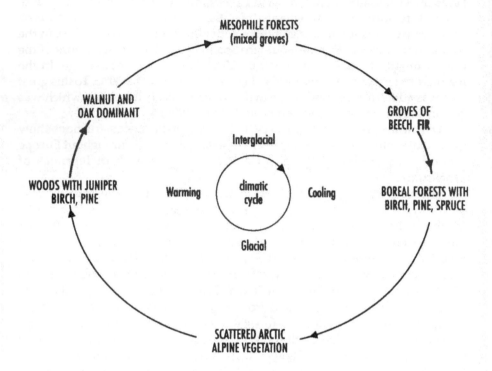

Fig. 17.4. Diagram of succession of different types of plant cover along a climatic cycle proposed by J. Iversen

regions around the Mediterranean. In particular, it has been shown that species survived in the three refuge zones located in the Iberian and Italian peninsulas as well as in the Balkans (Taberlet et al., 1998) (Fig. 17.5), with, however, marked differences among those zones in terms of species. Other refuge zones probably existed in eastern Europe or Asia. According to Taberlet et al. (1998), there was a recolonization route from the Iberian peninsula to Great Britain and to Scandinavia for species such as the brown bear and *Quercus* sp. There was also a recolonization route from the Balkans to southeastern France for species such as the beech and the grasshopper. The barrier of the Alps would have hampered the dispersal of many species that had found refuge in the Italian peninsula.

The comparison of colonization routes indicates the existence of "suture" zones, which are zones of meeting and hybridization of populations of a single species that recolonize a region from different refuge zones.

Oak, which tolerates the concurrence of other species, colonized Europe more rapidly than other tree species. The propagation of beech, the refuge zones of which were located close to the Black Sea and in southern Italy, was thus delayed because of its poor ability to compete with the oak. It began to spread around 6500 to 6000 years BP in the Apennines, and it was only about 3500

Fig. 17.5. Principal refuge zones and routes of post-glacial recolonization in Europe

years BP that it appeared in Spain and northwestern Europe. Spruce, which spread in the eastern Alps in the late glacial period, took more than 6000 years to reach the French Alps and the Jura and did not colonize the Massif Central till the 19th century, thanks to reforestation.

From the study of plant successions during the last four glacial cycles we can draw some general conclusions about plant successions. In all the cases, after a phase of pioneer colonization by pines and secondary colonization by birches, there was a more or less long phase of expansion of the oak-hazel group that corresponded to the thermal optimum. Then there developed deciduous trees and conifers that were less thermophilous or that dispersed more slowly, and finally the boreal forest. There are differences between the forest successions over the past few interglacial periods, which would have been due to the fact that the refuges, which were not at the same location as in the preceding glaciations, determined the different migratory routes during each recolonization.

France during the last glacial maximum

At the last glacial maximum, 18,000 years ago, the temperature at what is now Paris was 12 to 15°C lower. France escaped much of the ice that descended as far as Great Britain. But the soil was constantly frozen over nearly all the territory, and at great depths. In some regions, there may have been short periods of thawing during the cold summers. Water was stored in massive quantities in huge glaciers and the sea level was 120 m lower than at present. The Loire was dry, the English Channel was a valley that linked France and Great Britain, and the coasts lay much further west. La Rochelle was thus well inland, and the Marseilles harbour was a landscape of hills overhanging the Mediterranean Sea. It was in the heights of the hinterlands, on a plateau overlooking the sea, that the inhabitants of this epoch explored the Cosquer cave, which today lies 36 m under the sea.

The tundra and taiga landscape dominated in the northern part of France, while the southern regions were partly covered with wooded and grassy steppes. The large mammals (reindeer horses, bison, mammoth, cave bears, megaceros, etc.) rubbed shoulders with the penguins.

17.3.2. Mediterranean ecosystems

The history of the evolution of forest cover in the northern Mediterranean region is now quite well known (Quezel and Barbero, 1990). During the last glacial maximum (18,000 to 15,000 years BP), there remained only a grassy carpet between the frozen regions of the north and the Mediterranean in the south. In this tundra with low floral diversity, trees were eliminated by low summer temperatures and the short duration of summer. There was no longer a boreal forest of conifers or a deciduous temperate forest, or a Mediterranean forest (Rossignol-Strick, 1997). The woody and more or less thermophilous plants survived in the lower parts of river thalwegs of the Mediterranean, as well as on the southern slopes of the mountains located at the edge of the Mediterranean region (Pons, 1991). In any case, however, it was the more or less random and fragmentary assemblages of animal and plant species that were able to survive.

Around 9000 years BP, climatic conditions again became favourable in the Mediterranean basin with a rapid expansion of deciduous temperate trees, mainly oak. As the land began to open up between 7500 and 4700 years BP, holm oak (*Quercus ilex*) spread at the expense of pubescent oak (*Q. pubescens*). Landscapes dominated by sclerophylls (deciduous trees with persistent leaves such as holm oak and kermes oak) also appeared at this time. Then, the gradual emergence of the characteristics of the Mediterranean climate between 5000 and 2500 years BP (marked seasonality of precipitation and temperature, summer dryness) favoured the generalization of all the sclerophyll forests: the oak spread towards the north, cork oak (*Quercus suber*) progressed in the moors, and the Aleppo pine (*Pinus halepensis*), till then reduced to groves, took a significant place in forest formations.

17.3.3. Tropical rainforests

There is a sometimes irrational discourse concerning tropical forests and their protection that is linked partly to a misunderstanding of their past dynamics. To affirm, as some still do, that the tropical forests are the lungs of the earth is to make a scientific error or to deliberately ignore the results of research. In fact, the tropical forest, which has long been considered relatively stable since the end of the last glaciation, and which has often been presented as an example of wild and inviolate nature, has known periods of extension and recession, in relation with the major climate changes during the Holocene.

During the last glacial maximum, as a result of more severe climatic conditions (temperature drops of 2 to 6°C and decline in precipitation), the South American rainforest began to regress around 28,000 years BP and the African forest between 20,000 and 15,000 BP. In many regions, the savannahs overtook the rainforests, which occupied only reduced areas.

During the post-glacial period, the forest reinvasion was not synchronized in Africa and South America and it had a sometimes turbulent history. In Africa, there was a maximum extension of tree cover between 10,000 and 8000 years BP. After that period, there was a reduction in the rainfall in central Africa and, around 3000 to 2500 years BP, the savannahs spread at the cost of the forest in the south of the Congo and in regions in which the seasonal hydric deficits were most severe. Intense disturbances took place also in the rainforest of west Cameroon. For around a millennium, the general trend in central Africa was an extension of the forest with rates of progression of some tens of metres per century. Extrapolating from this, most of the enclosed savannahs will disappear from this region in the next five centuries. This forest invasion is linked to a return of more humid climatic conditions. In Cameroon, more rapid invasions have been observed over the past thirty years. This process seems to combine the progression of edges and the generalized invasion of pure savannah zones by the forest from groves or thickets (Schwartz, 1997).

Carbon 13 method and the study of forest-savannah changes

Carbon 13 (^{13}C) is one of three isotopes of carbon. Unlike ^{14}C, it is a stable isotope. In the atmosphere, $^{13}CO_2$ molecules coexist with $^{12}CO_2$ molecules in a proportion of 1.1% of total CO_2. Plants use both types of carbon during photosynthesis, but with slight differences linked to the type of photosynthesis. Plants with a C4 cycle (essentially Graminaceae and tropical Cyperaceae of an open environment) absorb more ^{13}C than C3 plants (practically all the other plants, including forest species).

The discrimination is small, but it is transmitted without great change in the organic matter of soils. This gives us an excellent marker that indicates the replacement of a forest by a savannah or the inverse, at time scales that are a function of the turnover rate of organic matter in the soil (Mariotti, 1997).

In South America, the Amazonian forest since the last deglaciation exhibits significant disturbances but the forest dynamics is not synchronized with that of Africa (Schwartz, 1997) (Fig. 17.6).

Between 10,000 and 8000 years BP, the forest grew in some places (southwest of Amazonia and central Brazil) but not in Guyana. Between 7000 and 4000 years BP, the forest receded to make way for the grassy formations in the north and southeast of Amazonia, as well as in central Brazil. For 4000 years, there has been a reinvasion of the forest on the sites from which it had disappeared and it reached its full extent only a thousand years ago on the Atlantic coast of Brazil. In other words, the Amazonian forest is quite young.

Studies carried out in Guyana have shown, moreover, that between 3000 and 2000 years BP the forest was more humid than at present. Between 1700 years BP and now, there were two periods of drought associated with disturbances of the forest system with major forest clearing that allowed the development of pioneer plants: one between 1700 and 1200 years BP and the other between 900 and 600 years BP. Similar phenomena took place at the same time in the west of the Amazonian basin, where traces of palaeo-fires between 2000 and 1400 years BP were observed. The Guyanese forest acquired its present characteristics only 300 years ago.

The comparison of tropical forest flora reveals great differences between the three continents (America, Africa, and Asia). These differences are partly attributed to the history of past climatic changes (Blanc, 1997). Malaysia harbours

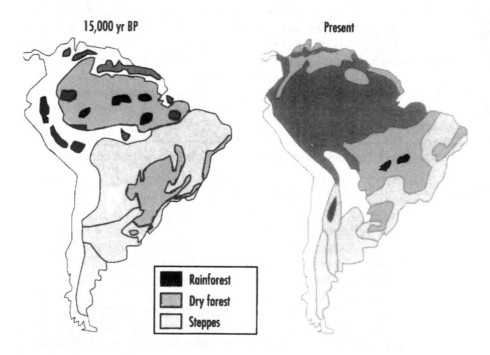

Fig. 17.6. Changes in distribution of major plant domains since the last glaciation, in South America (Legendre, 1991)

a flora twice as diversified as that of Guyana and it is richer in endemic species. These two floras also exhibit different characteristics as to their modes of dispersal as well as their rate of endemism. In Guyana, as in other forest zones of tropical America, around 80% of tree species are dispersed by mammals and birds. The proportions are comparable in Africa. In tropical Asia, on the other hand, the majority of tree species produce dry fruits from which the seeds fall on the soil and remain close to the mother tree. There is no system of rapid dispersal. The differences observed between Malaysia and Guyana may be explained by the more moderate phases of dryness in Malaysia, and by the mountainous topography of that country, which offers more possibilities for refuge zones that will attenuate the impact of climatic disturbances. The relative floral poverty of Guyana in comparison to Malaysia would be the result of severe forest regressions during the dry climatic phases as well as the existence of smaller and fewer refuges, because of a more monotonous relief.

For plants of the undergrowth, the fragmentation of the forest during dry phases of the Quaternary and the Holocene may have two consequences for the species richness:

— Forest fragmentation favours the speciation of species with seeds dispersed by vertebrates, which survive in isolated refuge zones.

— These same dry phases may have led to the disappearance of species with a low migration capacity and slow dispersal that diversify mostly during the humid phases.

Refuge zones

The hypothesis of the existence of refuges (Haffzer, 1982) arose from observation of the present distribution of plant and animal species in dense rainforest and the levels of differentiation attained. In a dry period, the forest is fragmented and only some islands can survive in places where the climatic conditions allow them to. The forest fragments that survive the dry phases thus serve as refuges for some of the species that earlier occupied the territory covered by the forest. If the fragmentation lasts long enough, it favours speciation by vicariance. These species then reinvade the zones in-between the refuges at rates varying with their migration capacity, which depends on their biological characteristics. In theory, the refuges will correspond to the present zones of high endemism and great biodiversity. This diversity decreases as the distance from refuge zones increases.

17.3.4. Freshwater ecosystems

Generally, the size of an aquatic habitat fluctuates as a function of the climatic conditions and over relatively short periods on the geological scale. With a few exceptions, the continental aquatic systems can be described as transition systems (Carpenter et al., 1992), in that they react constantly to climatic variations.

A good illustration is the Sahelian region, in which, during the middle of the glacial period around 20,000 years ago, the temperature was lower than at present and the climate was dry. The tropical rainforest was limited to some refuge zones, and there were good reasons to believe that the extension of aquatic systems was also highly reduced. With post-glacial warming, around 12,000 years BP, heavy rains allowed the aquatic ecosystems to spread northward. We find many traces of the existence of large lakes dating from this period in the present Sahara (Petit-Maire, 1980). At the peak of the rainy period, around 8000 to 9000 years BP, permanent rivers flowed from the north to the south. The hydrographic basin of Chad extended up to Ennedi and Tibesti, and that of the Niger up to Air (Lévêque, 1997). A little after this period, around 8000 years BP, the hydrological conditions deteriorated gradually and the rains diminished to the north of the Sahara. By 4500 years BP, the aridity was established, the Saharan lakes disappeared, and part of the hydrographic network dried up. As a result, the present hydrographic basins of Lake Chad and the Niger river cover only half of the area occupied 9000 years ago. A very impoverished fish fauna exists still in some parts of the Sahara. It is the vestige of that period of expansion (Lévêque, 1997) (Fig. 17.7).

The degradation of aquatic systems is still active now in Lake Chad, for example, which is a shallow basin in which the waters of surface tributaries concentrate. At present, it is fed essentially by the Chari and its area is small, a few thousands of km². During the glacial period, Lake Chad was dry, but during

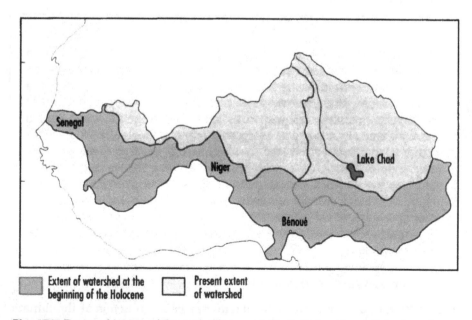

Extent of watershed at the beginning of the Holocene   Present extent of watershed

Fig. 17.7. Extent of basins of Senegal, Niger, and Chad during the period of maximum humidity at the beginning of the Holocene. All the basins were then active; now, only the southern part is still functional, the north being occupied by fossil valleys (Lévêque, 1997).

the rainy period of around 7000 to 6000 years BP, it occupied an area of more than 300,000 km². This aquatic system has thus seen great variations over relatively short time scales. At a still shorter scale, a century, significant fluctuations have been observed in the area of the lake, in response to fluctuations in rainfall. In the early 1960s the lake extended over more than 20,000 km², while in the early 1970s, after some years of relative drought, it occupied less than 5000 km². The consequences of this recession on the fauna and flora have been studied (Carmouze et al., 1983), and the information gained could be useful in predicting the future evolution of the system.

In the temperate regions, repeated glaciations wiped out the aquatic environments and thus the associated fauna, which explains why the fish communities are poorer than in the equatorial regions. We can for example compare Lake Geneva, which was frozen during the last glacial optimum (20,000 years), to the large lakes of East Africa (Tanganyika, Malawi), the existence of which is attested over several millions of years. Lake Geneva is a young lake in which the present fauna is the result of a recent recolonization during the Holocene warming, from refuge zones in which the aquatic fauna could survive at least partly. This relatively poor fauna comprised only 14 autochthonous species of fish. On the contrary, the lakes of East Africa are ancient, perennial over millions of years even though the levels have been shown to vary by several hundreds of metres during this period. These lakes harbour a fish and invertebrate fauna that is highly diversified and results from a long coevolution of the environment and the species. Simply put, the fauna of these lakes is diversified to better exploit all the resources of the ecosystem (Lévêque, 1997), while in Lake Geneva there is a heteroclitic set of species that is the consequence of a recent recolonisation. From the evidence, we cannot study the biological functioning of these lakes (production, food webs) without taking into account the history of environments. In the case of the East African lakes it is possible to develop theories about coevolution and speciation, adaptive radiation, and concepts of niche and competition for resources (Lévêque and Paugy, 1999). In the case of Lake Geneva, we can only observe that a certain number of species have succeeded in recolonizing the lake since the ice melted, but such a situation, partly based on chance, is not the best one to test ecological theories.

The same applies to fluvial systems. Climatic variability could lead to significant variations in their morphology, and sometimes even to their temporary disappearance. For fish to recolonize basins that were once dried up or frozen, or basins newly created by geological and climatic events, physical communication must have been established with basins that remained active and conserved a diversified fauna, in other words, refuge zones. For example, the fish fauna of Ireland is very poor and at present comprises only about twenty species, most of them introduced by man (Fitzmaurice, 1984). In fact, after the retreat of the ice during the Holocene, only eight anadromous migratory species (including salmon, shad, eel) were able to recolonize the Irish waters by natural routes. The situation is different in North America, where the Mississippi, which was the principal refuge zone for aquatic species during the recent glaciations, runs north-south. The result is that species are able to migrate to and shelter in the southern part of the river with the advance of the ice in the north, and then recolonize the hydrographic network in the other direction as it thaws. In Europe, the Danube serves as a refuge zone for fish, but it runs west-east, so that species cannot migrate south, as in North America. Consequently, the extinction of species was probably greater during the glaciations, which explains the poverty of the European ichthyologic fauna when compared to that of North America (Oberdorff et al., 1997).

The evolution of a river ecosystem over 20,000 years: the Somme basin

The reconstruction of palaeoenvironments involves only the flora and the fauna. The history of the Somme basin was reconstructed from the last glacial episode. The major trends were the following (Antoine, 1997).

• During the last glacial maximum, the climate was arid and the fluvial network was blocked up owing to a high sediment load that resulted from the erosion of watersheds due to gelifluction processes (the result of freeze-thaw alternations). The alluvial plain was exhausted and the network was characterized by braided channels, shallow and unstable.

• During the deglaciation, around 13,000 years BP (according to ^{14}C dating), there was an important incision of the previous deposits and a transition from a braided channel pattern to a stable multi-channel system. This phenomenon can result from an increase in water inputs. Then the plant cover of watersheds developed, stabilizing the slopes and reducing the inputs. Around 12,000 years BP, the dynamics became that of a classic alluvial plain with a system with a large main channel and meanders.

• Around 11,000 to 10,000 years BP there began a new phase of organic soil erosion under a climate that became colder and drier. The valley bottoms were the site of an important deposition of silts fed by the erosion of the drainage basin.

• The beginning of the Holocene corresponded to a new phase of incision that occurred also in the European hydrographic networks, in relation with the beginning of the global climatic warming of the early Holocene.

• During the first half of the Holocene, the whole valley filled in gradually. The development of the vegetation on the slopes induced a rise in the relative amount of organic matter in the sediments and a reduction of detritic inputs. This period was characterized by the development of peat bogs bordered by small lateral channels. The peaty valley, which worked as a sponge, was the site of high evapotranspiration linked to the development of the vegetation, reducing in this way the amount of water flowing in the channels.

• In the middle of the Holocene, a phase of less severe incision than the preceding was observed, followed by the deposition of organic silts in a large single meandering channel.

17.4. ECOLOGICAL IMPACT OF PRESENT GLOBAL WARMING

When we address the question of consequences of climatic change to come in the present ecosystems, we must avoid the trap of catastrophic language, laying undue stress on the destruction and damage caused to ecosystems by climatic evolution, or make value judgements that ignore history. For example, to speak as some do of the fragility of coastal ecosystems that will be endangered by the projected rise in sea level is to ignore, intentionally or otherwise, that the sea level has fluctuated greatly in the past also. Around 20,000 years ago it was 120 m below the present level, which means that the English channel was then dry. The coastal ecosystems were simply displaced as the sea level changed. Apart from the lower temperatures, the coastal ecosystems of Europe were probably very different during the glacial period from what we observe today. Obviously, the possible rise in sea level will have social and economic consequences that we cannot ignore, but this question is separate from our main question and it would be inappropriate to combine the two.

17.4.1. The general context

The task of the ecologist is to predict as well as possible the effect of climate modifications on species and ecosystems (Fig. 17.8). Such predictions may be stated in simple terms in four principal aspects (Hughes, 2000).

- *Effects on physiology.* Changes in the CO_2 concentration, temperature, and humidity may directly affect the metabolism and growth rates of plants, particularly processes such as photosynthesis and respiration.

- *Effects on distribution.* Species are expected to migrate towards the north or to higher altitudes in response to the displacement of climatic zones. An increase of 3°C in the average annual temperature will lead to a displacement of 200 to 400 km in latitude of isotherms in the temperate zone, or of 500 km in altitude.

- *Effects on phenology.* Changes occurring in the environmental factors that control the different phases of the life cycle of species can modify the phenological relationships between species and, consequently, the trophic relations (Harrington et al., 1999).

- *Adaptation.* The populations of species having short generation times and rapid growth rates may adapt *in situ* to changes in the conditions of their environment.

All these transformations, on the individual and species level, will inevitably modify the interactions between species with feedback effects on the geographic distribution and local abundance of those species. Some species will probably disappear and the composition of ecosystems will undoubtedly be modified.

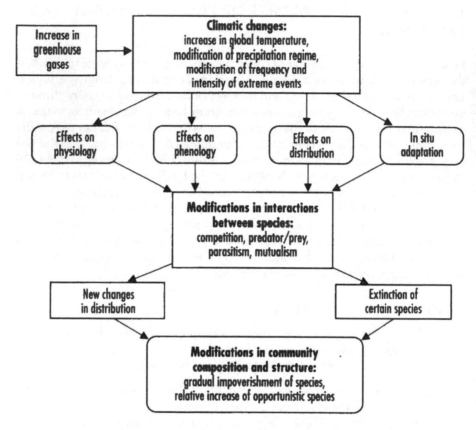

Fig. 17.8. Potential effects of climatic changes on the dynamics of communities. The increase in carbon dioxide concentration acts directly on the physiology of species that respond to climatic changes by adaptation or migration. From this come modifications in the interspecies relationships that could lead to extinction or changes in the distribution area. The final result is a change in the structure and composition of ecosystems (modified from Hughes, 2000).

17.4.2. Impact of CO_2 increase on species physiology and structure of ecosystems

The increase in atmospheric CO_2 affects the metabolism and structure of terrestrial ecosystems with feedback effects on the climatic system.

When the other resources present in the environment are available in sufficient quantity, various experiments carried out on isolated plants or plants grouped into simplified ecosystems have shown that a doubling of the atmospheric CO_2 level stimulates photosynthesis. For example, in young chestnut saplings planted in a greenhouse in an atmosphere containing 700 ppm of CO_2, or twice the atmospheric level, it was noted that the plant production increased by 20 to 30%. The saplings flowered in three years instead of the usual five years. The increase in plant biomass was concentrated in this case in the root system but other species tended to develop their above-ground parts. For grasslands maintained in an atmosphere twice as rich in CO_2, biomass above the soil increased on average by 14% but the individual responses were highly variable, ranging from negative responses to increases of 85% (Walter et al., 1999). This variability would be due to the influence of other factors, such as the availability of water and nutrients or the temperature.

Experimental results also show that the carbon fixed by photosynthesis in atmospheres rich in CO_2 is invested essentially below the soil surface (Jones et al., 1998). The result is a growth in organic matter present in the soil, which has an impact on the chain of decomposers. The microbial biomass of the soil has not been modified but the species composition of soil fungi has changed. There was also a modification of the species composition of Collembola, which are selective consumers of fungi. It was shown, moreover, that in herbaceous plants exposed to higher levels of CO_2 there was a reduction in the stomatal conductance that generated a reduction of transpiration losses and thus a maintenance of the soil humidity. This increase in water availability will be one of the factors responsible for a production increase in grasslands in which water is a limiting factor.

The Global Change Terrestrial Ecosystem (GCTE) launched in 1992 a programme of field experimentation in parcels of 0.5 m² to some hundreds of square metres distributed in various places in the world in order to evaluate the problem of a rise in CO_2. According to the results of this programme, the response of ecosystems to such a rise depends heavily on other factors such as the availability of water and nutrients or the temperature. For example, the alpine prairies or Arctic tundra respond least to the rise in CO_2 and even show complete acclimatization after some years of CO_2 increase. Besides, there is a strong correlation between the responses to CO_2 increase and the soil humidity in ecosystems in which the availability of water is limited, such as the Mediterranean grasslands. There is no stimulation when the soils are moist but there is an increase in the biomass during dry years or at the end of the wet season, when the soils dry out (Canadell, 1999).

An *in situ* experimental approach was used to evaluate the influence of doubling the atmospheric CO_2 and an increase of 3 to 5°C in temperature on a boreal forest system in Norway (Breemen et al., 1998). This ecosystem was composed of pine (*Pinus sylvestris*), birch (*Betula pubescens*), heather (*Calluna vulgaris*), and bilberry (*Vaccinium myrtillus*). Over a period of three years, the rates of mineralization of soil organic matter rose, supporting an increase in the availability of nitrogen in the soil and in the inorganic nitrogen export by runoff. The increase in nitrogen availability led to greater plant growth and better water use efficiency by plants.

The dilemma is to understand to what extent these results can be extrapolated to natural systems. We can investigate especially the direct effects at the level of certain processes (growth of plants, consequences on biogeochemical cycles) and indirect effects such as changes in the species composition. Experiments in a controlled environment cannot thus be generalized without caution because, first, there may be responses linked to the nature of the experimental communities used and, second, we do not know the feedback effects that could have taken place over the long term in nature, nor the effects of competition between species, which could result in a better growth.

Some observations made over the long term, however, offer the beginnings of an answer to the question whether climatic changes have already affected the functioning of ecosystems. At middle latitudes of the northern hemisphere, a relationship has been indicated between the air temperature at the soil surface and the variations, amplitude, and duration of the seasonal cycle of atmospheric CO_2. According to satellite data, photosynthetic activity of the terrestrial vegetation also appears to have increased between 1981 and 1991, largely because of the lengthening of the growing season, i.e., the period of photosynthetic activity, by around 12 days during the 1980s. It begins earlier in the spring and lasts later in the autumn (Myneni et al., 1997). Moreover, the amplitude of the atmospheric CO_2 cycle has increased by 7 to 14% since the early 1980s, which also leads to greater photosynthetic activity of the vegetation at high latitudes of the northern hemisphere. These results will confirm the hypothesis that higher temperatures have led to an increase in the biospheric activity at middle latitudes of the northern hemisphere.

Rise in CO_2 level, carbonate/bicarbonate equilibrium and the biology of corals

The rise in atmospheric CO_2 level is seen in an increase in the level of CO_2 dissolved in the water, with indirect effects including a lower pH and saturation of calcium carbonate in water. The result is a diminution in the calcification of corals and other marine calcifying organisms, depending on the processes described below, which are a good example of complex relationships between the atmosphere, aquatic systems, and the biology of marine species.

In the water, the dynamics of carbon dioxide concentrations follows complex rules that result from the existence of several chemical forms. The carbon dioxide may hydrate and form carbonic acid according to a reversible reaction:

$$\text{dissolved } CO_2 + H_2O \longleftrightarrow H_2CO_3 \text{ (carbonic acid)}$$

Carbonic acid is subsequently dissociated into bicarbonate ions (HCO_3^-) and carbonate ions (CO_3^{--}) as a function of the pH, i.e., the acidity of waters.

$$H_2CO_3 \longleftrightarrow H^+ + HCO_3^- \quad \text{and} \quad HCO_3^- \longleftrightarrow H^+ + CO_3^-$$

According to the laws of chemistry, the relative concentrations of free carbon dioxide, carbonate, and bicarbonate are based on a subtle state of equilibrium between the pH, the dissolved carbon dioxide level, and the derivative forms of carbonic acid. Calcium, abundant in many aquatic environments, plays a key role in these processes of equilibrium of derivative forms of CO_2. The presence of carbonic acid has the property of increasing considerably the dissolving capacity of water, which is then able to attack the calcium carbonate or calcite ($CaCO_3$), which is poorly soluble in water. This calcite is transformed into calcium bicarbonate ($Ca(HCO_3)_2$), which is much more soluble in water, according to a reversible reaction:

$$CaCO_3 + CO_2 + H_2O \longleftrightarrow Ca(HCO_3)_2) = Ca^{++} + 2\ HCO_3^-$$

When the CO_2 level rises, there is dissolution of calcite and formation of carbonates and bicarbonates. The waters are thus called "aggressive". On the other hand, when the dissolved CO_2 level falls, there is a transformation of soluble calcium bicarbonate into calcium carbonate, which precipitates. The result, for organisms that elaborate calcic structures, is that calcification diminishes with the increase in carbon dioxide level. Such a reduction is probably ongoing and could have amounted to 10% between 1880 and 1990 in the marine environment (Gattuso et al., 1999). This rate will probably accelerate with the increase in atmospheric CO_2: a reduction of around 20% over a century is predicted (Kleypas et al., 1999), which may cause problems for the health of the coral reefs and more generally for calcifying organisms living in aquatic environments.

17.4.3. Impact on the water cycle and aquatic ecosystems

The climate influences the key factors that control the hydrological balance and thus the dynamics of continental aquatic systems through changes in precipitation, soil humidity, and evapotranspiration. The low quantity of water found in the atmosphere (0.01 to 0.02% of the quantity of water contained on the earth) that falls in the form of rain has a major significance for ecosystems and human societies. It is indeed the availability of water that determines the distribution of the major biomes and agricultural productivity.

The distribution of rain is controlled by atmospheric circulation that transports water vapour. With warming, the increase in evaporation leads to greater average precipitation on the earth's surface However, statistical simulations of the evolution of the climate show that climatic warming will not affect the planet in a uniform manner. Regional modifications in the quantity and seasonal distribution of atmospheric precipitation (rain, snow, hail, etc.) as well as in the frequency of exceptional events will influence the scope and temporal distribution of surface flows and the supply of ground water, as well as the magnitude of floods and droughts. The flow of the Colorado river, for example, could diminish by half, while the flow of the Ganga and Brahmaputra rivers will increase at least by 50%. Other rivers may begin to flow again after having been dry for thousands of years. The water resources in arid and semi-arid regions are particularly sensitive to small changes in temperature and precipitation. A study carried out in the United States shows that an increase of 1 to 2°C of the temperature and a reduction of 10% of precipitation will reduce by half the availability of water in arid and semi-arid zones (UNEP, 1993).

Many models predict a reduction of precipitation in southern Europe and an increase of precipitation in winter at high latitudes. Because of this, since the beginning of the 20th century, precipitation has increased at high latitudes in the northern hemisphere, particularly in winter, while in the tropical and subtropical regions it has effectively decreased during the last decades, as for example in the Sahel and the eastern part of Indonesia. The warming will also modify the inputs of water from snow melt in the temperate regions. In a warmer world, there is more rain but less snow. The rivers will thus run at high water in winter and dry out in summer, a period in which they are presently fed by waters from snow melt.

The temperature rise will at the same time lead to modifications in evapotranspiration, soil humidity, and conditions of infiltration. The soils in North America, southern Europe, and many other regions will certainly become more dry in the next few decades. The higher temperatures will accelerate evaporation and transpiration by plants. However, studies have also shown that in a more humid period the more abundant snow cover could reduce the evaporation (Karl et al., 1997). It is clear that the predictive models still need to be improved.

It is likely that changes in the characteristics of the average climate will be accompanied by changes in the extreme meteorological events. Some authors foresee an increased frequency and intensification of tropical storms linked to the climate change.

The difficulties in predicting the future evolution of continental aquatic ecosystems lie mostly in our uncertainty about the regional consequences of climatic evolutions. Moreover, we do not know the possible feedback effects between modifications of the terrestrial environment and those of the aquatic environment, e.g., modification of plant cover, soil humidity.

17.4.4. Impact on terrestrial ecosystems

The earth's climate changes continually and terrestrial ecosystems constantly react to those changes. However, there is now a new aspect: the climatic changes are superimposed on the modifications of plant cover and landscape linked to human activities. In other words, the climatic evolution will exert an additional disturbance on many ecosystems, the functioning of which is already modified because of pollution, land use for agricultural or industrial purposes, and communication networks that break up the landscape.

Some studies show that the composition and structure of the vegetation will necessarily alter in response to some major characteristics of climate change (Walker et al., 1999).

- Biomes do not change or migrate all at once. The species react individually to changes as a function of their biological characteristics and their capacities to respond to disturbances. New communities may thus appear, and they may be transitory or perennial.

- Climatic warming is manifested in the growth of trees, which is accelerated over the past few decades. In France, comparisons of trees on some parcels over a hundred years show that the present oaks often grow taller than old ones. Over the past decade, the productivity of the forest in France increased from 80 to 90 million m³ of woods a year, a performance that cannot sufficiently be explained by forestation programmes.

- One possible response of species is to move. Studies of palaeoenvironments suggest that many plant species can shift quite quickly in the face of the climatic changes foreseen. The dispersal rates of plant species in response to the climatic changes of the Holocene were assessed from pollen data. They went from 50 to 2000 m/yr on average for woody species of Europe and North America. Plant species spread better if their seeds are adapted to transport by wind or by animals. Even though animals are in principle better adapted to move around, the distribution of some of them will be limited by the distribution of the plant species they depend on for food or habitat. However, in comparison to what has occurred in the past, the modifications of the landscape and the fragmentation of ecosystems resulting from human activities could constitute hindrances to movement and modify the migratory capacity of many species: fewer spaces are available to colonize and particularly hostile territories must be crossed to reach those space. It has been observed that the northern limits of distribution of 59 species of European birds were extended by around 19 km northward over the course of 20 years (1988-1991 compared to 1968-1972) (Thomas and Lennon, 1999).

- Modifications of interspecies relations consecutive to climatic changes probably play a non-negligible role in the constitution of new biomes. Some species may be doomed to extinction if they encounter new predators or parasites. Those that are more vulnerable to evolutions of the climate are the species that cannot move or adapt rapidly, and the highly specialized species linked to certain types of habitats. The great uncertainty, however, about which palaeocological research remains silent, is what relations of competition will be established between the autochthonous species and the many introduced species that now populate the ecosystems of all the continents.

- One of the consequences expected of climatic changes is the shifting of zones of vegetation and their associated fauna towards new regions, especially towards high latitudes. In Europe, for example, an increase of 5°C in the annual average temperature could initiate a shift 1000 km northward of the taiga and temperate deciduous forests (UNEP, 1993). It has been observed that the birch (*Betula pubescens*) seems to have responded rapidly to recent climatic warming, extending its area of distribution northward into the Swedish tundra. In Norway, conditions may become too hot for spruce, fir, and pine, so oak and beech may replace them. In that country, a temperature rise of 1.5 to 4.5°C will extend the area occupied by the temperate forest from 0.7 to 13%. In the arid regions of the Mediterranean, a temperature rise of 1.5 to 4.5°C will reduce the soil humidity and plant growth, and the desert may extend to the semi-arid zones of North Africa and the Middle East. The plant species and biomes could also spread in altitude as it grows warmer. It is known that in the Holocene, when the temperatures were 2°C higher than at present in North America, the distribution of Canadian hemlock (*Tsuga canadensis*) and white pine (*Pinus strobes*) extended 350 m higher in altitude than at present.

Are biocoenoses random assemblages?

There is no doubt that the climate is a controlling element of the distribution of species, as demonstrated by the vast biogeographic literature. Palaeoclimatological research has indicated the existence of significant fluctuations of climate with a periodicity of about a millennium during the Pleistocene, for which we have good chronological series. These oscillations were too rapid to allow a significant response in terms of macroevolution but could have had a perceptible effect on the ecology of biological communities. In particular, they support modifications in the areas of distribution of species, which react individually to climatic changes. Depending on the nature and amplitude of these changes, some species may come in contact with each other that did not do so earlier. Inversely, species that coexisted may become separated. Such individual adjustments of the distribution of species are common in the Pleistocene series, in terrestrial as well as marine environments. According to some authors, these observations confirm the hypothesis that biological communities are random assemblage and not groups with a strong internal coherence (Roy et al., 1996).

17.4.5. Rise in sea level

The present sea level is rising at a rate of around 2 mm/yr over the past few decades (Schneider, 1997). The question is whether this phenomenon will continue, or even be amplified because of climatic warming. The response has, of course, implications for coastal ecosystems as well as important socioeconomic consequences, which explains why many scientists are closely interested in the question. Just as with other complex phenomena, opinions do not agree. Experts of the IPCC estimate that the rise may be between 20 cm and 1 m in a century from now.

First of all, the climatic warming will probably increase the temperature of the oceans, and the volume of sea water, like any other physical body, increases when it is warm. It is estimated that this expansion alone will raise the sea level by 30 cm during the 21st century (Schneider, 1997). However, the warming of the climate could also cause the melting of part of the polar ice, especially Antarctica. The rise in the sea level will affect the coastal environment by reduction and erosion of shores and coastal lagoons, and saline intrusions in deltas and in coastal lagoons.

In addition to the rise in sea level, climatic change may cause an increase in the frequency and intensity of storms and thus of temporary and exceptional rises in the sea level above the tidal level predicted. In fact, when a depression passes, the combined effect of winds that accumulate the layers of surface water towards the coast and the lower atmospheric pressure contribute to a rise in the sea level. It is acknowledged that a warming of the ocean surface in tropical regions can lead to a rise in the frequency and force of cyclones and tropical storms (Pittock and Flather, 1993). In the Bay of Bengal, for example, cyclonic storms could have dramatic consequences for Bangladesh. However, such rises also exist in Europe, where there has been an increase in storms in the Atlantic and the North Sea during the last few decades. An increase in the height, duration, and frequency of ocean waves has also been observed along the German coast of the North Sea since the late 1960s (Toppe and Fuhrboter, 1994). These waves, which result in a temporal marine invasion of the coastal environment, can have a major economic impact in countries in which the coastal strip is a highly active area.

Long-term studies on the evolution of ecosystems as a function of climatic fluctuations are useful in understanding the relationships between climatic changes and ecosystem dynamics.

Observations taken in the northwest of the North Sea over 33 years (1955-1987) show that there is a remarkable similarity of variations in abundance of phytoplankton, zooplankton, herring (*Clupea harengus*), and black-legged kittiwake (*Rissa tridactyla*) (Aebischer et al., 1990). The long-term trend for these four trophic levels show a parallel and continuous decrease from 1955 to 1979-1980, followed by a revival after 1980. There is a strong correlation between these variations and that of the climate, estimated here by the annual frequency of western winds from 1955 to 1987. Even though it is difficult to interpret these observations in detail, the signal of climate is visible at all the trophic levels of this marine ecosystem.

Other observations also show that marine communities react to climatic fluctuations. In the northeast Pacific, long-term observations of over 80 years showed the existence of anomalies in the temperature of the ocean surface, on the inter-annual and inter-decade scale. The inter-annual anomalies appeared and disappeared in some months and are synchronous along the coast. There were 12 warm episodes in the California current since 1916, and 10 cold episodes, with anomalies of temperature that did not exceed 2.5 or 3°C, but the frequency of warm events increased from 1977 onward and seemed linked to the El Niño phenomenon. Significant and long-term responses of the marine ecosystems were associated with these climatic events. For example, in the California current, from the end of the 1970s, a reduction was observed in the abundance of zooplankton, sea birds, and fish catches, as well as changes in the structure of benthic communities. Also, northward movement of southern species was observed more frequently, and some of those species now dominate the communities (McGovan et al., 1998). Inversely, in the Gulf of Alaska, since the late 1970s, an increase has been observed in zooplankton and catches of several commercial species of fish. These biological changes are attributed to physical mechanisms controlling the primary productivity, notably those that influence the depth and speed of the vertical turnover.

17.4.6. Climate changes and sub-Antarctic terrestrial ecosystems

Sub-Antarctic islands such as the Kerguelen islands in the southern Indian Ocean are relatively simple environments in ecological terms. For example, they have a small number of species, low anthropization, and short food chains. The cold temperate oceanic climate with heavy precipitation and its low seasonal thermal amplitudes have a preponderant effect on the potential for implantation and development of the fauna and flora. The present climatic conditions are marked by a significant rise in the average annual temperatures (+1.3°C since the mid-1960s) as well as a severe water deficit: 600 to 800 mm since the 1970s as opposed to 1000 to 1200 mm in the 1950s. The result is a retreat of the ice fronts of the archipelago, some of which have retreated more than 4 km since the mid-1970s. The summer water deficit has been more marked for some years. The autochthonous vegetation is under stress: e.g., withering of leaves, drying of young saplings, disappearance of plant cover on soils with low water holding capacity. The particularly dry conditions of the austral summer in 1999 led to an unusual drying out of ponds and peat bogs, the disappearance of a large number of Kerguelen cabbage plants, and other phenomena. In fact, the autochthonous plants are physiologically adapted to permanent cold and cannot survive a rise of several degrees in the annual average temperature (Hennion and Bouchereau, 1998). On the other hand, the introduced plants were favoured by these changes in climate. For example, since the 1980s a significant increase was observed in introduced phanerogams (close to 60 at present) and the climatic changes observed these past few years have favoured the establishment of species that arrived spontaneously or were introduced accidentally. Some of these plants, such as *Stellaria alsine*, show invasive characters that locally threaten the autochthonous flora (Frenot and Goaguen, 1994). These allochthonous plants also help to diversify the food webs.

17.4.7. Effects on health

Extreme climatic phenomena such as torrential rain or long periods of drought have repercussions on the public health as well as on the health of species in general. More frequent precipitation facilitates the reproduction of species such as mosquitoes, the appearance of rodents, and the release of chemical products and microorganisms into the surface or ground waters.

It has been observed for some years that diseases associated with algal, bacterial, and viral toxins are affecting many marine species. Most of the new epidemics seem to be due not to new microorganisms but rather to pathogenic agents attacking species they had not attacked before, or at least species on which they were not recorded before (Harvell et al., 1999). Apart from diseases, there has been an increase in toxic algal blooms in the past decade, most of them due to dinoflagellates.

Massive mortalities due to epidemics have been observed in several marine taxa (Harvell et al., 1999). For widely studied groups, such as corals and mammals, the frequency of epidemics and the number of new diseases has increased. In 1998, for example, in the coral reefs throughout the world, the most severe phase of bleaching and mortality of the modern era was recorded. At the same time, the temperatures of the ocean surface were the highest ever recorded. Unlike bleaching phenomena observed before, which never exceeded 15 m depth, the bleaching of 1998 reached 50 m depth and many other organisms carrying zooxanthella were also affected. These events could not be attributed to just the natural variability or to local disturbances. Similarly, El Niño alone cannot explain the observations made throughout the world. It is probable that the bleaching of corals is the consequence of a constant increase in the temperature of sea water, under the effect of a general warming due partly to human activities. Such episodes of bleaching may become more frequent and more severe with the warming of the climate, and this global menace may add to the more local impacts of human activities on corals (Bryant et al., 1998; Wilkinson et al., 1999).

Bleaching of corals

When corals are physiologically disturbed, there is also a disturbance of their symbiotic relationship with the algae (zooxanthella) they harbour in their cells. The corals may lose part or even all the algae, which are the source of their colour and nutrition. This is what results in the *bleaching of the corals*. They survive if the disturbance lasts a short time, but they die if it lasts a long time. Moreover, corals subjected to sublethal stress are much more sensitive to infection by many pathogens. The physiological stress may have diminished the resistance and increased the frequency of opportunistic diseases. Epizootics may also lead to significant mortalities of corals (Hayes and Goreau, 1998).

The Global Reef Monitoring Network was created in 1998 under the responsibility of the Intergovernmental Oceanographic Commission of UNESCO, the UNEP, and the World Conservation Union. It is located at the International Centre for Living Aquatic Resource Management (ICLARM) at Manila. Its aim is to draw attention to the situation of coral reefs, the damage they suffer, and the causes of that damage through basic means and with the involvement of scientists and volunteers in a vast observation network. From mid-1997 to mid-1998, in parts of the world that are very far apart (Indian Ocean, the Antilles, South Asia, Southwest Asia, and East Asia, and the Middle East), there was widespread bleaching of hard and soft corals. For the most part, this phenomenon coincided with an exceptional Niño, but in many regions these events seemed unrelated. Although the degree of bleaching varied with the regions (severe in the Maldives, Sri Lanka, Singapore, and Tanzania, moderate to mild in Madagascar, the Great Barrier Reef, French Polynesia, the Galapagos, and the Bahamas), the general consensus was that the episode was the most serious ever recorded (Wilkinson and Hodgson, 1999). The question is whether the event of 1997-1998 was an exception or whether the phenomenon may recur in the context of the warming of waters and the atmosphere.

17.5 THE LIKELY EFFECTS OF CLIMATE CHANGE IN EUROPE

The effects of climate change in Europe will involve both losses and gains to the natural resource base. These effects will vary from region to region, but their significance will depend to a considerable extent on the non-climate drivers of environmental change, socio-economic development and policy evolution within Europe. Future climate change scenarios of impacts and adaptations to climate in Europe have been developed on the basis of present knowledge and expert review (Parry, 2000). Overall it is evident that the balance of impacts of climate change will be more negative in southern than in northern Europe. Primary sectors such as agriculture and forestry will be affected more than secondary and tertiary sectors such as manufacturing and retailing.

In very general terms we may expect:

- Effects of weather extremes on natural, social, and economic systems in ways that reveal sensitivities and vulnerabilities to climate change in these systems.

- Current and future pressures on water resources are likely to be exacerbated. Climate change is likely to increase the water resource differences between northern and southern Europe: we can expect an increase in annual stream flow in northern Europe and a decrease in the south. Flood hazard is likely to increase across most of Europe while the risk of summer drought may be exacerbated in southern Europe.

- Soil quality will deteriorate under warmer and drier climate scenarios in most regions of both northern and southern Europe. The magnitude of this effect will vary markedly between geographic locations and may be moderated by increasing precipitation.

- Most northern and western European ecosystems are temperature limited. Increasing temperature and CO_2 is expected to result in changing natural ecosystems, leading to a northward displacement of boreal forests, increasing the encroachment of trees and shrubs in the northern tundra, and broad-leaved trees in parts of northern Europe coniferous forests. In some nature reserves, the diversity of vegetation is under threat of rapid change. Fire disturbance may become more frequent in Mediterranean areas.

- The response of most animal communities depends on the fate of specific habitats. For example, loss of subarctic tundra and/or wetlands due to thermokarst, sea level rise, or shrub encroachment could threaten migratory bird populations

- Climate change is also likely to result in faunal shifts affecting freshwater and marine biodiversity.

- In coastal areas, the risk of flooding, erosion, and wetland loss will increase substantially with implications for human settlement, industry, tourism, and coastal habitats.

- Climate change poses a range of risks to human population health, including changes in the frequency of temperature extremes and the spread of vector-borne diseases.

For all these reasons, climate change has major implications for Europe's policies of development and environmental management.

Selected Bibliography

Académie des Sciences. 1993. *Ozone et Propriétés Oxydantes de la Troposphrè*. Rep. no. 30. Tec&Doc, Lavoisier, Paris.

Académie des Sciences. 1994. *L'effet de Serre*. Rep no. 31. Tec&Doc, Lavoisier, Paris.

Académie des Sciences. 1998. *L'Ozone Stratosphérique*. Rep no. 41. Tec&Doc, Lavoisier, Paris.

Acot, P. 1988. *Histoire de l'Écologie*. Presses Universitaires de France, Paris.

Agrawal, A.A., Laforsch, C., and Tollrian, R. 1999. Transgenerational induction of defences in animals and plants. *Nature*, 401: 60-63.

Allan, J.D. 1995. *Stream Ecology. Structure and Function of Running Waters*. Chapman & Hall, London.

Allen, T.F.H., and Starr, T.B. 1982. *Hierarchy: Perspectives forEcological Complexity*. University of Chicago Press, Chicago.

Amblard, C., Boisson, J.C., Bourdier, G., Fontvieille, D., Gayte, X. and Sime-Ngando, T. 1998. Ecologie microbienne en milieu aquatique: des virus aux protozoaires. *Revue des Sciences de l'Eau*, spec. No.: 145-162.

Amoros, C., and Petts, G.E. 1993. *Hydrosystèmes Fluviaux*. Collection d'écologie 24, Masson, Paris.

Andersson, G., Granéli, W., and Stenson, J. 1988. The influence of animals on phosphorus cycling in lake ecosystems. *Hydrobiologia*, 170: 267-284.

Arsac, J. 1993. *La Science et le Sens de la Vie*. Le temps des sciences, Fayard, Paris.

Auger, P., Baudry, J. and Fournier, F. (eds). *Hiérarchies et Échelles en Écologie*. Naturalia Publications.

Azam, F. 1998. Microbiological control of oceanic carbon flux: the plot thickens. *Science*, 280: 694-696.

Bailey, R.G. 1996. *Ecosystem Geography*. Springer Verlag, New York.

Bak, P. 1999. *Quand la Nature s'Organise. Avalanches, Tremblements de Terre et Autres Cataclysmes*. Flammarion, Paris.

Barbault, R. 1992. *Ecologie des Peuplements*. Masson, Paris.

Barbault, R. 1997. *Biodiversité*. Les Fondamentaux, Hachette, Paris.

Barbault, 1997. *Ecologie générale. Structure et Fonctionnement de la Biosphère*. Masson, Paris.

Barrue Pastor, M. and Muxart, T. 1992. Le géosystème: nature "naturelle" ou nature "anthropisée"? pp. 259-266 in M. Jollivet (éd.), *Les Passeurs de Frontières*. CNRS Éditions, Paris.

Begon, M., Harper, J.L., and Townsend, C.R. 1990. *Ecology: Individuals, Populations and Communities*. Blackwell Scientific Publications, Oxford.

Beier, C., and Rasmussen, L. 1994. Effects of whole-ecosystem manipulations on ecosystem internal processes. *Trends in Ecol. and Evol.* 9: 218-223.

Bergandi, D., and Blandin, P. 1998. Holism vs. Reductionism: do ecosystem ecology and landscape ecology clarify the debate? *Acta Biotheoretica*, 46: 185-206.

Berkes, F., Kislalioglu, M., Folke, C., and Gadgil, M. 1998. Exploring the basic ecological unit: ecosystem-like concepts in traditional societies. *Ecosystems*, 1: 409-415.

Bernard, J-J., Contini, D., Godet, G., and Gohau, G. 1995. *Le Temps en Géologie*. Collection Synapses, Hachette, Paris.

Berner, E.K., and Berner, R.A. 1996. *Global Environment. Water, Air and Geochemical Cycles*. Prentice Hall, New Jersey.

Besnier, J.M. 1996. *Les Théories de la Connaissance*. Dominos, Flammarion, Paris.

Blackburn, T.M., Brown, V.K., Doube, B.M., Greenwood, J.J.D., Lawton, J.H., and Stork, N.E. 1993. The relationship between abundance and body size in natural animal assemblages. *J. Anim. Ecol.*, 62: 519-528.

Blandin P., and Lamotte, M.1988. Recherche d'une entité écologique correspondant à l'étude des paysages: la notion d'écocomplexe. *Bull. Ecol.*, 19: 547-555.

Blasco, F., and Weill, A. (eds.) 1999. *Advances in Environmental and Ecological Modelling*. Elsevier, Paris.

Blasco, F. (ed.). 1997. *Tendances Nouvelles en Modélisation pour l'Environnement*. Elsevier, Paris.

Borman, F.H., and Likens, G.E. 1979. *Pattern and Process in a Forested Ecosystem*. Springer-Verlag, New York.

Boucher, G. 1997. Diversité spécifique et fonctionnement des écosystèmes: revue des hypothéses et perspectives de recherche en écologie marine. *Vie et Milieu*, 47(4): 307-316.

Boudreau, P.R., and Dickie, L.M. 1992. Biomass spectra of aquatic systems in relation to fisheries yield. *Can. J. Fish. Aquat. Sci.*, 49: 1528-1538.

Bourget, E., and Fortin, M. J. 1995. A commentary on current approaches in the aquatic science. *Hydrobiologia*, 300/301: 1-16.

Boutot, A. 1993. *L'Invention des Formes*. Éditions Odile Jacob, Paris.

Bravard, J.P., and Pettit, F. (eds.). 1997. *Les Cours d'Eau: Dynamique du Système Fluvial*. Armand Colin, Paris.

Brown, J.H. 1995. *Macroecology*. The University of Chicago Press, Chicago.

Bruin, J., Sabelis, M.W., and Dicke, M. 1999. Plants tap SOS signals from their infested neighbours? *Trends in Ecol. and Evol.*, 10(4): 167-170.

Burel, F., and Baudry, J. 1999. *Ecologie du Paysage. Concepts, Méthodes et Applications*. Éditions Tec&Doc, Lavoisier, Paris.

Campbell, I.D., McDonald, K., Flannigan, M.D., and Kringayark, J. 1999. Long-distance transport of pollen into the Arctic. *Nature*, 399: 29-30.

Carmouze, J.P., Durand, J.R., and Lévêque, C. (eds.). 1983. *Lake Chad: Ecology and Productivity of a Shallow Tropical Ecosystem*. Monographiae Biologica 53, W. Junk, The Hague.

Carpenter, S.R. 1996. Microcosm experiments have limited relevance for community and ecosystem ecology. *Ecology*, 77: 677-680.

Chalmers, A.F. 1987. *Qu'est-ce que la Science?* La Découverte, Paris.

Chapel, A., Fieux, M., Jacques, G., Jacques, J.M., Laval, K., Legrand, M., and Le Treut, H. 1996. *Océans et Atmosphère*. Collection Synapses, Hachette, Paris.

Chapin, F.S., Walker, H.H., Hobbs, R.J., Hooper, D.U., Lawton, J.H., Sala, O.E., and Tilman, D. 1997. Biotic control over the functioning of ecosystems. *Science*, 277: 500-504.

Charvolin, F. 1994. L'invention de la biosphère: les fondements d'une méthode. *Nature, Sciences, Sociétés*, 2(1): 21-28.

Cherrett, J.M. 1989. Key concepts: the results of a survey of our members opinions, pp. 1-16 in Cherrett, J.M. (ed.). *Ecological Concepts. The Contribution of Ecology to an Understanding of the Natural World*. Blackwell Scientific Publications.

Christensen, N.L., et al. 1996. The report of the Ecological Society of America Committee on the scientific basis for ecosystem management. *Ecol. Appl.*, 6: 665-691.

Cohen, J.E., Briand, F., and Newman, C.M. 1990. *Community Food-Webs: Data and Theory*. Springer-Verlag, New York.

Colinvaux, P. 1993. *Initiation à la Science de l'Écologie*. Points Sciences, Éditions du Seuil, Paris.

Combles, C. 1995. *Interactions Durables. Ecologie et Évolution du Parasitisme*. Masson, Paris.

Coque, R. 1993. *Géomorphologie*. Armand Colin, Paris.

Coquillard, P. and Hill, D.R.C. 1997. *Modélisation et Simulation d'Écosystèmes. Des Modèles Déterministes aux Simulations à Événements Discrets*. Masson, Paris.

Cosandey, C., and Robinson, M. 2000. *Hydrologie Continentale*. Armand Colin, Paris.

Cury, P., and Roy, C. 1991. *Pêcheries Ouest-Africaines. Variabilité, Instabilité et Changement*. ORSTOM, Paris.

Cushing, D.H. 1996. *Towards a Science of Recruitment in Fish Populations.* Excellence in Ecology 7, Ecology Institute, Oldendorf/luhe, Germany.

Daily, G. 1997. *Nature's Services.* Island, Washington, DC.

Dajoz, R. 2000. *Précis d'Écologie,* 7th ed. Dunod, Paris.

De Meester, L. 1996. Local genetic differentiation and adaptation in freshwater zooplankton populations: patterns and processes. *Ecoscience,* 3(4): 385-399.

Deléage, J.P. 1992. *Histoire de l'Écologie. Une Science de l'Homme et de la Nature.* La Découverte, Paris.

Deman, K., Hofmann, E., and Marchant, H. 1996. Marine biotic response to environmental change and feedbacks to climate. In: Houghton, J.T., Meira Filho, L.G., Callander, B.A., Harris, N., Kattenberg, A., and Maskell, K. (eds.), *The Science of Climate Change:* contribution of Working Group 1 to the Second Assessment Report of the Intergovernmental Panel on Climate Change (IPCC). Cambridge University Press, Cambridge, UK.

Dobrovolsky, V.V. 1994. *Biogeochemistry of the World's Land.* CRC Press, Boca Raton, Florida.

Dodson, S.I., Crowl, T.A., Pechkarsky, B.L., Kats, L.B., Covich, A.P., and Culp, J.M. 1994. Non-visual communication in freshwater benthos: an overview. *J.N. Amer. Benthol. Soc.,* 13: 268-282.

Drouin, J.M. 1993. *L'Écologie et Son Histoire.* Coll. Champs, Flammarion, Paris.

Duplessy, J.C., and Morel, P. 1990. *Gros Temps sur la Planéte.* Éditions Odile Jacob, Paris.

Durand, D. 1996. *La Systémique.* Que Sais-Je? Presses Universitaires de France, Paris.

Duvigneaud, P. 1980. *La Synthèse Écologique.* Doin, Paris.

Egerton, F.N. 1973. Changing concepts of the balance of nature. *Q. Rev. Biol.,* 48: 322-350.

Elton, C.S. 1927. *Animal Ecology.* Sidgwick & Jackson, London.

Fabiani, J.L. 1985. Sciences des écosystèmes et protection de la nature. pp. 75-93 in Cadoret A. (ed.), *Protection de la Nature. Histoire et Idéologie.* L'Harmattan, Paris.

Faurie, C., Ferra, C., Médori, P., and Dévaux, J. 1998. *Ecologie. Approche Scientifique et Pratique.* Tec&Doc, Lavoisier, Paris.

Field, C.B., Bethrenfeld, M.J., Randerson, J.T., and Falkowski, P. 1998. Primary production of the biosphere: integrating terrestrial and oceanic components. *Science,* 281: 237-240.

Folt, C.L., and Burns, C.W. 1999. Biological drivers of zooplankton patchiness. *Trends in Ecol. and Evol.,* 14(8): 300-305.

Ford, E.D. 2000. *Scientific Method for Ecological Research.* Cambridge University Press, Cambridge.

Forman, R.T.T., and Godron, M. 1986. *Landscape Ecology*. Wiley and Sons, New York.

Foucault, A. 1993. *Climate. Histoire et Avenir du Milieu Terrestre*. Fayard, Le Temps des Sciences.

Frochot, B., and Godreau, V. 1995. Intérêt écologique des carrières, terrils et mines. *Nature, Sciences, Sociétés*, Hors Série, 1995: 66-76.

Frontier, S., and Pichod-Viale, D. 1998. *Ecosystèmes. Structure, Fonctionnement, Évolution*. Dunod, Paris.

Frontier, S. (ed.). 1983. *Stratégies d'Échantillonnage en Écologie*. Collection d'écologie 17, Masson, Paris.

Fuhrman, J.A. 1999. Marine virus and their biogeochemical and ecological effects. *Nature*, 399: 541-548.

Gall, J.C. 1995. *Paléoécologie. Paysages et Environnements Disparus*. Masson, Paris.

Gleick, J. 1989. *La Théorie du Chaos: vers une Nouvelle Discipline*. Albin Michel, Paris.

Goldenfeld, N., and Kadanoff, L.P. 1999. Simple lessons from complexity. *Science*, 284: 87-89.

Golley, F. 1993. *A History of the Ecosystem Concept in Ecology: More than the Sum of the Parts*. Yale University Press, New Haven, Connecticut.

Grossman, D.H., Bourgeron, P., Busch, W.D.N., Cleland, D., Platts, W., Ray, G.C., Robins, C.R., and Roloff, G. 1999. Principles of ecological classification. pp. 353-393 in Szaro, R.C., Johnson, N.C., Sexton, W.T., and Malk, A.J. (eds), *Ecological Stewardship. A Common Reference for Ecosystem Management*. Elsevier.

Hairston, J.G., Sr. 1989. *Ecological Experiments: Purpose, Design and Execution*. Cambridge University Press, Cambridge.

Hanski, I., and Gilpin, M. 1997. *Metapopulation Biology, Ecology, Genetics and Evolution*. Academic Press.

Harmelin-Vivien, M.L., Harmelin, J.G., and Leboulleux, V. 1995. Microhabitat requirements for settlment of juvenile sparid fishes on Mediterranean rocky shores. *Hydrobiologia*, 300/301: 309-320.

Hartvigsen, G., Kinzig, A., and Peterson, G. 1998. Use and analysis of complex adaptive systems in ecosystem science: overview of special section. *Ecosystems*, 1: 427-430.

Harvell, C.D., Burkholder, J.M., Colwell, P.R., Grimes, D.J., Hofmann, E.E., Lipp, E.K., Osterhaus, A.D.M.E., Overstreet, R.M., Porter, J.W., Smith, G.W., and Vasta, G.R. 1999. Emerging marine diseases—climate links and anthropogenic factors. *Science*, 285: 1505-1510.

Hastings, A., Hom, C.L., Ellner, S., Turchin, P., and Godfray, H.C.J. 1993. Chaos in ecology: is mother nature a strage attractor? *Annu. Rev. Ecol. Syst.*, 24: 1-33.

Hayden, B.P. 1998. Ecosystem feedbacks on climate at the landscape scale. *Phil. Trans. R. Soc. Lond.*, B, 353: 5-18.

Hector, A., et al. 1999. Plant diversity and productivity experiments in European grasslands. *Science*, 5442: 1123-1127.

Hildrew, A.G. 1996. Whole river ecology: spatial scale and heterogeneity in the ecology of running waters. *Arch. Hydrobiol.*, suppl. 113, Large Rivers 10: 25-43.

Holling, C.S. 1992. Cross-scale morphology, geometry, and dynamics of ecosystems. *Ecol. Monogr.*, 62: 447-502.

Hong, S., Candelone, J.P., Patterson, C.C., and Boutron, C. 1996. History of ancient copper smelting pollution during roman and medieval time recorded in Greenland ice. *Science*, 272: 246-249.

Hooper, D., and Vitousek, P.M. 1997. The effects of plant composition and diversity on ecosystem processes. *Oecologia*, 110: 449-460.

Hugueny, B. 1989. West African rivers as biogeographic islands: species richness of fish communities. *Oecologia*, 79: 235-243.

Huston, A.M. 1994. *Biological Diversity. The Coexistence of Species on Changing Landscapes.* Cambridge University Press, Cambridge.

Hutchinson, G.E. 1961. The paradox of the plankton. *Amer. Nat.*, 95: 137-147.

Hutchinson, G.E. 1965. *The Ecological Theater and the Evolutionary Play.* Yale University Press, New Haven.

Hynes, H.B.N. 1975. The stream and its valley. *Vehr. Int. Ver. Theror. Ang. Limnol.*, 19: 1-15.

Ipcc 1996. Houghton, J.T., Meira Filho, L.G., Callander, B.A., Harris, N., Kattenberg, A. and Maskell, K. (eds.). 1996. *The Science of Climate Change: contribution of Working Group 1 to the Second Assessment Report of the Intergovernmental Panel on Climate Change (IPCC).* Cambridge UniversityPress, Cambridge, UK.

Jacques, G. 1996. *Le Cycle de l'Eau.* Les Fondamentaux, Hachette, Paris.

Jaeger, J.J. 1996. *Les Mondes Fossiles.* Sciences. Odile Jacob, Paris.

Jaupart, C. 2000. Les sciences de la terre en quête d'une méthode. *La Recherche*, 330: 60-65.

Jax, K. 1998. Holocoen and ecosystem—on the origin and historical consequences of two concepts: *J. Hist. Biol.*, 31: 113-142.

Johnson, L. 1994. Pattern and process in ecological systems: a step in the development of a general ecological theory. *Can. J. Fish. Aquat. Sci.*, 51: 226-246.

Jones, C.G., Lawton, J.H., and Schchak, M. 1994. Organisms as ecosystem engineers. *Oikos*, 69: 373-386.

Jorgensen, S.E. 1997. *Integration of Ecosystem Theory: a Pattern.* Kluwer Academic Publishers, Dordrecht.

Joussaume, S., and Guiot, J. 1999. Reconstruire les chauds et froids de l' Europe. *La Recherche*, 321: 54-59.

Joussaume, S. 1993. *Climat d'Hier à Demain*. Science au présent, CNRS Éditions/CEA.

Jouzel, J., and Lorius, C. 1999. Evolution du climat: du passé récent vers le futur. *C.R. Acad. Sci., Paris, Sciences de la terre et des planètes*, 328: 229-239.

Jupin, H. 1996. *Le Cycle du Carbone*. Les Fondamentaux, Hachette, Paris.

Karl, D.M. 1999. A sea of change: biogeochemical variability in the North Pacific subtropical gyre. *Ecosystems*, 2: 181-214.

Kawanabe, H., and Iwasaki, K. 1993. Introduction: flexibility and synergism of biological relationships in natural communities, pp. 1-10 in Kawanabe, H., Cohen, J.E., and Iwasaki, K. (eds.), *Mutualism and Community Organization. Behavioural, Theoretical, and Food-web Approaches*. Oxford University Press, Oxford.

Klinj, F. (ed.). 1994. *Ecological Classification for Environmental Management*. Kluwer Academic Publishers, Dordrecht.

Kratz, T.K., Webster, K.E., Bowser, C.J., Magnuson, J.J., and Benson, B.J. 1997. The influence of landscape position on lakes in northern Wisconsin. *Freshwater Biol.*, 37: 209-217.

Kuhn, T.S. 1983. *La Structure des Révolutions Scientifiques*. Flammarion, Paris.

Lamotte, M., and Bourlière, F. (sci. eds.). 1967. *Problèmes de Productivité Biologique*. Masson, Paris.

Lamotte, M., and Bourlière, F. (sci. eds.). 1978. *Problèmes d'Écologie: Structure et Fonctionnement des Écosystèmes Terrestres*. Masson, Paris.

Lamotte, M., and Bourlière, F. (sci. eds.). 1983. *Problemes d'Écologie: Structure et Fonctionnement des Écosystémes Limniques*. Masson, Paris.

Lamotte, M. 1985. *Fondements Rationnels de I' Amenagement d'un Territoire*. Masson, Paris.

Larsson, P., and Dodson, S.I. 1993. Invited review—chemical communication in planktonic animals. *Arch Hydrobiol.*, 129: 129-155.

Lawton, J.H., and Brown, V.K. 1994. Redundancy in ecosystems. pp. 255-270 in Schulze, E.D., and Mooney, H.A. (eds.). *Biodiversity and Ecosystem Function*. Springer-Verlag, Berlin.

Lawton, J.H. 1994. What do species do in ecosystems. *Oikos*, 71: 367-374.

Lawton, J.H. 1996. The ecotron facility at Sdilwood Park: the value of "Big Bottle" experiments. *Ecology*, 77(3): 665-669.

Lawton, J.H. 1999. Are there general laws in ecology? *Oikos*, 84: 177-192.

Lazlo, P. 1999. *La Découverte Scientifique*. Que Sais-Je? Presses Universitaires de France, Paris.

Le Tacon, F., and Selosse, M.-A. 1997. Le rôle des mycorhizes dans la colonisation des continents et la diversification des écosystèmes terrestres. *Rev. For. Fr.*, spec. no.: 15-24.

Lebel, T. 1991. Le transfert d'échelle en hydrologie: concept ou confusion? pp. 147-155 in Mullon, C. (ed.), SEMINFOR 4. *Le Transfert d'Échelle.* ORSTOM, Paris.

Lefeuvre, I.C., and Barnaud, G. 1988. Ecologie du paysage: mythe ou réalité? *Bull. Ecol.,* 19(4): 493-522.

Lefeuvre, J.C. 1991. La recherche en écologie en France. Heur et malheur d'une discipline en difficulté. *Courrier de la Cellule Environnement de l'INRA,* 13: 17-26.

Legay, J.M., and Barbault, R. (eds.). 1995. *Le Révolution technologique en Écologie.* Masson, Paris.

Legay, J.M. 1997. *L'Expérience et la Modèle. Un Discours sur la Méthode.* Sciences en question, INRA, Paris.

Lenton, T.M. 1998. Gaia and natural selection. *Nature,* 394: 439-447.

Lepart, J., and Escarre, J. 1983. La succession vegétale, mécanismes et modèles: Analyse bibliographique. *Bull. Ecol.,* 14(3): 133-178.

Lévêque, C., and PaugY, D. (sci. eds.). 1999. *Les Poissons des Eaux Continentales Africaines. Diversité Écologie, Utilisation Par l'Homme.* Éditions de l'IRD, Paris.

Lévêque, C. 1995. Role and consequences of fish diversity in the functioning of African freshwater ecosystems: a review. *Aquatic Living Resources,* 8: 59-78.

Lévêque, C. 1996. *Ecosystèmes Aquatiques.* Les Fondamentaux, Hachette, Paris.

Lévêque, C. 1997. *Biodiversity Dynamics and Conservation: the Freshwater Fish of Tropical Africa.* Cambridge University Press, Cambridge.

Lévêque, C. 1997. *La Biodiversité.* Que Sais-Je? Presses Universitaires de France, Paris.

Levin, S.A. 1998. Ecosystems and the biosphere as complex adaptive systems. *Ecosystems,* 1: 431-436.

Levin, S.A. 1999. *Fragile Dominion. Complexity and the Commons.* Helix Books, Perseus Books.

Likens, G.E. 1985. An experimental approach for the study of ecosystems. The fifth Tansley lecture. *J. Ecol.* 73: 381-396.

Likens, G.E. 1995. *The Ecosystem Approach: Its Use and Abuse.* Oldenborf/luhe (Germany), Ecology Institute.

Lindeman, R. 1942. The trophic-dynamic aspect of ecology. *Ecology,* 23: 399-418.

Linne, C. 1972. *L'équilibre de la Nature.* Vrin, Paris. Text translated and annotated by B. Jasmin. Introd. and notes by C. Limoges.

Lovelock, J. 1990. *Les Âges de Gaïa.* Robert Laffont, Paris.

Lubchenko, J. et al., 1991. The sustainable biosphere initiative: an ecological research agenda. *Ecology,* 72(2): 371-412.

Lurcat, F. 1999. *Le Chaos.* Que Sais-Je? Presses Universitaires de France, Paris.

MaCann, K., Hastings, A., and Huxel, G.R. 1998. Weak interactions and the balance of nature. *Nature,* 395: 794-798.

MacArthur, R.H., and Wilson, E.O. 1967. *The Theory of Island Biogeography.* Princeton Univesity Press, Princeton, New Jersey.

Magnuson, J.J., and Kratz, T.K. 2000. Lakes in the landscape: approaches to regional limnology. *Verh. Internat. Verein. Limnol.,* 27: 74-87.

Magnuson, J.J. 1990. Long-term eclogical research and the invisible present. *BioScience,* 40(7): 495-501.

Magny, M. 1995. *Une Histoire du Climat. Des Derniers Mammouths au Siècle de l'Automobile.* Éditions Errance.

Mandelbrot, B.B. 1967. How long is the coast of Britain? Statistical self similarity and fractional dimension. *Science,* 155: 636-638.

Margalef, R. 1968. *Perspectives in Ecological Theory.* University of Chicago Press.

May, R.M. 1972. Will a large complex system be stable? *Nature,* 238: 413-414.

May, R.M. 1974. *Stability and Complexity in Model Ecosystems.* Princeton University Press.

Mayr, E. 1982. *The Growth of Biological Thought, Diversity, Evolution. and Inheritance.* The Belknap Press of Harvard University Press, Cambridge, Massachusetts.

Mayr, E. 1988. *Towards a New Philosophy of Biology: Observations of an Evolutionist.* Harvard University Press, Cambridge.

McArthur, R.H. 1955. Fluctuations of animal populations and a measure of community stability. *Ecology,* 36: 533-536.

McIntosh, R.P. 1991. Concept and terminology of homegeneity ande heterogeneity in ecology. pp. 24-46 in Kolosa, J., and Pickett, S.T.A. (eds.), *Ecological Heterogeneity.* Ecological Studies 86, Springer Verlag.

McKenzie, R., Connor, B., and Bodeker, G. 1999. Increased summertime UV radiation in New Zealand in response to ozone loss. *Science,* 285: 1709-1711.

Mills, L.S., Soulé, M.E., and Doak, D.F. 1993. The keystone species concept in ecology and conservation. *BioScience,* 43: 219-224.

Milne, B.T. 1997. Applications of fractal geometry in wildlife biography. pp. 32-69 in Bissonette, J.A. (ed.), *Wildlife and Landscape Eclogy: Effects of Pattern and Scale.* Springer Verlag, New York.

Minster, J.F. 1994. *Les Océans.* Dominos, Flammarion.

Moles, A.A. 1990. *Les Sciences de l'Imprécis.* Les Seuil, Paris.

Molinier, M., Leprun, J.C., and Audry, P. 1991. Effet d'échelle observé sur le ruissellement dans le Nordeste brésilien. pp. 95-104 in Mullon, C. (ed.), SEMINFOR 4. *LeTransfert d'Échelle.* ORSTOM, Paris.

Moore, P. 1998. Nutrient consequences of salmon spread. *Nature,* 396: 314-315.

Naem, S., Hahn, D.R., and Schuurman, G. 2000. Producer-decomposer co-dependency influences biodiversity effects. *Nature,* 403: 462-764.

Naem, S., Thompson, L.J., Lawler, J.H., Lawton, J.H., and Woodfin, R.M. 1995. Biodiversity and ecosystem funtioning: empirical evidence from experimental microcosms. *Phil. Trnas. R. Lond.,* 347: 249-262.

Naiman, R.J., and Décamps, H. (eds.). 1990. *The Ecology and Management of Aquatic-Terrestrial Ecotones.* Man and Biosphere series, vol. 4, UNESCO, Paris.

Nicolis, G., and Prigogine, I. 1989. *A la Rencontre du Complexe.* Presses Universitaires de France, Paris.

Nicolis, G. 1992. Dynamical systems, biological complexity and global change. pp. 21-32 in Solbrig, O.T., van Emden, H.M., and van Oordt, P.G.W.J. (eds.), *Biodiversity and Global Change.* Monograph no. 8, IUBS.

Northcote, T.G. 1988. Fish in the structure and function of freshwater ecosystems: a "top-down" view. *Can J. Fish. Aquat. Sci.,* 45: 361-379.

O'Neill, R.V., DeAngelis, D.L., Waide, J.B., and Allen, T.F.H. 1986. *A Hierachical Concept of Ecosystems.* Princeton University Press, Princeton, New Jersey.

Oberdforff, T., Guégan, J.F., and Hugueny, B. 1995. Global scale patterns of fish species rechness in rivers. *Ecography,* 18: 345-352.

Oberdorff, T., Hugueny, B., and Guégan, J.F. 1997. Is there an influence of historical events on contemporary fish species richness in rivers? Comparisons between Western Europe and North America. *J. Biogeogr.,* 24(4): 461-467.

Odum, E.P. 1971. *Fundamentals of Ecology,* 3d ed. W.B. Saunders Company, Philadelphia.

Odum, H.T. 1957. Trophic structure and productivity of Silver Springs, Florida. *Ecol. Mongor.,* 27: 55-112.

Odum, H.T. 1983. *System Ecology.* Wiley, New York.

Palmer, M.A., et al. 1997. Biodiversity and ecosystem processes in freshwater sediments. *Ambio,* 26(8): 571-577.

Parry, M.L. (ed.), 2000. *Assessment of Potential Effects and Adaptations for Climate Change in Europe: The Europe ACACIA Project.* Jackson Environment Institute, University of East Anglia, Norwich, UK, 320 pp.

Patten, B.C. 1978. Systems approach to the conceptof the environment. *Ohio J. Sci.,* 78: 206-222.

Patten, B.C., and Odum, E.P. 1981. The cybernetic nature of ecosystems. *Am. Nat.,* 118: 886-895.

Patten, B.C., Straskraba, M., and Jorgensen, S.E. 1997. Ecosystem emerging: 1. Conservation. *Ecol. Modelling,* 96: 221-284.

Pauly, D., Christensen, V., Dalsgaard, J., Froese, R., and Torres, F. 1998. Fishing down marine food webs. *Science,* 279: 860-863.

Pavé, A. 1994. *Modélisation en Biologie et en Écologie*. Aléas, Lyon.

Pérès, J.M. 1961. *Océanographie Biologique et Biologie Marine. I—La Vie Benthique*. Presses Universitaires de France, Paris.

Person, L., Diehl, S., Johansson, L., Andersson, G., and Hamrin, S., 1992. Trophic interactions in temperate lake ecosystems—a test of foodchain theory. *Am. Nat*. 140: 59-84.

Peters, R.H. 1991. *A Critique for Ecology*. Cambridge Univesity Press, Cambridge.

Pett, J.R., Jouzel, J., Raynaud, D., Barkov, N.I., Barnola, J.M., Basile, I., Benders, M., Chappellaz, J., Davis, M., Delaygue, G., Delmotte, M., Kotlyakov, V.M., Legrand, M., Lipenkov, V.Y., Louius, C., Pépin, L., Ritz, C., Saltzman, E., and Stievenard, M. 1999. Climate and atmospheric history of the past 420,000 years from the Vostok ice core, Antarctica. *Nature*, 399: 429-435.

Pickett, S.T.A., and Cadenasso, M.L. 1995. Landscape ecology: spatial heterogencity in ecological systems. *Science*, 269: 331-334.

Pickett, S.T.A., and White, P.S. 1985. *The Ecology of Natural Disturbance and Patch Dynamics*. Academic Press, Orlando, Florida.

Pickett, S.T.A. 1999. The culture of synthesis: habits of mind in novel ecological integration. *Oikos*, 87: 479-487.

Pickett, S.T.A., Kolasa, J., and Jones, C.G. 1994. *Ecological Understanding. The Nature of Theory and the Theory of Nature*. Academic Press.

Pinay, G., and Décamps, H. 1988. The role of riparian woods in regulating nitrogen fluxes between the alluvial aquifer and surface water: a conceptual model. *Regulated Rivers: Res. and Mgmt*. 2: 507-516.

Pinel-Alloul, B. 1995. Spatial heterogeneity as a multiscale characteristic of zooplankton community. *Hydrobiologia*, 300/301: 17-42.

Polis, G. 1994. Food webs, trophic cascades and community structure. *Austral. J. Ecol.*, 19(2): 121-136.

Polis, G.A., Holt, R.D., Menge, B.A., and Winemiller, K.O. 1996. Time, space and life history: influences on food webs. pp. 435-460, in Polis, G.A., and Winemiller, K.O. (eds.), *Food-Webs, Integration of Patterns and Dynamics*. Chapman and Hall, New York.

Popper, K. 1962. *Conjectures and Refutations: the Growth of Scientific knowledge*. Basic Books, New York.

Popper, K. 1985. *Conjectures et Réfutations*. Payot, Paris.

Post, E., Peterson, R.O., Stenseth, N.C., and McLaren, B.E. 1999. Ecosystem consequences of wolf behavioural response to climate. *Nature*, 401: 905-907.

Prance, G. 1982. *Biological Diversification in the Tropics*. Columbia University Press, New York.

Prigogine, I., and Stengers, I. 1986. *La Nouvelle Alliance. Métamorphose de la Science*. Gallimard, Paris.

Prigogine, I. 1994. *Les Lois du Chaos*. Coll. Champs, Flammarion, Paris.

Rasmussen, L., and Wright, R.F. 1998. Large-scale ecosystem experiements: ecological research and European environmental policy. *Forest Ecol. and Mgmt.*, 101: 353-363.

Rasool, I. 1993. *Système Terre.* Coll. Dominos, Flammarion, Paris.

Reice, S.R. 1994. Nonequilibrium determinants of biological community structure. *Amer. Scientist*, 82: 424-435.

Reille, M. 1990. *Leçons de Palynologie et d'Analyse Polinique.* Éditions du CNRS.

Resh, V.H., Brown, A.V., Covich, A.P., Guriz, M.E., Li, H.W., Minshall, G.W., Reice, S.R., Sheldon, A.L., Wallace, J.B., and Wissmar, R.C. 1988. The role of disturbance in stream ecology. *J. N. Amer. Benthol. Soc.*, 7: 433-455.

Rieu, M., and Perrier, E. 1994. Modélisation fractale de la structure des sols. *C.R. Acad. Agric. Fr.*, 80(6): 21-39.

Rind, D. 1999. Complexity and climate. *Science*, 284: 105-107.

Ronsay, J. de. 1975. *Le Macroscope. Vers une Vision Globale.* Éditions du Seuil, Paris.

Roughgarden, J. 1998. *Primer of Ecological Theory.* Prentice Hall, New Jersey.

Roughgarden, J., May, R.M., and Levin, S.A. (eds.). 1989. *Perspectives in Ecological Theory.* Princeton Univsity Press, Princeton, New Jersey.

Ruelle, D. 1991. *Hasard et Chaos.* Odile Jacob, Paris.

Sadourny, R. 1994. *Le Climat de la Terre.* Coll. Dominos, Flammarion, Paris.

Sala, E., Boudoureque, C.F., and Harmelin-Vivien, M. 1998. Fishing, trophic cascades and the structure of algal assemblages: evaluation of an old but untested paradigm. *Oikos*, 83: 425-439.

Saugier, B. 1996. *Végétation et Atmosphère.* Dominos, Flammarion, Paris.

Schindler, D.W. 1990. Experimental perturbations of whole lakes as tests of hypotheses concerning ecosystem structure and function. *Oikos*, 57: 25-41.

Schindler, D.W. 1998. Replication versus realism: the need for ecosystem-scale experiments. *Ecosystems*, 1: 323-334.

Schmidt-Lainé, C., and Pavé, A. 2000. Environnement: modélisation et modèles pour comprendre, agir ou décider dans un contexte interdisciplinaire. *Nature, Sciences, Sociétés.*

Scholes, R.J., Schulze, E.D., Pitelka, L.F., and Hall, D.O. 1999. Biogeochemistry of terrestrial ecosystems. pp. 271-303 in Walker, B., Steffen, W., Canadell, J., and Ingram, J. (eds.). *The Terrestrial Biosphere and Global Change. Implications for Natural and Managed Ecosystems.* Synthesis Volume. International Geosphere-Biosphere Programme Book Series 4, Cambridge University Press.

Schultz, E.D. 1995. Flux control at the ecosystem level. *Trends in Ecol. and Evol.* 10(1): 40-43.

Schulze, E.D., and Mooney, H.A. (eds.). 1993. *Biodiversity and Ecosystem Function.* Springer Verlag.

Sime-Ngando, T. 1997. Importance des virus dans la structure et le fonction-nement des réseaux trophiques. *Année Biologique*, XXXVI(3): 181-210.

Smith, R.F.J. 1992. Alarm signals in fishes. *Rev. Fish Biol. Fish.*, 2: 33-63.

Smith, S.E., and Read, D.J. 1997. *Mycorrhizal Symbiosis*. Academic Press, London, 600 p.

Snelgrove, V.R., et al. 1997. The importance of marine sediment biodiversity in ecosystem processes. *Ambio*, 26(8): 578-583.

Solbrig, 1991. *Biodiversity. scientific issues and collaborative research proposals*. MAB Digest 9, UNESCO, Paris.

Soranno, P.A., Webster, K.E., Riera, J.L., Kratz, T.K., Baron, J.S., Bukaveckas, P.A., Kling, G.W., White, D.S., Caine, N., Lathrop, R.C., and Leavitt, P.R. 1999. Spatial variation among lakes within landscapes: ecological organization along lake chains. *Ecosystems*, 2: 395-410.

Southwood, T.R.E., 1987. The concept and nature of the community. pp. 3-27 in Gee, J.H.R., and Giller, P.S. (eds.), *Organization of Communities, Past and Present*. Blackwell, Oxford.

Statzner, B. 1997. Complexity of theoretical concepts in ecology and predictive power: patterns observed in stream organisms. pp. 211-218 in Ladolt, P., and Sartori, M. (eds.), *Ephemeroptera and Plecoptera: Biology-Ecology-Systematics*. MTL, Fribourg.

Statzner, B., Gore, J.A., and Resh, V.H. 1988. Hydraulic stream ecology: observed patterns and lpotential applications. *J. N. Amer. Benthol. Soc.*, 7(4): 307-360.

Statzner, B., Resh, V.H., and Dolédec, S. 1994. Ecology of the Upper Rhône River: a test of habitat template theories. *Freshwater Biology*, special issue, 31(3): 253-554.

Steele, J.H., and Schmacher, M. 2000. Ecosystem structure before fishing. *Fisheries Research*, 44: 201-205.

Stengers, I., and Schlanger, J. 1991. *Les Concepts Scientifiques*. Gallimard, Coll. Folio/Essais, Paris.

Stone, L., and Weisburd, S.J. 1992. Positive feedback in aquatic ecosystems. *Trends in Ecol. and Evol.*, 7(8): 263-267.

Straskraba, M., Jorgensen, S.E., and Pattern, B.C. 1999. Ecosystem emerging: 2. Dissipation. *Ecol. Modelling*, 117: 3-39.

Swanson, F.J., and Sparks, R.E. 1990. Long-term ecological Research and the invisible place. The local to global spatial scales of the Long-Term Ecological Research Program. *BioScience*, 40(7): 502-508.

Taberlet, P., Fumagalli, L., Wust-Sancy, A.G., and Cosson, J.F. 1998. Comparative phylogeography and postglacial colonization routes in Europe. *Mol. Ecol.*, 7: 453-464.

Tansley, A.G. 1935. The use and abuse of vegetational concepts and terms. *Ecology*, 16(3): 284-307.

Thieneman, A. 1926. *Limnologie*. Breslau.

Thom, R. 1980. *Les Buts de la Science, Modèles Mathématiques de la Morphogenèse*, 2nd ed. Bourgeois, Paris.

Thuillier, P. 1988. *Les Passions du Savoir. Essais sur les Dimensions Culturelles de la Science*. Le Temps des Sciences, Fayard, Paris.

Tilman, D., and Downing, J.A. 1994. Biodiversity and stability in grasslands. *Nature*, 367: 363-365.

Tilman, D. 1996. Biodiversity: population versus ecosystem stability. *Ecology*, 77: 350-363.

Tilman, D., Knops, J., Wedin, D., Reich, P., Ritchie, P., and Siemann, E. 1997. The influence of functional diversity and composition on ecosystem processes. *Science*, 277: 1300-1302.

Town, W.M. 1990. Climate change and fish communities: a conceptual framework. *Trans. Amer. Fisheries Soc.*, 119: 337-352.

Townsend, C.R., and Hildrew, A.G. 1994. Species traits in relation to a habitat templet for river systems. *Freshwater Biol.*, 31: 265-275.

Townsend, C.R. 1989. The patch dynamics concept of stream community ecology. *J. N. Amer. Benthol. Soc.*, 8: 36-50.

Turekian, K.K. 1996. *Global Environmental Change. Past, Present, Future*. Prentice Hall, New Jersey.

Ulanowicz, R.E. 1997. *Ecology, the Ascendent Perspective*. Complexity in Ecological Systems Series, Columbia University Press, New York.

Underwood, A.J. 1990. Experiments in ecology and management: their logics, functions and interpretations. *Aust. J. Ecol.*, 15: 363-389.

Van Der, Heijden, M.G.A., Klironomos, J.N., Ursic, M., Moutoglis, P., Strettwolf-Engei, R., Boller, T., Wiemken, A., and Sanders, I.R. 1998. Mycorrhizal fungal diversity determines plant biodiversity, ecosystem variability and productivity. *Nature*, 396: 69-72.

Van Der Leun, J.C., Tang, X., and Tevini, M. 1995. Environmental effects of ozone depletion: 1994 assessment. *Ambio*, 24(3): 138-143.

Vannote, R.L., Minshall, G.W., Cummins, K.W., Sedell, J.R., and Cushing, C.E. 1980. The river continuum concept. *Can. J. Fish. Aquat. Sci.*, 37: 130-137.

Vaughan, D.G., and Drake, C.S.M. 1996. Recent atmospheric warming and retreat of ice shelves on the Antarctic Peninsula. *Nature*, 379: 328-331.

Vernadsky, W. 1997. *La Biosphère*. Latitudes, Diderot Editeur, Paris.

Vivien, F.D. 1997. L'économie et l'écologie entre science et idéologie. *Nature, Sciences, Sociétés*, 5(4): 12-22.

Vogt, K.A., Gordon, J.C., Wargo, J.P., and Vogt, D., et al. 1997. *Ecosystems. Balancing Science with Management*. Springer-Verlag, New York.

Voituriez, B. 1992. *Les Climats de la Terre*. Explora, Presses Pocket.

Wagensberg, J. 1997. *L'Âme de la Méduse. Idées sur la Complexité du Monde.* Éditions du Seuil, Paris.

Walker, B.H., Kinzig, A., and Langridge, J. 1999. Plant atribute diversity, resilience, and ecosystem function: the nature and significance of dominant and minor species. *Ecosystems,* 2: 95-113.

Walker, B.H., and Steffen, W.L. 1999. The nature of global change, pp. 1-18 in Walker, B., Steffen, W., Canadell, J., and Ingram, J. 1999. *The Terrestrial Biosphere and Global Change.* Synthesis volume. International Geosphere-Biosphere Programme Book series, 4. Cambridge University Press.

Walker, B.H., Steffen, W.L., and Langridge, J. 1999. Interactive and integrated effects of global change on terrestrial ecosystems. pp. 329-375 in Walker, B., Steffen, W., Canadell, J., and Ingram, J. (eds.). *The Terrestrial Biosphere and Global Change.* Synthesis volume. International Geosphere-Biosphere Programme Book series, 4. Cambridge University Press.

Wcislo, W.T. 1989. Behavioral environments and evolutionary change. *Annu. Rev. Ecol. and Systematics,* 20: 137-169.

Werner, B.T. 1999. Complexity in natural landform patterns. *Science,* 284: 102-104.

West-Eberhard, M.J. 1989. Phenotypic plasticity and the origins of diversity. *Annu. Rev. Ecol. and Systematics,* 20: 249-278.

Westbroek, P. 1998. *Vive la Terre. Physiologie d'une Planète.* Éditions du Seuil, Paris.

Whittaker, R.H. 1975. *Communities and Ecosystems.* MacMillan, New York.

Wiens, J.A. 1989. Spatial scaling in ecology. *Functional Ecology,* 3: 385-397.

Wilson, E.O. 2000. *L'unicité du savoir. De la biologie à l'art, une même connaissance.* Robert Lafont, Paris.

Witte, F., and Oijen, M.J.P. van. 1990. Taxonomy, ecology and fishery of Lake Victoria haplochromine trophic groups. *Zool. Vehr., Leiden,* 262: 1-47.

Witte, F., Goldschmidt, T., Goudswaard, P.C., Ligtvoet, W., van Oijen, M.P.J., and Wanink, J.H. 1992. Species extinction and concomitant ecological changes in Lake Victoria. *Neth. J. Zool.,* 42: 214-232.

Worster, D. 1992. *Les Pionniers de l'Écologie.* Sang de la Terre, Paris.

Wright, D.H., Curie, D.J., and Maurer, B.A. 1993. Energy supply and patterns of species richness on local and regional scales. pp. 66-74 in Ricklefs, R.E., and Schluter, D. (eds.), *Species Diversity in Ecological Communities. Historical and Geographical Perspectives.* University Chicago Press, Chicago.

Index

A

Acid rain 123, 348, 364, 407, 408
Adaptive radiation 292, 293, 437
Adaptive strategy 38
Adaptive system 86, 105, 251, 252
Adaptive systems theory 86, 240
Aerobic environment 340, 345, 346
Aerosols 348, 364, 370, 392, 402, 413, 422, 428, 429
Albedo effect 348, 408, 409, 414
Allopatric speciation 168
Ammonia 338, 340, 342, 395, 399
Amplification 89, 95
Anabolism 102, 251
Anaerobic environment 340, 345, 346
Analogy 16, 26, 30, 32, 44-46, 49, 66, 95, 154, 249, 266, 295, 314, 381
Annual variability 185, 188, 389
Aquatic ecosystem 75, 96, 125, 187, 189, 255, 273- 275, 278, 283, 314, 322, 324, 333, 337, 369, 406, 438, 444, 446
Atmosphere 1, 5, 25, 30, 87, 90, 108-113, 125, 157, 174, 187, 195, 237, 239, 255, 303, 307, 329, 331-339, 340, 342, 345, 346, 348, 353, 354, 356, 357, 358-381, 384, 386-414, 416, 420, 421, 427, 433, 439-442
Atmospheric circulation 375, 376, 388, 389, 411, 442
Atmospheric transport 117
Attractor 237, 238
Autecology 20
Autotroph 184, 249, 251- 253, 256, 260, 264, 266-268, 271, 272, 274, 278, 279, 283, 306, 311, 329, 332, 335, 340, 346, 355
Auxiliary energy 114, 127, 167

B

Balance of nature 203, 205-212, 228, 251
Benthic communities 136, 137, 273, 295, 319, 446
Benthos 264, 274, 282, 309, 320, 321
Bioavailability 254, 340, 349
Biocoenosis 7, 19, 22-24, 28, 136, 137, 152, 165, 174, 187, 231, 233, 243, 430, 431, 435
Biocomplexity 91
Biodiversity, changes in 68, 75, 91
Biodiversity, concept 283
Biogeochemical cycle 5, 6, 34, 36, 131, 255, 298, 305, 314, 320, 327-331, 338, 348, 353, 355, 358, 363-366, 422, 440
Biogeochemical flows 10, 179, 227, 284, 287
Biogeochemistry 7, 30, 40, 130, 331, 362
Biogeocoenosis 24, 25
Biogeographic zone 173, 218, 219
Biological carbon pumpe 332, 349
Biological diversity 9, 10, 73, 89, 201, 227, 230, 234, 235, 284-287, 296, 298, 300-304, 310-317, 365
Biological nitrogen fixation 302, 336, 349
Biomass 19, 22, 29, 31, 67, 94, 101-105, 113, 166, 174, 177, 178, 187, 221, 222, 225, 226, 229, 237, 249, 251-

Milton Keynes UK
Ingram Content Group UK Ltd.
UKHW021908071024
449327UK00022B/1640